Airplane Flying Handbook

Foreword by David Soucie

U.S. Department of Transportation
FEDERAL AVIATION ADMINISTRATION
Flight Standards Service

Skyhorse Publishing

First published in 2016
First Skyhorse Publishing edition 2017
Foreword © 2014 by David Soucie

All rights to any and all materials in copyright owned by the publisher are strictly reserved by the publisher. All inquiries should be addressed to Skyhorse Publishing, 307 West 36th Street, 11th Floor, New York, NY 10018.

Skyhorse Publishing books may be purchased in bulk at special discounts for sales promotion, corporate gifts, fund-raising, or educational purposes. Special editions can also be created to specifications. For details, contact the Special Sales Department, Skyhorse Publishing, 307 West 36th Street, 11th Floor, New York, NY 10018 or info@skyhorsepublishing.com.

Skyhorse® and Skyhorse Publishing® are registered trademarks of Skyhorse Publishing, Inc.®, a Delaware corporation.

Visit our website at www.skyhorsepublishing.com.

10 9 8 7

Library of Congress Cataloging-in-Publication Data is available on file.

Cover design by Federal Aviation Administration

Print ISBN: 978-1-5107-1283-6
Ebook ISBN: 978-1-5107-1284-3

Printed in China

The Art Of Flying

Freedom from the bounds of gravity, reserved for the Gods in eras past, has somehow evaded the fate of other wondrous artful discoveries that eventually became commonplace. Photography, railways, telephones, automobiles, television, Internet, and countless other modern discoveries born or art and invention seem to have fallen victim to the ordinary. Ancient history also provides lessons of art relegated to science.

Greek artists had visions of glorious structures to honor and shelter the gods. Magnificent coliseums, housing hundreds of thousands of people, were built to provide a view of gladiators ruling over man and beast. These structures, while beautiful to behold, ultimately failed, structurally crumbling under their own weight and taking hundreds of lives with them. Mourning citizens demanded that Caesar take action to assure their safety. This compelled the advancement of scientific design, architecture, and government regulation. Workers with only rote skills replaced the sculptor's talent. Repetitive and routine designs replaced creative invention. Today, architecture and structural engineering has come into balance with art and beauty. Magnificent, modern architectural achievement around the world, safely sheltering hundreds of thousands, stand in beautiful testament to the balance of art, function, and safety.

Flight was also born of visionaries. They conceived of glorious flying machines suspended by lighter-than-air balloons that would someday grace the skies, transporting passengers over impassable land routes to new and exciting places.

These early visionaries gave way to inventors with grand ideas of leaving the earth behind in heavier-than-air flying machines. From within the depths of da Vinci's artistic mind, innovation, and creativity, these visions materialized. He discovered the principle of flight was balance. Lift must balance with weight, thrust must balance with drag. His works of art then evolved into practical and viable transportation. Like the marvelous ancient structures, flying machines also crumbled under their own weight, claiming the lives of those inside. Before commercial air travel was viable, citizens demanded the government take action to assure their safety. This compelled the advancement of aeronautical design and systems engineering. Similar to the evolution of architecture, aeronautics and engineering have come into balance with innovation and creativity. Magnificen flying machines with incredible complexity gracefully fly over our heads, carrying hundreds of people, proudly displaying their balance of beauty, function, and safety. With every balance, however, there is a new imbalance.

As modern technology encroaches on the controls in the cockpit, the art and intrigue of flying persists. Flying has not yet relegated itself to rote repetitive skills as have other arts. Technology is exceeding its role of making flight safer and creating a new imbalance. Will flying continue to transcend the commonplace or will technology push it into routine?

Perhaps the wonder and fascination of technology and automated flight started to crumble under its own weight? Hundreds of people have recently lost their lives as technology was allowed to take the controls. Will citizens call upon government regulation to be more restrictive, to take action to protect the passengers, crew, and their families from more technological failures? Or is it in the hands and minds of the flyers—artists of aviation—to find balance

I am hopeful the art of flying will find the perfect balance with technology and automated flight. Only time will tell. Only the minds of flyers whose artful hands are on the controls will be able to find this balance—the hands of flyers like you, who find beauty and art in being free from the bounds of earth. Flyers who won't succumb to the lure of technology, automation, and rote skills that would otherwise commandeer the art of flight

The joy of piercing clouds, soaring over majestic mountains and vast oceans like gods, can never be replaced with technology. Flying will continue to defy its presumed destiny of becoming routine. It will attain yet another natural balance: a perfect balance of technology and art, function, and safety.

—David Soucie

Preface

The Airplane Flying Handbook provides basic knowledge that is essential for pilots. This handbook introduces basic pilot skills and knowledge that are essential for piloting airplanes. It provides information on transition to other airplanes and the operation of various airplane systems. It is developed by the Flight Standards Service, Airman Testing Standards Branch, in cooperation with various aviation educators and industry. This handbook is developed to assist student pilots learning to fly airplanes. It is also beneficial to pilots who wish to improve their flying proficiency and aeronautical knowledge, those pilots preparing for additional certificates or ratings, and flight instructors engaged in the instruction of both student and certificated pilots. It introduces the future pilot to the realm of flight and provides information and guidance in the performance of procedures and maneuvers required for pilot certification. Topics such as navigation and communication, meteorology, use of flight information publications, regulations, and aeronautical decision making are available in other Federal Aviation Administration (FAA) publications.

Occasionally the word "must" or similar language is used where the desired action is deemed critical. The use of such language is not intended to add to, interpret, or relieve a duty imposed by Title 14 of the Code of Federal Regulations (14 CFR).

It is essential for persons using this handbook to become familiar with and apply the pertinent parts of 14 CFR and the Aeronautical Information Manual (AIM). The AIM is available online at www.faa.gov. The current Flight Standards Service airman training and testing material and learning statements for all airman certificates and ratings can be obtained from www.faa.gov.

This handbook supersedes FAA-H-8083-3A, Airplane Flying Handbook, dated 2004.

This handbook is available for download, in PDF format, from www.faa.gov.

This handbook is published by the United States Department of Transportation, Federal Aviation Administration, Airman Testing Standards Branch, AFS-630, P.O. Box 25082, Oklahoma City, OK 73125.

Comments regarding this publication should be sent, in email form, to the following address:

AFS630comments@faa.gov

John S. Duncan
Director, Flight Standards Service

Acknowledgments

The Airplane Flying Handbook was produced by the Federal Aviation Administration (FAA) with the assistance of Safety Research Corporation of America. The FAA wishes to acknowledge the following contributors:

    Mr. Shane Torgerson for imagery of the Sedona Airport (Chapter 1)

    Mr. Robert Frola for imagery of an Evektor-Aerotechnik EV-97 SportStar Max (Chapter 16)

Additional appreciation is extended to the General Aviation Joint Steering Committee (GA JSC) and the Aviation Rulemaking Advisory Committee's (ARAC) Airman Certification Standards (ACS) Working Group for their technical support and input.

Table of Contents

Preface...iii

Acknowledgments...v

Table of Contents vii

Chapter 1
Introduction to Flight Training 1-1
Introduction..1-1
Role of the FAA..1-2
Flight Standards Service 1-5
Role of the Pilot Examiner......................................1-6
Role of the Flight Instructor....................................1-7
Sources of Flight Training 1-8
    Practical Test Standards (PTS) and Airman
    Certification Standards (ACS)..................................1-10
    Safety of Flight Practices 1-11
        Collision Avoidance 1-11
        Runway Incursion Avoidance...........................1-12
        Stall Awareness 1-12
        Use of Checklists 1-13
        Positive Transfer of Controls............................1-15
Chapter Summary 1-15

Chapter 2
Ground Operations...2-1
Introduction..2-1
Preflight Assessment of the Aircraft..........................2-2
    Visual Preflight Assessment.....................................2-3
    Outer Wing Surfaces and Tail Section 2-5
    Fuel and Oil...2-6
    Landing Gear, Tires, and Brakes...............................2-8
    Engine and Propeller 2-9
Risk and Resource Management.................................2-9
    Risk Management...2-10
        Identifying the Hazard 2-10
        Risk...2-10
        Risk Assessment 2-10
        Risk Identification 2-10

        Risk Mitigation 2-10
        Resource Management.......................................2-11
Ground Operations...2-11
Engine Starting...2-12
Hand Propping 2-13
Taxiing 2-14
Before-Takeoff Check 2-17
Takeoff Checks 2-18
After-Landing 2-18
Clear of Runway and Stopped 2-18
Parking 2-19
Engine Shutdown 2-19
Post-Flight..2-19
    Securing and Servicing..2-19
Chapter Summary 2-19

Chapter 3
Basic Flight Maneuvers 3-1
Introduction..3-1
The Four Fundamentals 3-2
Effect and Use of the Flight Controls 3-2
    Feel of the Airplane 3-4
Attitude Flying 3-4
Integrated Flight Instruction 3-5
Straight-and-Level Flight...3-6
    Straight Flight...3-7
    Level Flight 3-8
Trim Control 3-10
Level Turns 3-10
    Turn Radius 3-12
    Establishing a Turn...3-13
Climbs and Climbing Turns.......................................3-16
    Establishing a Climb 3-17
    Climbing Turns 3-18
Descents and Descending Turns 3-19
Glides..3-20
    Gliding Turns 3-21
Chapter Summary 3-23

Chapter 4
Maintaining Aircraft Control: Upset Prevention and Recovery Training 4-1

Introduction...4-1
  Defining an Airplane Upset....................................4-2
  Coordinated Flight...4-2
  Angle of Attack 4-2
Slow Flight..4-3
  Performing the Slow Flight Maneuver.....................4-4
Stalls...4-5
  Stall Recognition...4-5
    Angle of Attack Indicators 4-6
  Stall Characteristics...4-6
  Fundamentals of Stall Recovery..............................4-7
  Stall Training..4-8
  Approaches to Stalls (Impending Stalls),
  Power-On or Power-Off...4-8
  Full Stalls, Power-Off 4-8
  Full Stalls, Power-On 4-9
  Secondary Stall...4-10
  Accelerated Stalls..4-10
  Cross-Control Stall..4-11
  Elevator Trim Stall..4-12
  Common Errors...4-13
  Spin Awareness...4-13
  Spin Procedures..4-14
    Entry Phase 4-14
    Incipient Phase 4-14
    Developed Phase..4-15
    Recovery Phase..4-15
  Intentional Spins...4-16
  Weight and Balance Requirements Related
  to Spins..4-17
  Common Errors...4-17
Upset Prevention and Recovery 4-17
  Unusual Attitudes Versus Upsets............................4-17
  Environmental Factors 4-18
  Mechanical Factors..4-18
  Human Factors..4-18
    VMC to IMC..4-18
    IMC...4-18
    Diversion of Attention 4-18
    Task Saturation 4-18
    Sensory Overload/Deprivation 4-18
    Spatial Disorientation 4-19
    Startle Response...4-19
    Surprise Response...4-19
  Upset Prevention and Recovery Training (UPRT)...4-19
  UPRT Core Concepts...4-20

Academic Material (Knowledge and
  Risk Management) 4-20
    Prevention Through ADM and
    Risk Management 4-21
    Prevention through Proportional
    Counter-Response..4-21
    Recovery 4-22
  Common Errors...4-22
  Roles of FSTDs and Airplanes in UPRT 4-22
  Airplane-Based UPRT..4-22
  All-Attitude/All-Envelope Flight
  Training Methods 4-23
  FSTD–based UPRT...4-23
  Spiral Dive..4-23
  UPRT Summary...4-24
Chapter Summary 4-24

Chapter 5
Takeoffs and Departure Climbs 5-1

Introduction...5-1
  Terms and Definitions..5-2
Prior to Takeoff...5-2
Normal Takeoff..5-3
  Takeoff Roll 5-3
  Lift-Off...5-4
  Initial Climb 5-5
Crosswind Takeoff...5-6
  Takeoff Roll 5-6
  Lift-Off...5-8
  Initial Climb 5-8
Ground Effect on Takeoff...5-9
Short-Field Takeoff and Maximum
Performance Climb 5-10
  Takeoff Roll 5-10
  Lift-Off...5-10
  Initial Climb 5-11
Soft/Rough-Field Takeoff and Climb............................5-11
  Takeoff Roll 5-12
  Lift-Off...5-12
  Initial Climb 5-12
Rejected Takeoff/Engine Failure 5-12
Noise Abatement...5-13
Chapter Summary 5-13

Chapter 6
Ground Reference Maneuvers 6-1

Introduction...6-1
Maneuvering by Reference to Ground Objects 6-2
Drift and Ground Track Control 6-3
  Correcting Drift During Straight-and-Level Flight....6-3

Constant Radius During Turning Flight.....................6-4
Tracking Over and Parallel to a Straight Line...........6-6
Rectangular Course...6-6
Turns Around a Point...6-8
S-Turns..6-10
Elementary Eights..6-11
Eights Along a Road 6-11
Eights Across A Road 6-13
Eights Around Pylons..6-13
Eights-on-Pylons..6-14
Chapter Summary 6-18

Chapter 7
Airport Traffic Patterns 7-1
Introduction...7-1
Airport Traffic Patterns and Operations 7-2
Standard Airport Traffic Patterns................................7-2
Non-Towered Airports...7-5
Safety Considerations 7-5
Chapter Summary 7-6

Chapter 8
Approaches and Landings 8-1
Introduction...8-1
Normal Approach and Landing 8-2
Base Leg..8-2
Final Approach..8-3
Use of Flaps...8-4
Estimating Height and Movement...............................8-5
Round Out (Flare) 8-6
Touchdown...8-7
After-Landing Roll..8-8
Stabilized Approach Concept.....................................8-9
Intentional Slips 8-11
Go-Arounds (Rejected Landings)...............................8-12
Power..8-13
Attitude..8-13
Configuration 8-13
Ground Effect..8-14
Crosswind Approach and Landing..............................8-14
Crosswind Final Approach..8-14
Crosswind Round Out (Flare) 8-15
Crosswind Touchdown...8-15
Crosswind After-Landing Roll....................................8-16
Maximum Safe Crosswind Velocities..........................8-17
Turbulent Air Approach and Landing..........................8-18
Short-Field Approach and Landing 8-18
Soft-Field Approach and Landing 8-21
Power-Off Accuracy Approaches...............................8-22
90° Power-Off Approach 8-22
180° Power-Off Approach 8-23

360° Power-Off Approach 8-25
Emergency Approaches and Landings (Simulated).....8-26
Faulty Approaches and Landings 8-27
Low Final Approach..8-27
High Final Approach...8-28
Slow Final Approach...8-28
Use of Power 8-29
High Round Out 8-29
Late or Rapid Round Out 8-30
Floating During Round Out.......................................8-30
Ballooning During Round Out 8-30
Bouncing During Touchdown 8-31
Porpoising...8-32
Wheel Barrowing 8-33
Hard Landing...8-33
Touchdown in a Drift or Crab 8-34
Ground Loop 8-34
Wing Rising After Touchdown 8-35
Hydroplaning...8-35
Dynamic Hydroplaning 8-35
Reverted Rubber Hydroplaning.................................8-35
Viscous Hydroplaning 8-36
Chapter Summary 8-36

Chapter 9
Performance Maneuvers..9-1
Introduction...9-1
Steep Turns...9-2
Steep Spiral 9-4
Chandelle 9-5
Lazy Eight..9-6
Chapter Summary 9-8

Chapter 10
Night Operations 10-1
Introduction..10-1
Night Vision..10-2
Night Illusions...10-3
Pilot Equipment 10-4
Airplane Equipment and Lighting 10-4
Airport and Navigation Lighting Aids.........................10-5
Training for Night Flight 10-6
Preparation and Preflight 10-6
Starting, Taxiing, and Runup 10-6
Takeoff and Climb 10-7
Orientation and Navigation......................................10-7
Approaches and Landings..10-8
Night Emergencies 10-9
Chapter Summary 10-9

Chapter 11
Transition to Complex Airplanes 11-1

Introduction..11-1
   Function of Flaps...11-2
   Flap Effectiveness 11-3
   Operational Procedures 11-3
Controllable-Pitch Propeller 11-4
   Constant-Speed Propeller................................11-4
   Takeoff, Climb, and Cruise 11-6
   Blade Angle Control..11-7
   Governing Range...11-7
   Constant-Speed Propeller Operation 11-7
Turbocharging...11-8
   Ground Boosting Versus Altitude Turbocharging ...11-9
   Operating Characteristics 11-9
   Heat Management...11-10
   Turbocharger Failure 11-10
      Over-Boost Condition.................................11-10
      Low Manifold Pressure 11-11
Retractable Landing Gear 11-11
   Landing Gear Systems...................................11-11
   Controls and Position Indicators 11-11
   Landing Gear Safety Devices.........................11-11
   Emergency Gear Extension Systems...............11-12
   Operational Procedures 11-12
      Preflight 11-12
      Takeoff and Climb 11-13
      Approach and Landing 11-15
Transition Training 11-16
Chapter Summary 11-16

Chapter 12
Transition to Multiengine Airplanes 12-1

Introduction..12-1
General..12-2
Terms and Definitions...12-2
Operation of Systems 12-3
   Propellers...12-3
   Propeller Synchronization 12-6
   Fuel Crossfeed 12-6
   Combustion Heater..12-6
   Flight Director/Autopilot..................................12-6
   Yaw Damper..12-7
   Alternator/Generator 12-7
   Nose Baggage Compartment...........................12-7
   Anti-Icing/Deicing..12-8
Performance and Limitations.................................12-9
Weight and Balance 12-11
Ground Operation 12-12
Normal and Crosswind Takeoff and Climb..........12-13
Level Off and Cruise...12-14

Normal Approach and Landing 12-14
Crosswind Approach and Landing........................12-16
Short-Field Takeoff and Climb............................12-17
Short-Field Approach and Landing 12-17
Go-Around 12-18
Rejected Takeoff..12-19
Engine Failure After Lift-Off...............................12-19
   Landing Gear Down 12-19
   Landing Gear Control Selected Up, Single-
   Engine Climb Performance Inadequate 12-20
   Landing Gear Control Selected Up, Single-
   Engine Climb Performance Adequate....................12-20
      Control 12-20
      Configuration...12-21
      Climb 12-21
      Checklist 12-21
Engine Failure During Flight................................12-22
Engine Inoperative Approach and Landing 12-23
Engine Inoperative Flight Principles 12-23
Slow Flight..12-26
Stalls..12-26
   Power-Off Approach to Stall (Approach
   and Landing)..12-26
   Power-On Approach to Stall (Takeoff
   and Departure) 12-27
   Full Stall 12-27
   Accelerated Approach to Stall.........................12-27
   Spin Awareness 12-28

Chapter 13
Transition to Tailwheel Airplanes 13-1

Introduction..13-1
Landing Gear 13-2
   Instability..13-2
Angle of Attack...13-2
Taxiing 13-2
Weathervaning 13-3
Visibility 13-3
Directional Control 13-3
Normal Takeoff Roll 13-3
Liftoff...13-4
Crosswind Takeoff..13-4
Short-Field Takeoff...13-4
Soft-Field Takeoff...13-4
Landing 13-5
Touchdown 13-5
   Three-Point Landing.......................................13-5
   Wheel Landing 13-6
   Crosswinds 13-6
After-Landing Roll..13-6
Crosswind After-Landing Roll 13-7

Short-Field Landing...13-7
Soft-Field Landing...13-8
Ground Loop...13-8
Chapter Summary 13-8

Chapter 14
Transition to Turbopropeller-
Powered Airplanes 14-1
Introduction..14-1
Gas Turbine Engine 14-2
Turboprop Engines...14-2
Turboprop Engine Types 14-3
    Fixed Shaft 14-3
    Split Shaft/ Free Turbine Engine 14-5
Reverse Thrust and Beta Range Operations 14-7
Turboprop Airplane Electrical Systems......................14-8
Operational Considerations...14-9
Training Considerations..14-11
    Ground Training...14-12
    Flight Training...14-12
Chapter Summary 14-13

Chapter 15
Transition to Jet-Powered Airplanes 15-1
Introduction..15-1
Jet Engine Basics 15-2
Operating the Jet Engine 15-3
    Jet Engine Ignition...15-4
    Continuous Ignition 15-4
    Fuel Heaters...15-4
    Setting Power 15-4
    Thrust To Thrust Lever Relationship 15-5
    Variation of Thrust with RPM 15-5
    Slow Acceleration of the Jet Engine 15-6
Jet Engine Efficiency 15-6
Absence of Propeller Effect 15-6
Absence of Propeller Slipstream...........................15-6
Absence of Propeller Drag 15-7
Speed Margins 15-7
Recovery From Overspeed Conditions......................15-9
Mach Buffet Boundaries 15-9
Low Speed Flight...15-10
Stalls...15-11
Drag Devices..15-14
Thrust Reversers 15-15
Pilot Sensations in Jet Flying...............................15-17
Jet Airplane Takeoff and Climb...........................15-18
    Minimum Equipment List and Configuration
    Deviation List 15-18
    V-Speeds 15-20
    Pre-Takeoff Procedures..................................15-20

Takeoff Roll 15-21
Rejected Takeoff 15-22
Rotation and Lift-Off...15-24
Initial Climb 15-24
Jet Airplane Approach and Landing 15-25
    Landing Requirements 15-25
    Landing Speeds 15-25
    Significant Differences...................................15-26
    Stabilized Approach 15-27
    Approach Speed 15-27
    Glidepath Control 15-28
    The Flare...15-28
    Touchdown and Rollout 15-29
Key Points 15-30
Chapter Summary 15-31

Chapter 16
Transition to Light Sport Airplanes (LSA).......16-1
Introduction..16-1
Light Sport Airplane (LSA) Background 16-2
LSA Synopsis...16-3
Sport Pilot Certificate 16-3
Transition Training Considerations 16-4
    Flight School 16-4
    Flight Instructors 16-4
LSA Maintenance 16-5
Airframe and Systems..16-5
    Construction 16-5
    Engines 16-6
    Instrumentation..16-6
Weather Considerations 16-6
Flight Environment 16-7
    Preflight...16-7
    Inside of the Airplane 16-8
    Outside of the Airplane 16-9
    Before Start and Starting Engine....................16-10
    Taxi..16-10
    Takeoff and Climb..16-11
    Cruise 16-11
    Approach and Landing 16-12
    Emergencies..16-12
    Postflight..16-12
Key Points 16-12
Chapter Summary 16-13

Chapter 17
Emergency Procedures 17-1
Emergency Situations 17-1
Emergency Landings 17-2
    Types of Emergency Landings........................17-2
    Psychological Hazards 17-2

Basic Safety Concepts..17-2
   General 17-2
   Attitude and Sink Rate Control 17-4
   Terrain Selection 17-4
   Airplane Configuration...17-4
   Approach 17-5
Terrain Types 17-5
   Confined Areas 17-5
   Trees (Forest) 17-5
   Water (Ditching) and Snow...................................17-6
Engine Failure After Takeoff (Single-Engine) 17-6
Emergency Descents 17-6
In-Flight Fire 17-7
   Engine Fire 17-8
   Electrical Fires..17-8
   Cabin Fire..17-8
Flight Control Malfunction/Failure.............................17-9
   Total Flap Failure 17-9
   Asymmetric (Split) Flap 17-9
   Loss of Elevator Control 17-9
   Landing Gear Malfunction 17-10
Systems Malfunctions 17-11
   Electrical System 17-11
   Pitot-Static System 17-12
Abnormal Engine Instrument Indication 17-13
Door Opening In-Flight 17-13
Inadvertent VFR Flight Into IMC 17-15
   Recognition 17-15
   Maintaining Airplane Control 17-15
   Attitude Control..17-16
   Turns...17-16
   Climbs 17-17
   Descents..17-17
   Combined Maneuvers...17-17
   Transition to Visual Flight 17-18
Chapter Summary 17-18

Glossary **G-1**

Index **I-1**

Addendum to Chapter 12 ....**Add.-1**

Chapter 1

Introduction to Flight Training

Introduction

The overall purpose of primary and intermediate flight training, as outlined in this handbook, is the acquisition and honing of basic airmanship skills. *[Figure 1-1]* Airmanship is a broad term that includes a sound knowledge of and experience with the principles of flight, the knowledge, experience, and ability to operate an airplane with competence and precision both on the ground and in the air, and the application of sound judgment that results in optimal operational safety and efficiency. *[Figure 1-2]* Learning to fly an airplane has often been likened to learning to drive an automobile. This analogy is misleading. Since an airplane operates in a three-dimensional environment, it requires a depth of knowledge and type of motor skill development that is more sensitive to this situation, such as:

- Coordination—the ability to use the hands and feet together subconsciously and in the proper relationship to produce desired results in the airplane.

- Timing—the application of muscular coordination at the proper instant to make flight, and all maneuvers, a constant, smooth process.

- Control touch—the ability to sense the action of the airplane and knowledge to determine its probable actions immediately regarding attitude and speed variations by sensing the varying pressures and resistance of the control surfaces transmitted through the flight controls.

- Speed sense—the ability to sense and react to reasonable variations of airspeed.

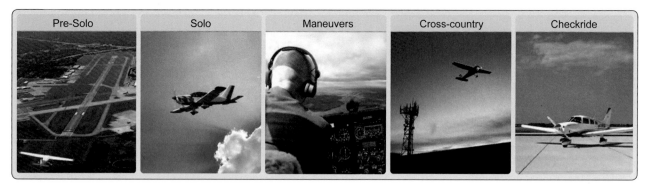

| Pre-Solo | Solo | Maneuvers | Cross-country | Checkride |

Figure 1-1. *Primary and intermediate flight training teaches basic airmanship skills and creates a good foundation for student pilots.*

An accomplished pilot demonstrates the knowledge and ability to assess a situation quickly and accurately and determine the correct procedure to be followed under the existing circumstance. He or she is also able to analyze accurately the probable results of a given set of circumstances or of a proposed procedure; to exercise care and due regard for safety; to gauge accurately the performance of the airplane; to recognize personal limitations and limitations of the airplane and avoid approaching the critical points of each; and the ability to identify, assess, and mitigate risk. The development of airmanship skills requires effort and dedication on the part of both the student pilot and the flight instructor, beginning with the very first training flight where proper habit formation begins with the student being introduced to good operating practices.

Every airplane has its own particular flight characteristics. The purpose of primary and intermediate flight training; however,

is not to learn how to fly a particular make and model airplane. The underlying purpose of flight training is to develop the knowledge, experience, skills, and safe habits that establish a foundation and are easily transferable to any airplane. The pilot who has acquired necessary skills during training, and develops these skills by flying training-type airplanes with precision and safe flying habits, is able to easily transition to more complex and higher performance airplanes. It should also be remembered that the goal of flight training is a safe and competent pilot; passing required practical tests for pilot certification is only incidental to this goal.

Role of the FAA

The Federal Aviation Administration (FAA) is empowered by the U.S. Congress to promote aviation safety by prescribing safety standards for civil aviation. Standards are established

Figure 1-2. *Good airmanship skills include sound knowledge of the principles of flight and the ability to operate an airplane with competence and precision.*

for the certification of airmen and aircraft, as well as outlining operating rules. This is accomplished through the Code of Federal Regulations (CFR), formerly referred to as Federal Aviation Regulations (FAR). Title 14 of the CFR (14 CFR) is titled Aeronautics and Space with Chapter 1 dedicated to the FAA. Subchapters are broken down by category with numbered parts detailing specific information. *[Figure 1-3]* For ease of

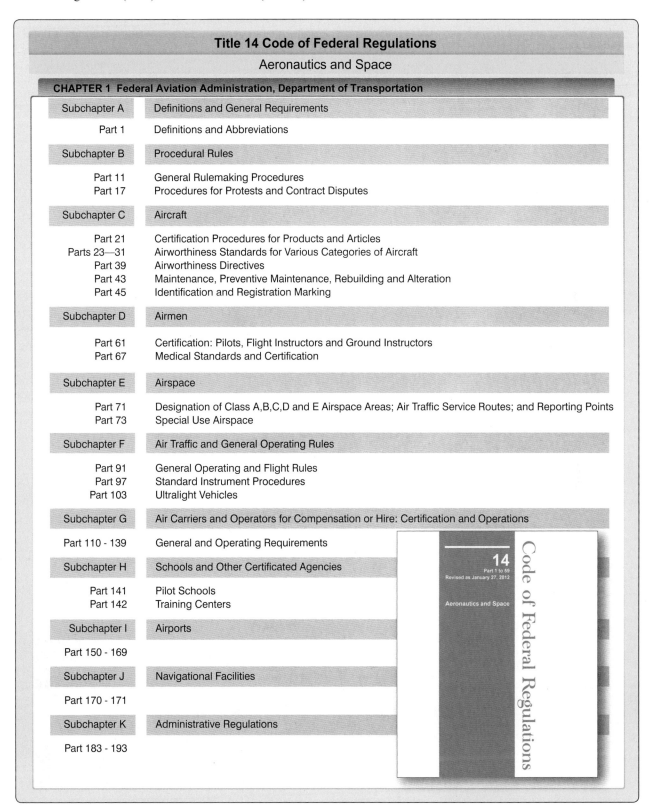

Figure 1-3. *Title 14 CFR, Chapter 1, Aeronautics and Space and subchapters.*

reference since the parts are numerical, the abbreviated pattern 14 CFR part ___ is used (e.g., 14 CFR part 91).

While the various subchapters and parts of 14 CFR provide general to specific guidance regarding aviation operations within the U.S., the topic of aircraft certification and airworthiness is spread through several interconnected parts of 14 CFR.

- 14 CFR part 21 prescribes procedural requirements for issuing airworthiness certificates and airworthiness approvals for aircraft and aircraft parts. A standard airworthiness certificate, FAA Form 8100-2, is required to be displayed in the aircraft. *[Figure 1-4]* It is issued for aircraft type certificated in the normal, utility, acrobatic, commuter or transport category, and for manned free balloons. A standard airworthiness certificate remains valid as long as the aircraft meets its approved type design, is in a condition for safe operation and maintenance, and preventative maintenance and alterations are performed in accordance with 14 CFR parts 21, 43, and 91.

- 14 CFR part 39 is the authority for the FAA to issue Airworthiness Directives (ADs) when an unsafe condition exists in a product, aircraft, or part, and the condition is likely to exist or develop in other products of the same type design.

- 14 CFR part 45 identifies the requirements for the identification of aircraft, engines, propellers, certain replacement and modification parts, and the nationality and registration marking required on U.S.-registered aircraft.

- 14 CFR part 43 prescribes rules governing the maintenance, preventive maintenance, rebuilding, and alteration of any aircraft having a U.S. airworthiness certificate. It also applies to the airframe, aircraft engines, propellers, appliances, and component parts of such aircraft.

- 14 CFR part 91 outlines aircraft certifications and equipment requirements for the operation of aircraft in U.S. airspace. It also prescribes rules governing maintenance, preventive maintenance, and alterations. Also found in 14 CFR part 91 is the requirement to maintain records of maintenance, preventive maintenance, and alterations, as well as records of the 100-hour, annual, progressive, and other required or approved inspections.

While 14 CFR part 91 outlines the minimum equipment required for flight, the Airplane Flight Manual/Pilot's Operating Handbook (AFM/POH) lists the equipment required for the airplane to be airworthy. The equipment list found in the AFM/POH is developed during the airplane certification process. This list identifies those items that are required for airworthiness, optional equipment installed in addition to the required equipment, and any supplemental items or appliances.

UNITED STATES OF AMERICA
DEPARTMENT OF TRANSPORTATION-FEDERAL AVIATION ADMINISTRATION
STANDARD AIRWORTHINESS CERTIFICATE

1 NATIONALITY AND REGISTRATION MARKS	2 MANUFACTURER AND MODEL	3 AIRCRAFT SERIAL NUMBER	4 CATEGORY
N12345	Douglas DC-6A	43219	Transport

5 AUTHORITY AND BASIS FOR ISSUANCE
This airworthiness certificate is issued pursuant to 49 U.S.C. § 44704 and certifies that, as of the date of issuance, the aircraft to which issued has been inspected and found to conform to the type certificate therefore, to be in condition for safe operation, and has been shown to meet the requirements of the applicable comprehensive and detailed airworthiness code as provided by Annex 8 to the Convention on International Civil Aviation, except as noted herein. Exceptions:

None

6 TERMS AND CONDITIONS
Unless sooner surrendered, suspended, revoked, or a termination date is otherwise established by the FAA, this airworthiness certificate is effective as long as the maintenance, preventative maintenance, and alterations are performed in accordance with Parts 21, 43, and 91 of the Federal Aviation Regulations, as appropriate, and the aircraft is registered in the United States.

DATE OF ISSUANCE	FAA REPRESENTATIVE	DESIGNATION NUMBER
01/20/2000	E.R. White E.R. White	NE-XX

Any iteration, reproduction, or misuse of this certificate may be punishable by a fine not exceeding $1,000 or imprisonment not exceeding 3 years or both.
THIS CERTIFICATE MUST BE DISPLAYED IN THE AIRCRAFT IN ACCORDANCE WITH APPLICABLE FEDERAL AVIATION REGULATIONS.
FAA Form 8100-2 (04-11) Supersedes Previous Edition

Figure 1-4. *FAA Form 8100-2, Standard Airworthiness Certificate.*

Figure 1-5 shows an example of some of the required equipment, standard or supplemental (not required but commonly found in the airplane) and optional equipment list for an aircraft. It is originally issued by the manufacturer and is required to be maintained by the Type Certificate Data Sheet (TCDS). An aircraft and its installed components and parts must continually meet the requirements of the original Type Certificate or approved altered conditions to be airworthy.

- 14 CFR part 61 pertains to the certification of pilots, flight instructors, and ground instructors. It prescribes the eligibility, aeronautical knowledge, flight proficiency training, and testing requirements for each type of pilot certificate issued.

- 14 CFR part 67 prescribes the medical standards and certification procedures for issuing medical certificates for airmen and for remaining eligible for a medical certificate.

- 14 CFR part 91 contains general operating and flight rules. The section is broad in scope and provides general guidance in the areas of general flight rules, visual flight rules (VFR), instrument flight rules (IFR), and as previously discussed aircraft maintenance, and preventive maintenance and alterations.

Flight Standards Service

Within the FAA, the Flight Standards Service (AFS) sets the aviation standards for airmen and aircraft operations

Sym:
Items in this listing are coded by a symbol indicating the status of the item. These codes are:
  C Required item for FAA Certification.
  S Standard equipment. Most standard equipment is applicable to all airplanes. Some equipment may be replaced by optional equipment.
  O Optional equipment. Optional equipment may be installed in addition to or to replace standard equipment.

Qty: The quantity of the listed item in the airplane. A hyphen (-) in this column indicates that the equipment was not installed.

ATA Item	Description	SYM	QTY	Part Number	Unit Weight	Arm
34-08	GPS 1 Antenna	C	1	12744-001	0.4	136.2
34-09	GPS 2 Antenna	S	1	12744-001	0.4	110.3
34-10	Transponder Antenna	C	1	12739-001	0.1	105.0
34-11	VOR/LOC Antenna	C	1	12742-001	0.4	331.0
34-12	Turn coordinator, modified	C	1	11891-001	1.8	118.0
34-13	GMA 340 audio panel	S	1	12717-050	1.5	121.5
34-14	GNS 420 (GPS/COM/NAV)	O	1	12718-004	5.0	121.0
34-15	GNS 420 (GPS/COM/NAV)	C	1	12718-051	5.0	121.0
34-16	GNS 420 (GPS/COM/NAV)	O	1	12718-051	5.0	122.4
	EMax engine monitoring					
34-17	• Data acquisition unit	O	1	16692-001	2.0	118.0
34-18	• Monitor cabin harness	O	1	16695-005	2.0	108.0
	Sky watch option					
34-19	• Sky watch inverter	O	1	14484-001	0.5	118.0
34-20	• Sky watch antenna nsti	O	1	14480-001	2.3	150.5
34-21	• Sky watch track box	O	1	14477-050	10.0	140.0
	Stormscope option					
34-22	• Processor	O	1	12745-050	1.7	199.0
34-23	• Antenna	O	1	12745-070	0.9	191.0
	Transponder option					
34-24	• Mode A/C transponder	C	1	13587-001	1.6	124.9
34-25	• Mode S transponder	O	-	15966-050	2.6	121.0
	TAWS option					
34-26	• KGP 560 processor	O	1	15963-001	1.3	117.0
	XM satellite option					
34-27	• XM WX/radio receiver	O	1	16121-001	1.7	114.0
34-28	• XM radio remote control	O	1	16665-501	0.2	149.3
61	**Propeller**					
61-01	• Hartzell propeller installation	C	1	15319-00X	79.8	48.0
61-02	• McCauley propeller installation	O	1	15825-00X	78.0	50.0
61-03	• Propeller governor	C	1	15524-001	3.2	61.7
71	**Power plant**					
71-01	• Upper cowl	C	1	20181-003	10.5	78.4
71-02	• Lower cowl LH	C	1	20182-005	5.4	78.4
71-03	• Lower cowl RH	C	1	20439-005	5.4	78.4
71-03	• Engine baffling installation	C	1	15460-001	10.7	78.4

Figure 1-5. *Example of some of the required, standard or supplemental and optional equipment for an aircraft.*

in the United States and for American airmen and aircraft around the world. The AFS is headquartered in Washington, D.C., and is broadly organized into divisions based on work function (Air Transportation, Aircraft Maintenance, Flight Technology, Training, Certification and Surveillance, a Regulatory Support Division based in Oklahoma City, OK, and a General Aviation and Commercial Division). Regional Flight Standards division managers, one at each of the FAA's nine regional offices, coordinate AFS activities within their respective regions.

The interface between AFS and the aviation community/general public is the local Flight Standards District Office (FSDO). The approximately ninety FSDOs are strategically located across the United States, each office having jurisdiction over a specific geographic area. *[Figure 1-6]* The individual FSDO is responsible for all air activity occurring within its geographic boundaries. The individual FSDOs are responsible for the certification and surveillance of air carriers, air operators, flight schools/training centers, airmen (pilots, flight instructors, mechanics and other certificate holders). Additional duties that are tasked to FSDO inspectors is accident investigation and enforcement actions. NOTE: Accident investigation and enforcement actions are a smaller part of a field inspectors job than surveillance and certification.

Each FSDO is staffed by Aviation Safety Inspectors (ASIs) whose specialties include operations, maintenance, and avionics. General Aviation ASIs are highly qualified and experienced aviators. Once accepted for the position, an inspector must satisfactorily complete indoctrination training conducted at the FAA Academy that includes airman evaluation and pilot testing techniques and procedures. Thereafter, the inspector must complete recurrent training on a regular basis. Among other duties, the FSDO inspector is responsible for administering FAA practical tests for pilot and flight instructor certificates and associated ratings. All questions concerning pilot certification (and/or requests for other aviation information or services) should be directed to the FSDO having jurisdiction in the particular geographic area. For specific FSDO locations and telephone numbers, refer to www.faa.gov.

Role of the Pilot Examiner

Pilot and flight instructor certificates are issued by the FAA upon satisfactory completion of required knowledge and practical tests. The administration of these tests is an FAA responsibility that the issuance of pilot and instructor certificates can be carried out at the FSDO level. In order to satisfy the public need for pilot testing and certification services, the FAA delegates certain responsibilities, as

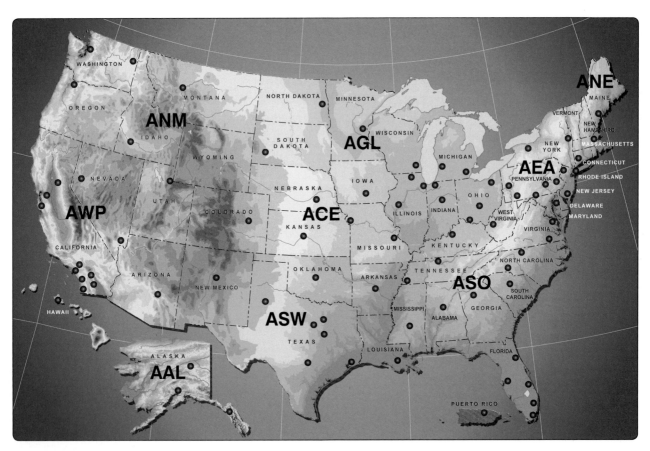

Figure 1-6. *Flight Standards District Office locations across the United States.*

the need arises, to private individuals who are not FAA employees. A Designated Pilot Examiner (DPE) is a private citizen who is designated as a representative of the FAA Administrator to perform specific (but limited) pilot certification tasks on behalf of the FAA and may charge a reasonable fee for doing so. Generally, a DPE's authority is limited to accepting applications and conducting practical tests leading to the issuance of specific pilot certificates and/or ratings. A DPE operates under the direct supervision of the FSDO that holds the examiner's designation file. A FSDO inspector is assigned to monitor the DPE's certification activities. Normally, the DPE is authorized to conduct these activities only within the designating FSDO's jurisdictional area.

The FAA selects only highly qualified individuals to be DPEs. These individuals must have good industry reputations for professionalism, high integrity, a demonstrated willingness to serve the public, and adhere to FAA policies and procedures in certification matters. A DPE is expected to administer practical tests with the same degree of professionalism, using the same methods, procedures, and standards as an FAA ASI. It should be remembered, however, that a DPE is not an FAA ASI. A DPE cannot initiate enforcement action, investigate accidents, or perform surveillance activities on behalf of the FAA. However, the majority of FAA practical tests at the recreational, private, and commercial pilot level are administered by FAA DPEs.

Role of the Flight Instructor

The flight instructor is the cornerstone of aviation safety. The FAA has adopted an operational training concept that places the full responsibility for student training on the authorized flight instructor. In this role, the instructor assumes the total responsibility for training the student pilot in all the knowledge areas and skills necessary to operate safely and competently as a certificated pilot in the National Airspace System (NAS). This training includes airmanship skills, pilot judgment and decision-making, hazard identification, risk analysis, and good operating practices. (See Risk Management Handbook, FAA-H-8083-2). [Figure 1-7]

An FAA Certificated Flight Instructor (CFI) has to meet broad flying experience requirements, pass rigid knowledge and practical tests, and demonstrate the ability to apply recommended teaching techniques before being certificated. In addition, the flight instructor's certificate must be renewed every 24 months by showing continued success in training pilots or by satisfactorily completing a flight instructor's refresher course or a practical test designed to upgrade aeronautical knowledge, pilot proficiency, and teaching techniques.

A pilot training program is dependent on the quality of the ground and flight instruction the student pilot receives. A good flight instructor has a thorough understanding of the learning process, knowledge of the fundamentals of instruction, and the ability to communicate effectively with the student pilot.

A good flight instructor uses a syllabus and insists on correct techniques and procedures from the beginning of training so that the student will develop proper habit patterns. The syllabus should embody the "building block" method of instruction in which the student progresses from the known to the unknown. The course of instruction should be laid out so that each new maneuver embodies the principles involved in the performance of those previously undertaken. Consequently, through each new subject introduced, the student not only learns a new principle or technique, but broadens his or her application of those previously learned and has his or her deficiencies in the previous maneuvers emphasized and made obvious. [Figure 1-8]

The flying habits of the flight instructor, both during flight instruction and as observed by students when conducting other pilot operations, have a vital effect on safety. Students consider their flight instructor to be a paragon of flying proficiency whose flying habits they, consciously or unconsciously, attempt to imitate. For this reason, a good flight instructor meticulously observes the safety practices taught to the students. Additionally, a good flight instructor carefully observes all regulations and recognized safety practices during all flight operations.

Figure 1-7. *The flight instructor is responsible for teaching and training students to become safe and competent certificated pilots.*

Lesson _____ **Stalls** _____	**Student** _____ **Date** _____

Objective
- To familiarize the student with the stall warnings and handling characteristics of the airplane as it approaches a stall. To develop the student's skill in recognition and recovery from stalls.

Content
- Configuration of airplane for power-on and power-off stalls.
- Observation of airplane attitude, stall warnings, and handling characteristics as it approaches a stall.
- Control of airplane attitude, altitude, and heading.
- Initiation of stall recovery procedures.

Schedule
- Preflight Discussion..:10
- Instructor Demonstrations...:25
- Student Practice ....:45
- Postflight Critique ....:10

Equipment
- Chalkboard or notebook for preflight discussion.

Instructor's actions
- Preflight—discuss lesson objective.
- Inflight—demonstrate elements. Demonstrate power-on and power-off stalls and recovery procedures. Coach student practice.
- Postflight—critique student performance and assign study material.

Student's actions
- Preflight—discuss lesson objective and resolve questions.
- Inflight—review previous maneuvers including slow flight. Perform each new maneuver as directed.
- Postflight—ask pertinent questions.

Completion standards
- Student should demonstrate competency in controlling the airplane at airspeeds approaching a stall. Student should recognize and take prompt corrective action to recover from power-on and power-off stalls.

This is a typical lesson plan for flight training which emphasizes stall recognition and recovery procedures.

Figure 1-8. *Sample lesson plan for stall training and recovery procedures.*

Generally, the student pilot who enrolls in a pilot training program is prepared to commit considerable time, effort, and expense in pursuit of a pilot certificate. The student may tend to judge the effectiveness of the flight instructor and the overall success of the pilot training program solely in terms of being able to pass the requisite FAA-practical test. A good flight instructor is able to communicate to the student that evaluation through practical tests is a mere sampling of pilot ability that is compressed into a short period of time. The flight instructor's role is to train the "total" pilot.

Sources of Flight Training

The major sources of flight training in the United States include FAA-approved pilot schools and training centers, non-certificated (14 CFR part 61) flying schools, and independent flight instructors. FAA-approved schools are those flight schools certified by the FAA as pilot schools under 14 CFR part 141. *[Figure 1-9]*

Application for certification is voluntary, and the school must meet stringent requirements for personnel, equipment, maintenance, and facilities. The school must operate in accordance with an established curriculum that includes a training course outline (TCO) approved by the FAA. The TCO must contain student enrollment prerequisites, detailed description of each lesson including standards and objectives, expected accomplishments and standards for each stage of training, and a description of the checks and tests used to measure a student's accomplishments. FAA-approved pilot school certificates must be renewed every 2 years.

Renewal is contingent upon proof of continued high quality instruction and a minimum level of instructional activity. Training at an FAA-certificated pilot school is structured and because of this structured environment, the graduates of these pilot schools are allowed to meet the certification experience requirements of 14 CFR part 61 with less flight time. Many FAA-certificated pilot schools have DPEs on staff to administer FAA practical tests. Some schools have been granted examining authority by the FAA. A school with examining authority for a particular course(s) has the authority to recommend its graduates for pilot certificates or ratings

UNITED STATES OF AMERICA
DEPARTMENT OF TRANSPORTATION
FEDERAL AVIATION ADMINISTRATION

Air Agency Certificate

Number

(Enter certificate number from
original certification)

This certificate is issued to

(Enter name of school)

whose business address is

(Enter address of main base of operations)

upon finding that its organization complies in all respects with the requirements of the Federal Aviation Regulations relating to the establishment of an Air Agency, and is empowered to operate an approved (Enter the words, Pilot School)

with the following ratings:

(Enter all authorized ratings; after the ratings with both
examining authorities, enter the words, (Knowledge and Flight Tests)

This certificate, unless canceled, suspended, or revoked, shall continue in effect (Enter expiration date of original certificate)

By direction of the Administrator

Date issued: (Enter date of original certification)

(Enter date of amendment)   (Have district office manager sign)

This Certificate is not Transferable, AND ANY MAJOR CHANGE IN THE BASIC FACILITIES, OR IN THE LOCATION THEREOF, SHALL BE IMMEDIATELY REPORTED TO THE APPROPRIATE REGIONAL OFFICE OF THE FEDERAL AVIATION ADMINISTRATION

Any alteration of this certificate is punishable by a fine of not exceeding $1,000, an imprisonment not exceeding 3 years, or both

FAA Form 8000-4 (1-67)   SUPERSEDES FAA FORM 390

Figure 1-9. *FAA Form 8000-4, Air Agency Certificate.*

without further testing by the FAA. A list of FAA-certificated pilot schools and their training courses can be found at http://av-info.faa.gov/pilotschool.asp.

FAA-approved training centers are certificated under 14 CFR part 142. Training centers, like certificated pilot schools, operate in a structured environment with approved courses and curricula and stringent standards for personnel, equipment, facilities, operating procedures, and record keeping. Training centers certificated under 14 CFR part 142, however, specialize in the use of flight simulation (flight simulators and flight training devices) in their training courses.

There are a number of flying schools in the United States that are not certificated by the FAA. These schools operate under the provisions of 14 CFR part 61. Many of these non-certificated flying schools offer excellent training and meet or exceed the standards required of FAA-approved pilot schools. Flight instructors employed by non-certificated flying schools, as well as independent flight instructors, must meet the same basic 14 CFR part 61 flight instructor requirements for certification and renewal as those flight instructors employed by FAA-certificated pilot schools. In the end, any training program is dependent upon the quality of the ground and flight instruction a student pilot receives.

Practical Test Standards (PTS) and Airman Certification Standards (ACS

Practical tests for FAA pilot certificates and associated ratings are administered by FAA inspectors and DPEs in accordance with FAA-developed Practical Test Standards (PTS) and Airman Certification Standards (ACS). *[Figure 1-10]* 14 CFR part 61 specifies the areas of operation in which knowledge and skill must be demonstrated by the applicant. The CFRs provide the flexibility to permit the FAA to publish PTS and ACS containing the areas of operation and specific tasks in which competence must be demonstrated. The FAA requires that all practical tests be conducted in accordance with the appropriate PTS and ACS and the policies set forth in the introduction section of the PTS and ACS.

It must be emphasized that the PTS and ACS are testing documents rather than teaching documents. Although the pilot applicant should be familiar with these books and refer to the standards it contains during training, the PTS and ACS is not intended to be used as a training syllabus. It contains the standards to which maneuvers/procedures on FAA practical tests must be performed and the FAA policies governing the administration of practical tests. An appropriately rated flight instructor is responsible for training a pilot applicant to acceptable standards in all subject matter areas, procedures, and maneuvers included in, and

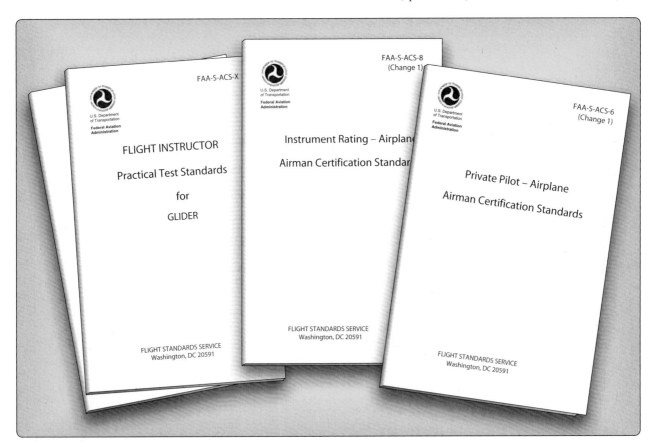

Figure 1-10. *Airman Certification Standards (ACS) developed by the FAA.*

encompassed by, the tasks within each area of operation in the appropriate PTS and ACS. Flight instructors and pilot applicants should always remember that safe, competent piloting requires a commitment to learning, planning, and risk management that goes beyond rote performance of maneuvers. Descriptions of tasks and information on how to perform maneuvers and procedures are contained in reference and teaching documents, such as this handbook. A list of reference documents is contained in the introduction section of each PTS and ACS. It is necessary that the latest version of the PTS and ACS, with all recent changes, be referenced for training. All recent versions and changes to the FAA PTS and ACS may be viewed or downloaded at www.faa.gov.

Safety of Flight Practices

In the interest of safety and good habit pattern formation, there are certain basic flight safety practices and procedures that must be emphasized by the flight instructor, and adhered to by both instructor and student, beginning with the very first dual instruction flight. These include, but are not limited to, collision avoidance procedures including proper scanning techniques and clearing procedures, runway incursion avoidance, stall awareness, positive transfer of controls, and flight deck workload management.

Collision Avoidance

All pilots must be alert to the potential for midair collision and impending loss of separation. The general operating and flight rules in 14 CFR part 91 set forth the concept of "See and Avoid." This concept requires that vigilance shall be maintained at all times by each person operating an aircraft regardless of whether the operation is conducted under IFR or VFR. Pilots should also keep in mind their responsibility for continuously maintaining a vigilant lookout regardless of the type of aircraft being flown and the purpose of the flight. Most midair collision accidents and reported near midair collision incidents occur in good VFR weather conditions and during the hours of daylight. Most of these accident/incidents occur within 5 miles of an airport and/or near navigation aids. *[Figure 1-11]*

The "See and Avoid" concept relies on knowledge of the limitations of the human eye and the use of proper visual scanning techniques to help compensate for these limitations. Pilots should remain constantly alert to all traffic movement within their field of vision, as well as periodically scanning the entire visual field outside of their aircraft to ensure detection of conflicting traffic. Remember that the performance capabilities of many aircraft, in both speed and rates of climb/descent, result in high closure rates limiting the time available for detection, decision, and evasive action. *[Figure 1-12]*

The probability of spotting a potential collision threat increases with the time spent looking outside, but certain techniques

Figure 1-11. *Most midair collision accidents occur in good weather.*

may be used to increase the effectiveness of the scan time. The human eyes tend to focus somewhere, even in a featureless sky. In order to be most effective, the pilot should shift glances and refocus at intervals. Most pilots do this in the process of scanning the instrument panel, but it is also important to focus outside to set up the visual system for effective target acquisition. Pilots should also realize that their eyes may require several seconds to refocus when switching views between items on the instrument panel and distant objects.

Proper scanning requires the constant sharing of attention with other piloting tasks, thus it is easily degraded by such

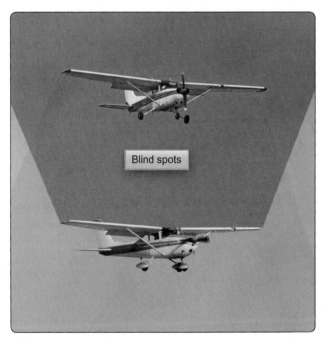

Figure 1-12. *Proper scanning techniques can mitigate midair collisions. Pilots must be aware of potential blind spots and attempt to clear the entire area that they are maneuvering in.*

psychological and physiological conditions, such as fatigue, boredom, illness, anxiety, or preoccupation.

Effective scanning is accomplished with a series of short, regularly-spaced eye movements that bring successive areas of the sky into the central visual field. Each movement should not exceed 10 degrees, and each area should be observed for at least 1 second to enable detection. Although horizontal back-and-forth eye movements seem preferred by most pilots, each pilot should develop a scanning pattern that is most comfortable to them and adhere to it to assure optimum scanning.

Peripheral vision can be most useful in spotting collision threats from other aircraft. Each time a scan is stopped and the eyes are refocused, the peripheral vision takes on more importance because it is through this element that movement is detected. Apparent movement is almost always the first perception of a collision threat and probably the most important because it is the discovery of a threat that triggers the events leading to proper evasive action. It is essential to remember, however, that if another aircraft appears to have no relative motion, it is likely to be on a collision course with you. If the other aircraft shows no lateral or vertical motion, but is increasing in size, take immediate evasive action.

The importance of, and the proper techniques for, visual scanning should be taught to a student pilot at the very beginning of flight training. The competent flight instructor should be familiar with the visual scanning and collision avoidance information contained in AC 90-48, Pilots' Role in Collision Avoidance, and the Aeronautical Information Manual (AIM).

There are many different types of clearing procedures. Most are centered around the use of clearing turns. The essential idea of the clearing turn is to be certain that the next maneuver is not going to proceed into another airplane's flightpath. Some pilot training programs have hard and fast rules, such as requiring two 90° turns in opposite directions before executing any training maneuver. Other types of clearing procedures may be developed by individual flight instructors. Whatever the preferred method, the flight instructor should teach the beginning student an effective clearing procedure and insist on its use. The student pilot should execute the appropriate clearing procedure before all turns and before executing any training maneuver. Proper clearing procedures, combined with proper visual scanning techniques, are the most effective strategy for collision avoidance.

Runway Incursion Avoidance

A runway incursion is any occurrence at an airport involving an aircraft, vehicle, person, or object on the ground that creates a collision hazard or results in a loss of separation with an aircraft taking off, landing, or intending to land. The three major areas contributing to runway incursions are communications, airport knowledge, and flightdeck procedures for maintaining orientation. *[Figure 1-13]*

Taxi operations require constant vigilance by the entire flight crew, not just the pilot taxiing the airplane. During flight training, the instructor should emphasize the importance of vigilance during taxi operations. Both the student pilot and the flight instructor need to be continually aware of the movement and location of other aircraft and ground vehicles on the airport movement area. Many flight training activities are conducted at non-tower controlled airports. The absence of an operating airport control tower creates a need for increased vigilance on the part of pilots operating at those airports. *[Figure 1-14]*

Planning, clear communications, and enhanced situational awareness during airport surface operations reduces the potential for surface incidents. Safe aircraft operations can be accomplished and incidents eliminated if the pilot is properly trained early on and throughout their flying career on standard taxi operating procedures and practices. This requires the development of the formalized teaching of safe operating practices during taxi operations. The flight instructor is the key to this teaching. The flight instructor should instill in the student an awareness of the potential for runway incursion, and should emphasize the runway incursion avoidance procedures. For more detailed information and a list of additional references, refer to Chapter 14 of the Pilot's Handbook of Aeronautical Knowledge.

Stall Awareness

14 CFR part 61 requires that a student pilot receive and log flight training in stalls and stall recoveries prior to solo flight. *[Figure 1-15]* During this training, the flight instructor should emphasize that the direct cause of every stall is an excessive angle of attack (AOA). The student pilot should fully understand that there are several flight maneuvers that may produce an increase in the wing's AOA, but the stall does not occur until the AOA becomes excessive. This critical AOA varies from 16°–20° depending on the airplane design. *[Figure 1-16]*

The flight instructor must emphasize that low speed is not necessary to produce a stall. The wing can be brought to an excessive AOA at any speed. High pitch attitude is not an absolute indication of proximity to a stall. Some airplanes are capable of vertical flight with a corresponding low AOA. Most airplanes are quite capable of stalling at a level or near level pitch attitude.

Figure 1-13. *Three major areas contributing to runway incursions are communications with air traffic control (ATC), airport knowledge, and flight deck procedures.*

The key to stall awareness is the pilot's ability to visualize the wing's AOA in any particular circumstance, and thereby be able to estimate his or her margin of safety above stall. This is a learned skill that must be acquired early in flight training and carried through the pilot's entire flying career.

Figure 1-14. *Sedona Airport is one of the many airports that operate without a control tower.*

The pilot must understand and appreciate factors such as airspeed, pitch attitude, load factor, relative wind, power setting, and aircraft configuration in order to develop a reasonably accurate mental picture of the wing's AOA at any particular time. It is essential to safety of flight that pilots take into consideration this visualization of the wing's AOA prior to entering any flight maneuver. Chapter 3, Basic Flight Maneuvers, discusses stalls in greater detail.

Use of Checklists

Checklists have been the foundation of pilot standardization and flight deck safety for years. *[Figure 1-17]* The checklist is a memory aid and helps to ensure that critical items necessary for the safe operation of aircraft are not overlooked or forgotten. Checklists need not be "do lists." In other words, the proper actions can be accomplished, and then the checklist used to quickly ensure all necessary tasks or actions have been completed. Emphasis on the "check" in checklist. However,

Figure 1-15. *All student pilots must receive and log flight training in stalls and stall recoveries prior to their first solo flight.*

checklists are of no value if the pilot is not committed to using them. Without discipline and dedication to using the appropriate checklists at the appropriate times, the odds are on the side of error. Pilots who fail to take the use of checklists seriously become complacent and begin to rely solely on memory.

The importance of consistent use of checklists cannot be overstated in pilot training. A major objective in primary flight training is to establish habit patterns that will serve pilots well throughout their entire flying career. The flight instructor must promote a positive attitude toward the use of checklists, and the student pilot must realize its importance.

At a minimum, prepared checklists should be used for the following phases of flight. *[Figure 1-18]*

- Preflight Inspection
- Before Engine Start
- Engine Starting
- Before Taxiing
- Before Takeoff
- After Takeoff
- Cruise
- Descent
- Before Landing
- After Landing
- Engine Shutdown and Securing

Tilt with respect to
horizontal plane

Turbulent wake

0°

Separation point

Separation point moves
slightly forward

5°

Maximum lift

Separation point
jumps forward

16° Stall angle

Separated flow region expands
and reduces AFT

20°

Large turbulent wake
(reduced lift end large pressure drag)

Figure 1-16. *Stalls occur when the airfoils angle of attack reaches the critical point which can vary between 16° and 20°.*

Figure 1-17. *Checklists have been the foundation of pilot standardization and flight safety for many years.*

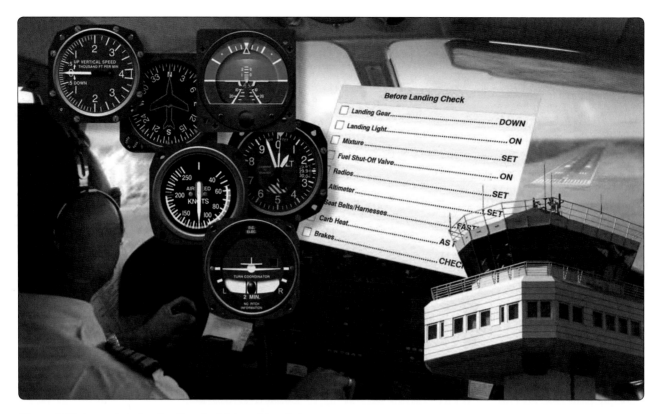

Figure 1-18. *A sample before landing checklist used by pilots.*

Positive Transfer of Controls

During flight training, there must always be a clear understanding between the student and flight instructor of who has control of the aircraft. Prior to any flight, a briefing should be conducted that includes the procedures for the exchange of flight controls. The following three-step process for the exchange of flight controls is highly recommended.

When a flight instructor wishes the student to take control of the aircraft, he or she should say to the student, "You have the flight controls." The student should acknowledge immediately by saying, "I have the flight controls." The flight instructor should then confirm by again saying, "You have the flight controls." Part of the procedure should be a visual check to ensure that the other person actually has the flight controls. When returning the controls to the flight instructor, the student should follow the same procedure the instructor used when giving control to the student. The student should

stay on the controls until the instructor says: "I have the flight controls." There should never be any doubt as to who is flying the airplane at any one time. Numerous accidents have occurred due to a lack of communication or misunderstanding as to who actually had control of the aircraft, particularly between students and flight instructors. Establishing the above procedure during initial training ensures the formation of a very beneficial habit pattern.

Chapter Summary

This chapter discussed some of the concepts and goals of primary and intermediate flight training. It identified and provided an explanation of regulatory requirements and the roles of the various entities involved. It also offered recommended techniques to be practiced and refined to develop the knowledge, proficiency, and safe habits of a competent pilot.

Chapter 2

Ground Operations

Introduction

All pilots must ensure that they place a strong emphasis on ground operations as this is where safe flight begins and ends. At no time should a pilot hastily consider ground operations without proper and effective thoroughness. This phase of flight provides the first opportunity for a pilot to safely assess the various factors of flight operations including the regulatory requirements, an evaluation of the airplane's condition, and the pilot's readiness for their pilot in command (PIC) responsibilities.

Flying an airplane presents many new responsibilities that are not required for other forms of transportation. Focus is often overly placed on the flying portion itself with less emphasis placed on ground operations; it must be stressed that a pilot should allow themselves adequate time to properly prepare for flight and maintain effective situational awareness at all times until the airplane is safely and securely returned to its tie-down or hangar.

This chapter covers the essential elements for the regulatory basis of flight including an airplane's airworthiness requirements, important inspection items when conducting a preflight visual inspection, managing risk and resources, and proper and effective airplane surface movements including the use of the Airplane Flight Manual/Pilot's Operating Handbook (AFM/POH) and airplane checklists.

Preflight Assessment of the Aircraft

The visual preflight assessment is an important step in mitigating airplane flight hazards. The purpose of the preflight assessment is to ensure that the airplane meets regulatory airworthiness standards and is in a safe mechanical condition prior to flight. The term "airworthy" means that the aircraft and its component parts meet the airplane's type design or is in a properly altered configuration and is in a condition for safe operation. The inspection has two parts and involves the pilot inspecting the airplane's airworthiness status and a visual preflight inspection of the airplane following the AFM/POH to determine the required items for inspection. *[Figures 2-1 through 2-3]* The owner/operator is primarily responsible for maintenance, but the pilot is (solely) responsible for determining the airworthiness (and/or safety) of the airplane for flight.

Each airplane has a set of logbooks that include airframe and engine and, in some cases, propeller and appliance logbooks, which are used to record maintenance, alteration, and inspections performed on a specific airframe, engine, propeller, or appliance. It is important that the logbooks

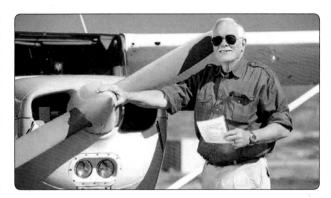

Figure 2-2. *A visual inspection of the aircraft before flight is an important step in mitigating airplane flight hazards.*

be kept accurate and secure but available for inspection. Airplane logbooks are not required, nor is it advisable, to be kept in the airplane. It should be a matter of procedure by the pilot to inspect the airplane logbooks or a summary of the airworthy status prior to flight to ensure that the airplane records of maintenance, alteration, and inspections are current and correct. *[Figure 2-4]* The following is required:

- Annual inspection within the preceding 12-calendar months (Title 14 of the Code of Federal Regulations (14 CFR) part 91, section 91.409(a))

- 100-hour inspection, if the aircraft is operated for hire (14 CFR part 91, section 91.409(b))

- Transponder certification within the preceding 24-calendar months (14 CFR part 91, section 91.413)

- Static system and encoder certification, within the preceding 24-calendar months, required for instrument flight rules (IFR) flight in controlled airspace (14 CFR part 91, section 91.411)

- 30-day VHF omnidirectional range (VOR) equipment check required for IFR flight (14 CFR part 91, section 91.171)

- Emergency locator transmitter (ELT) inspection within the last 12 months (14 CFR part 91, section 91.207(d))

- ELT battery due (14 CFR part 91, section 91.207(c))

- Current status of life limited parts per Type Certificate Data Sheets (TCDS) (14 CFR part 91, section 91.417)

- Status, compliance, logbook entries for airworthiness directives (ADs) (14 CFR part 91, section 91.417(a)(2)(v))

- Federal Aviation Administration (FAA) Form 337, Major Repair or Alteration (14 CFR part 91, section 91.417)

- Inoperative equipment (14 CFR part 91, section 91.213)

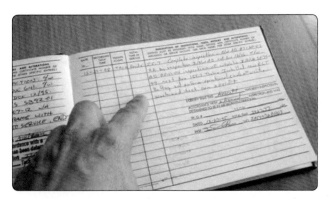

Figure 2-1. *Pilots must view the aircraft's maintenance logbook prior to flight to ensure the aircraft is safe to fly.*

Figure 2-3. *Airplane Flight Manuals (AFM) and the Pilot Operating Handbook (POH) for each individual aircraft explain the required items for inspection.*

A review determines if the required maintenance and inspections have been performed on the airplane. Any discrepancies must be addressed prior to flight. Once the pilot has determined that the airplane's logbooks provide factual assurance that the aircraft meets its airworthy requirements, it is appropriate to visually inspect the airplane. The visual preflight inspection of the airplane should begin while approaching the airplane on the ramp. The pilot should make note of the general appearance of the airplane, looking for discrepancies such as misalignment of the landing gear and airplane structure. The pilot should also take note of any distortions of the wings, fuselage, and tail, as well as skin damage and any staining, dripping, or puddles of fuel or oils.

It must be determined by the pilot that the following documents are, as appropriate, on board, attached, or affixed to the airplane:

- Original Airworthiness Certificate (14 CFR part 91, section 91.203)
- Original Registration Certificate (14 CFR part 91, section 91.203)
- Radio station license for flights outside the United States or airplanes greater than 12,500 pounds (Federal Communications Commission (FCC) rule)
- Operating limitations, which may be in the form of an FAA-approved AFM/POH, placards, instrument markings, or any combination thereof (14 CFR part 91, section 91.9)
- Official weight and balance
- Compass deviation card (14 CFR part 23, section 23.1547)
- External data plate (14 CFR part 45, section 45.11)

Visual Preflight Assessment

The inspection should start with the cabin door. If the door is hard to open or close, does not fit snugly, or the door latches do not engage or disengage smoothly, the surrounding structure, such as the door post, should be inspected for misalignment which could indicate structural damage. The visual preflight inspection should continue to the interior of the cabin or cockpit where carpeting should be inspected to ensure that it is serviceable, dry, and properly affixed; seats belts and shoulder harnesses should be inspected to ensure that they are free from fraying, latch properly, and are securely attached to their mounting fittings; seats should be inspected to ensure that the seats properly latch into the seat rails through the seat lock pins and that seat rail holes are not abnormally worn to an oval shape; *[Figure 2-5]* the windshield and windows should be inspected to ensure that they are clean and free from cracks, and crazing. A dirty, scratched, and/or a severely crazed window can result in near zero visibility due to light refraction at certain angles from the sun.

AFM/POH must be the reference for conducting the visual preflight inspection, and each manufacturer has a specified sequence for conducting the actions. In general, the following items are likely to be included in the AFM/POH preflight inspection:

- Master, alternator, and magneto switches are OFF
- Control column locks are REMOVED
- Landing gear control is DOWN
- Fuel selectors should be checked for proper operation in all positions, including the OFF position. Stiff fuel selectors or where the tank position is not legible or lacking detents are unacceptable.
- Trim wheels, which include elevator and may include rudder and aileron, are set for takeoff position.
- Avionics master OFF
- Circuit breakers checked IN
- Flight instruments must read correctly. Airspeed zero; altimeter when properly set to the current barometric setting should indicate the field elevation within 75 feet for IFR flight; the magnetic compass should indicate the airplane's direction accurately; and the compass correction card should be legible and complete. For conventional wet magnetic compasses, the instrument face must be clear and the instrument case full of fluid. A cloudy instrument face, bubbles in the fluid, or a partially filled case renders the compass unusable. The vertical speed indictor (VSI) should read zero. If the VSI does not show a zero reading, a small screwdriver can be used to zero the instrument.

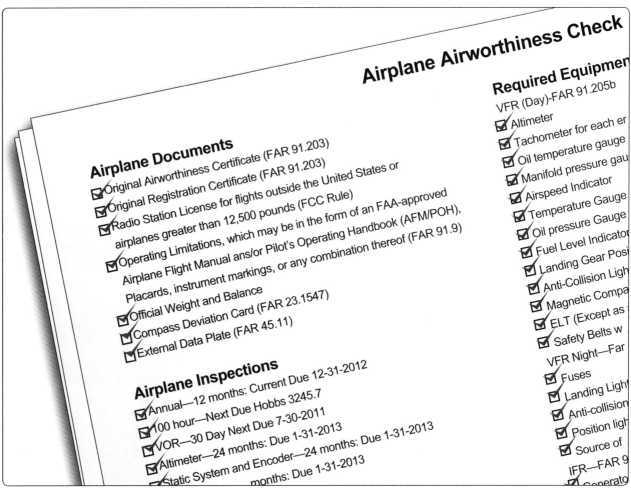

Figure 2-4. *A sample airworthiness checklist used by pilots to inspect an aircraft.*

The VSI is the only flight instrument that a pilot has the prerogative to adjust. All others must be adjusted by an FAA-certificated repairman or mechanic.

- Mechanical air-driven gyro instruments must be inspected for signs of hazing on the instrument face, which may indicate leaks.

Ensure that seats properly latch into the seat rails through the seat lock pins and that seat rail holes are not abnormally worn to an oval shape.

Figure 2-5. *Seats should be inspected to ensure that they are properly latched into the seat rails and checked for damage.*

- If the airplane has retractable gear, landing gear down and locked lights are checked green.

- Check the landing gear switch is DOWN, then turn the master switch to the ON position and fuel qualities must be noted on the fuel quantity gauges and compared to a visual inspection of the tank level. If so equipped, fuel pumps may be placed in the ON position to verify fuel pressure in the proper operating range.

- Other items may include checking that lights for both the interior and exterior airplane positions are operating and any annunciator panel checks.

Advanced avionics aircraft have specific requirements for testing Integrated Flight Deck (IFD) "glass-panel" avionics and supporting systems prior to flight. IFD's are complex electronic systems typically integrating flight control, navigation and communication, weather, terrain, and traffic subsystems with the purpose to enhance a pilot's situational awareness (SA), aeronautical decision-making (ADM), and single-pilot resource management (SRM) capability. Ground-based inspections may include verification that the flight deck reference guide is in the aircraft and assessable, system driven removal of "Xs" over engine indicators, pitot/static and attitude displays, testing of low level alarms, annunciator panels, setting of fuel levels, and verification that the avionics cooling fans, if equipped, are functional. *[Figure 2-6]* The AFM/POH specifies how these preflight inspections are to take place. Since an advanced avionics aircraft preflight checklist may be extensive, pilots should allow extra time for these aircraft to ensure that all items are properly addressed.

Outer Wing Surfaces and Tail Section

Generally, the AFM/POH specifies a sequence for the pilot to inspect the aircraft which may sequence from the cabin entry access opening and then in a counterclockwise direction until the aircraft has been completely inspected. Besides the AFM/POH preflight assessment, the pilot must also develop awareness for potential areas of concern, such as signs of deterioration or distortion of the structure, whether metal or composite, as well as loose or missing rivets or screws.

Besides all items specified in the AFM/POH that must be inspected, the pilot should also develop an awareness for critical areas, such as spar lines, wing, horizontal, and vertical attach points including wing struts and landing gear attachment areas. The airplane skin should be inspected in

Figure 2-7. *Example of rivet heads where black oxide film has formed due to the rivet becoming loose in its hole.*

these areas as load-related stresses are concentrated along spar lines and attach points. Spar lines are lateral rivet lines that extend from one side of the wing to the other, horizontal stabilizer, or vertical stabilizer. Pilots should pay close attention to spar lines looking for distortion, ripples, bubbles, dents, creases, or waves as any structural deformity may be an indication of internal damage or failure. Inspect around rivet heads looking for cracked paint or a black-oxide film that forms when a rivet works free in its hole. *[Figure 2-7]*

Additional areas that should be scrutinized are the leading edges of the wing, horizontal and vertical stabilizer. These areas may be impact damaged by rocks, ice, birds, and or

Figure 2-6. *Ground-based inspections include verification that "Xs" on the instrument display are displayed until the sensor activates.*

hangar rash incidents—dents and dings may render the structure unairworthy. Some leading edge surfaces have aerodynamic devices, such as stall fences, slots, or vortex generators, and deicing equipment, such as weeping wings and boots. If these items exist on the airplane which the pilot intends to fly, knowledge of an acceptable level of proper condition must be gained so that an adequate preflight inspection may take place.

On metal airplanes, wingtips, fairings, and non-structural covers may be fabricated out of thin fiberglass or plastic. These items are frequently affected by cracks radiating from screw holes or concentrated radiuses. Often, if any of these items are cracked, it is practice to "stop-drill" the crack to prevent crack progression. *[Figure 2-8]* Extra care should be exercised to ensure that these devices are in good condition without cracks that may render them unairworthy. Cracks that have continued beyond a stop drilled location or any new adjacent cracks that have formed may lead to in-flight failure.

Inspecting composite airplanes can be more challenging as the airplanes generally have no rivets or screws to aid the pilot in identifying spar lines and wing attach points; however, delamination of spar to skin or other structural problems may be identified by bubbles, fine hair-line cracks, or changes in sound when gently tapping on the structure with a fingertip. Anything out of place should be addressed by discussing the issue with a properly rated aircraft mechanic.

Fuel and Oil

While there are various formulations of aviation gasoline (AVGAS), only three grades are conventional: 80/87, 100LL, and 100/130. 100LL is the most widely available in the United States. AVGAS is dyed with a faint color for grade identification: 80/87 is dyed red; 100LL is dyed blue; and 100/130 is dyed green. All AVGAS grades have a familiar gasoline scent and texture. 100LL with its blue dye

Figure 2-8. *Cracks radiating from screw holes that have been stop-drilled to prevent crack progression.*

is sometimes difficult to identify unless a fuel sample is held up against a white background in reasonable white lighting.

Aircraft piston engines certificated for grade 80/87 run satisfactorily on 100LL if approved as an alternate. The reverse is not true. Fuel of a lower grade should never be substituted for a required higher grade. Detonation will severely damage the engine in a very short period of time. Detonation, as the name suggests, is an explosion of the fuel-air mixture inside the cylinder. During detonation, the fuel/air charge (or pockets within the charge) explodes rather than burning smoothly. Because of this explosion, the charge exerts a much higher force on the piston and cylinder, leading to increased noise, vibration, and cylinder head temperatures. The violence of detonation also causes a reduction in power. Mild detonation may increase engine wear, though some engines can operate with mild detonation regularly. However, severe detonation can cause engine failure in minutes. Because of the noise that it makes, detonation is called "engine knock" or "pinging" in cars.

When approved for the specific airplane to be flown, automobile gasoline is sometimes used as a substitute fuel in certain airplanes. Its use is acceptable only when the particular airplane has been issued a Supplemental Type Certificate (STC) to both the airframe and engine.

Jet fuel is a kerosene-based fuel for turbine engines and a new generation of diesel-powered airplanes. Jet fuel has a stubborn, distinctive, non-gasoline odor and is oily to the touch. Jet fuel is clear or straw colored, although it may appear dyed when mixed with AVGAS. Jet fuel has disastrous consequences when introduced into AVGAS burning reciprocating airplane engines. A reciprocating engine operating on jet fuel may start, run, and power the airplane for a time long enough for the airplane to become airborne only to have the engine fail catastrophically after takeoff.

Jet fuel refueling trucks and dispensing equipment are marked with JET-A placards in white characters on a black background. Because of the dire consequences associated with misfueling, fuel nozzles are specific to the type of fuel. AVGAS fuel filler nozzles are straight with a constant diameter. *[Figure 2-9]* However, jet fuel filler nozzles are flared at the end to prevent insertion into AVGAS fuel tanks. *[Figure 2-10]*

Using the proper, approved grade of fuel is critical for safe, reliable engine operation. Without the proper fuel quantity, grade, and quality, the engine(s) will likely cease to operate. Therefore, it is imperative that the pilot visually verify that the airplane has the correct quantity for the intended flight plus adequate and legal reserves, as well as inspect that the fuel is of the proper grade and that the quality of the fuel is

Figure 2-9. *An AVGAS fuel filler nozzle is straight with a constant diameter.*

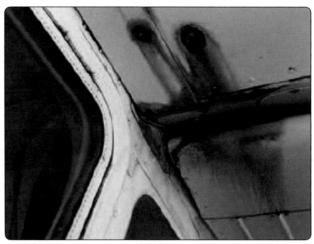

Figure 2-11. *Evidence of fuel leakage can be found along rivet lines.*

acceptable. The pilot should always ensure that the fuel caps have been securely replaced following each fueling.

Many airplanes are very sensitive to its attitude when attempting to fuel for maximum capacity. Nosewheel or main landing gear strut extension, both high as well as low, and the slope of the ramp can significantly alter the attitude of the aircraft and therefore the fuel capacity. Always positively confirm the fuel quantity indicated on the fuel gauges by visually inspecting the level of each tank.

The pilot should be aware that fuel stains anywhere on the wing or any location where a fuel tank is mounted warrants further investigation—no matter how old the stains appear to be. Fuel stains are a sign of probable fuel leakage. On airplanes equipped with wet-wing fuel tanks, evidence of fuel leakage can be found along rivet lines. *[Figure 2-11]*

Checking for water and other sediment contamination is a key preflight item. Water tends to accumulate in fuel tanks from condensation, particularly in partially filled tanks. Because water is heavier than fuel, it tends to collect in the low points of the fuel system. Water can also be introduced into the

Figure 2-10. *A jet fuel filler nozzle is flared at the end to prevent an inadvertent insertion into an AVGAS fuel tank.*

fuel system from deteriorated gas cap seals exposed to rain or from the supplier's storage tanks and delivery vehicles. Sediment contamination can arise from dust and dirt entering the tanks during refueling or from deteriorating rubber fuel tanks or tank sealant. Deteriorating rubber from seals and sealant may show up in the fuel sample as small dark specks.

The best preventive measure is to minimize the opportunity for water to condense in the tanks. If possible, the fuel tanks should be completely filled with the proper grade of fuel after each flight, or at least filled after the last flight of the day. The more fuel that is in the tanks, the less room inside the tank exists for condensation to occur. Keeping fuel tanks filled is also the best way to slow the aging of rubber fuel tanks and tank sealant.

Sufficient fuel should be drained from the fuel strainer quick drain and from each fuel tank sump to check for fuel grade/color, water, dirt, and odor. If water is present, it is usually in bubble or bead-like droplets, different in color (usually clear, sometimes muddy yellow to brown with specks of dirt), in the bottom of the sample jar. In extreme water contamination cases, consider the possibility that the entire fuel sample, particularly if a small sample was taken, is water. If water is found in the first fuel sample, continue sampling until no water and contamination appears. Significant and/or consistent water, sediment or contaminations are grounds for further investigation by qualified maintenance personnel. Each fuel tank sump should be drained during preflight and after refueling. The order of sumping the fuel system is often very important. Check the AFM/POH for specific procedures and order to be followed.

Checking the fuel tank vent is an important part of a preflight assessment. If outside air is unable to enter the tank as fuel is drawn into the engine, the eventual result is fuel starvation and engine failure. During the preflight assessment, the pilot should look for signs of vent damage and blockage. Some airplanes utilize vented fuel caps, fuel vent tubes, or recessed areas under the wings where vents are located. The pilot should use a flashlight to look at the fuel vent to ensure that it is free from damage and clear of obstructions. If there is a rush of air when the fuel tank cap is cracked, there could be a serious problem with the vent system.

Aviation oils are available in various single/multi-grades and mineral/synthetic-based formulations. It is important to always use the approved and recommended oil for the engine. The oil not only acts as a lubricant but also as a medium to transfer heat as a result of engine operation and to suspend dirt, combustion byproducts, and wear particles between oil changes. Therefore, the proper level of oil is required to ensure lubrication, effective heat transfer, and the suspension of various contaminators. The oil level should be checked during each preflight, rechecked with each refueling, and maintained to not have the oil level fall below the minimum required during engine operation.

During the preflight assessment, if the engine is cold, oil levels on the oil dipstick show higher levels than if the engine was warm and recently shutdown after a flight. When removing the oil dipstick, care should be taken to keep the dipstick from coming in contact with dirty or grimy areas. The dipstick should be inspected to verify the oil level. Typically, piston airplane engines have oil reservoirs with capacities between four and eight quarts, with six quarts being common. Besides the level of oil, the oil's color provides an insight as to its operating condition. Oils darken in color as the oil operating hours increase—this is common and expected as the oil traps contaminators; however, oils that rapidly darken in the first few hours of use after an oil change may indicate engine cylinder problems. Piston airplane engines consume a small amount of oil during normal operation. The amount of consumption varies on many factors; however, if consumption increases or suddenly changes, qualified maintenance personnel should investigate.

It is suggested that the critical aspect of fuel and oil not be left to line service personnel without oversight of the pilot responsible for flight. While line personnel are aviation professionals, it is the pilot who is responsible for the safe outcome of their flight. During refueling or when oil is added to an engine, the pilot must monitor and ensure that the correct quantity, quality, and grade of fuel and oil is added and that all fuel and oil caps have been securely replaced.

Landing Gear, Tires, and Brakes
The landing gear, tires, and brakes allow the airplane to maneuver from and return to the ramp, taxiway, and runway environment in a precise and controlled manner. The landing gear, tires, and brakes must be inspected to ensure that the airplane can be positively controlled on the ground. Landing gear on airplanes varies from simple fixed gear to complex retractable gear systems.

Fixed landing gear is a gear system in which the landing gear struts, tires, and brakes are exposed and lend themselves to relatively simple inspection. However, more complex airplanes may have retractable landing gear with multiple tires per landing gear strut, landing gear doors, over-center locks, springs, and electrical squat switches. Regardless of the system, it is imperative that the pilot follow the AFM/POH in inspecting that the landing gear is ready for operation.

On many fixed-gear airplanes, inspection of the landing gear system can be hindered by wheel pants, which are covers used to reduce aerodynamic drag. It is still the pilot's responsibility to inspect the airplane properly. A flashlight helps the pilot in peering into covered areas. On low-wing airplanes, covered or retraceable landing gear presents additional effort required to crouch below the wing to properly inspect the landing gear.

The following provides guidelines for inspecting the landing gear system; however, the AFM/POH must be the pilot's reference for the appropriate procedures.

- The pilot, when approaching the airplane, should look at the landing gear struts and the adjacent ground for leaking hydraulic fluid that may be coming from struts, hydraulic lines from landing gear retraction pumps, or from the braking system. Landing gear should be relatively free from grease, oil, and fluid without any undue amounts. Any amount of leaking fluid is unacceptable. In addition, an overview of the landing gear provides an opportunity to verify landing gear alignment and height consistency.

- All landing gear shock struts should also be checked to ensure that they are properly inflated, clean, and free from hydraulic fluid and damage. All axles, links, collars, over-center locks, push rods, forks, and fasteners should be inspected to ensure that they are free from cracks, corrosion, rust, and determined to be airworthy.

- Tires should be inspected for proper inflation, an acceptable level of remaining tread, and normal wear pattern. Abnormal wear patterns, sidewall cracks, and damage, such as cuts, bulges, imbedded foreign objects, and visible cords, render the tire unairworthy.

- Wheel hubs should be inspected to ensure that they are free from cracks, corrosion, and rust, that all fasteners are secure, and that the air valve stem is straight, capped, and in good condition.

- Brakes and brake systems should be checked to ensure that they are free from rust and corrosion and that all fasteners and safety wires are secure. Brake pads should have a proper amount of material remaining and should be secure. All brake lines should be secure, dry, and free of signs of hydraulic leaks, and devoid of abrasions and deep cracking.

- On tricycle gear airplanes, a shimmy damper is used to damp oscillations of the nose gear and must be inspected to ensure that they are securely attached, are free of hydraulic fluid leaks, and are in overall good condition. Some shimmy dampeners do not use hydraulic fluid and instead use an elastomeric compound as the dampening medium. Nose gear links, collars, steering rods, and forks should be inspected to ensure the security of fasteners, minimal free play between torque links, crack-free components, and for proper servicing and general condition.

- On some conventional gear airplanes, those airplanes with a tailwheel or skid, the main landing gear may have bungee cords to help in absorbing landing loads and shocks. The bungee cords must be inspected for security and condition.

- Where the landing gear transitions into the airplane's structure, the pilot should inspect the attachment points and the airplane skin in the adjacent area—the pilot needs to inspect for wrinkled or other damaged skin, loose bolts, and rivets and verify that the area is free from corrosion.

Engine and Propeller

Properly managing the risks associated with flying requires that the pilot of the airplane identify and mitigate any potential hazards prior to flight to prevent, to the furthest extent possible, a hazard becoming a realized risk. The engine and propeller make up the propulsion system of the airplane—failure of this critical system requires a well-trained and competent pilot to respond with significant time constraints to what is likely to become a major emergency.

The pilot must ensure that the engine, propeller, and associated systems are functioning properly prior to operation. This starts with an overview of the cowling that surrounds the airplane's engines looking for loose, worn, missing, or damaged fasteners, rivets, and latches that secure the cowling around the engine and to the airframe. The pilot should be vigilant as fasteners and rivets can be numerous and surround the cowling requiring a visual inspection from above, the sides, and the bottom to ensure that all areas have been inspected. Like other areas on the airframe, rivets should be closely inspected for looseness by looking for signs of a black oxide film around the rivet head. Pay attention to chipped or flaking paint around rivets and other fasteners as this may be a sign of a lack of security. Any cowling security issues must be referred to a competent and rated airplane maintenance mechanic.

From the cowling, a general inspection of the propeller spinner, if so equipped, should be completed. Not all airplane/propeller combinations have a spinner, so adherence to the AFM/POH checklist is required. Spinners are subjected to great stresses and should be inspected to be free from dents, cracks, corrosion, and in proper alignment. Cracks may not only occur at locations where fasteners are used but also on the rear facing spinner plate. In conditions where ice or snow may have entered the spinner around the propeller openings, the pilot should inspect the area to ensure that the spinner is internally free from ice. The engine/propeller/spinner is balanced around the crankshaft and a small amount of ice or snow can produce damaging vibrations. Cracks, missing fasteners, or dents results in a spinner that is unairworthy.

The propeller should be checked for blade erosion, nicks, cracks, pitting, corrosion, and security. On controllable pitch propellers, the propeller hub should be checked for oil leaks that tend to stream directionally from the propeller hub toward the tip. On airplanes so equipped, the alternator/generator drive belts should be checked for proper tension and signs of wear.

When inspecting inside the cowling, the pilot should look for signs of fuel dye, which may indicate a fuel leak. The pilot should check for oil leaks, deterioration of oil and hydraulic lines, and to make certain that the oil cap, filter, oil cooler, and drain plug are secure. This may be difficult to inspect without the aid of a flashlight, so even during day operations, a flashlight is handy when peering into the cowling. The inside of the cowling should be inspected for oil or fuel stains. The pilot should also check for loose or foreign objects inside the cowling, such as bird nests, shop rags, and/or tools. All visible wires and lines should be checked for security and condition. The exhaust system should be checked for white stains caused by exhaust leaks at the cylinder head or cracks in the exhaust stacks. The heat muffs, which provide cabin heating on some airplanes, should also be checked for general condition and signs of cracks or leaks.

The air filter should be checked to ensure that it is free from substantial dirt or restrictions, such as bugs, birds, or other causes of airflow restrictions. In addition, air filters elements are made from various materials and, in all cases, the element should be free from decomposition and properly serviced.

Risk and Resource Management

Ground operations also include the pilot's assessment of the risk factors that contribute to safety of flight and the pilot's management of the resources, which may be leveraged to maximize the flight's successes. The Risk Management Handbook (FAA-H-8083-2) should be reviewed for a comprehensive discussion of this topic, but presented below are a summary of key points.

Approximately 85 percent of all aviation accidents have been determined by the National Transportation Safety Board (NTSB) to have been caused by "failure of the pilot to..." As such, a reduction of these failures is the fundamental cornerstone to risk and resource management. The risks involved with flying an airplane are very different from those experienced in daily activities, such as driving to work. Managing risks and resources requires a conscious effort that goes beyond the stick and rudder skills required to pilot the airplane.

Risk Management

Risk management is a formalized structured process for identifying and mitigating hazards and assessing the consequences and benefits of the accepted risk. A hazard is a condition, event, object, or circumstance that could lead to or contribute to an unplanned or undesired event, such as an incident or accident. It is a source of potential danger. Some examples of hazards are:

- Marginal weather or environmental conditions

- Lack of pilot qualification, currency, or proficiency for the intended flight

Identifying the Hazard

Hazard identification is the critical first step of the risk management process. If pilots do not recognize and properly identify a hazard and choose to continue, the consequences of the risk involved is not managed or mitigated. In the previous examples, the hazard identification process results in the following assessment:

- Marginal weather or environmental conditions is an identified hazard because it may result in the pilot having a skill level that is not adequate for managing the weather conditions or requiring airplane performance that is unavailable.

- The lack of pilot training is an identified hazard because the pilot does not have experience to either meet the legal requirements or the minimum necessary skills to safely conduct the flight.

Risk

Risk is the future impact of a hazard that is not controlled or eliminated. It can be viewed as future uncertainty created by the hazard.

- If the weather or environmental conditions are not properly assessed, such as in a case where an airplane

may encounter inadvertent instrument conditions, loss of airplane control may result.

- If the pilot's lack of training is not properly assessed, the pilot may be placed in flight regimes that exceed the pilot's stick and rudder capability.

Risk Assessment

Risk assessment determines the degree of risk and whether the degree of risk is worth the outcome of the planned activity. Once the planned activity is started, the pilot must consider whether or not to continue. A pilot must always have viable alternatives available in the event the original flight plan cannot be accomplished. Thus, hazard and risk are the two defining elements of risk management. A hazard can be a real or perceived condition, event, or circumstance that a pilot encounters. Risk assessment is a quantitative value weighted to a task, action, or event. When armed with the predicted risk assessment of an activity, pilots are able to manage and mitigate their risk.

In the example where marginal weather is the identified hazard, it is relatively simple to understand that the risk associated with flight and that the consequences of loss of control in inadvertent meteorological conditions (IMC) are likely to be severe for a pilot without certification, proficiency, competency, and currency in instrument flight. A risk assessment in this example would determine that the risk is unacceptable and as a result, mitigation of the risk is required. Proper risk mitigation would require that flight be cancelled or delayed until weather conditions were not conducive for inadvertent flight into instrument meteorological conditions.

Risk Identification

Identifying hazards and associated risk is key to preventing risk and accidents. If a pilot fails to search for risk, it is likely that he or she will neither see it nor appreciate it for what it represents. Unfortunately, in aviation, pilots seldom have the opportunity to learn from their small errors in judgment because even small mistakes in aviation are often fatal. In order to identify risk, the use of standard procedures is of great assistance. Several procedures are discussed in detail in the Risk Management Handbook (FAA-H-8083-2).

Risk Mitigation

Risk assessment is only part of the equation. After determining the level of risk, the pilot needs to mitigate the risk. For example, the VFR pilot flying from point A to point B (50 miles) in marginal flight conditions has several ways to reduce risk:

- Wait for the weather to improve to good VFR conditions.

- Take a pilot who is more experienced or who is certified as an instrument flight rules (IFR) pilot.

- Delay the flight.

- Cancel the flight.

- Drive.

Resource Management

Crew resource management (CRM) and single-pilot resource management (SRM) is the ability for the crew or pilot to manage all available resources effectively to ensure that the outcome of the flight is successful. In general aviation, SRM is more often than CRM. The focus of SRM is on the single-pilot operation. SRM integrates the following:

- Situational Awareness

- Human Resource Management

- Task Management

- Aeronautical Decision-making (ADM)

Situational Awareness

Situational awareness is the accurate perception of operational and environmental factors that affect the flight. It is a logical analysis based upon the airplane, external support, environment, and the pilot. It is awareness on what is happening in and around the flight.

Human Resource Management

Human Resource Management requires an effective use of all available resources: human, equipment, and information.

Human resources include the essential personnel routinely working with the pilot to ensure safety of flight. These people include, but are not limited to: weather briefers, flight line personnel, maintenance personnel, crew members, pilots, and air traffic personnel. Pilots need to effectively communicate with these people. This is accomplished by using the key components of the communication process: inquiry, advocacy, and assertion. Pilots must recognize the need to seek enough information from these resources to make a valid decision. After the necessary information has been gathered, the pilot's decision must be passed on to those concerned, such as air traffic controllers, crew members, and passengers. The pilot may have to request assistance from others and be assertive to safely resolve some situations.

Equipment in many of today's aircraft includes automated flight and navigation systems. These automatic systems, while providing relief from many routine cabin or cockpit tasks, present a different set of problems for pilots. The automation intended to reduce pilot workload essentially removes the pilot from the process of managing the aircraft, thereby reducing situational awareness and leading to complacency. Information from these systems needs to be continually monitored to ensure proper situational awareness. It is essential that pilots be aware not only of equipment capabilities, but also equipment limitations in order to manage those systems effectively and safely.

Information workloads and automated systems, such as autopilots, need to be properly managed to ensure a safe flight. The pilot who effectively manages his or her workload completes as many of these tasks as early as possible to preclude the possibility of becoming overloaded by last minute changes and communication priorities in the later, more critical stages of the approach. Routine tasks delayed until the last minute can contribute to the pilot becoming overloaded and stressed, resulting in erosion of performance. By planning ahead, a pilot can effectively reduce workload during critical phases of flight.

Task Management

Pilots have a limited capacity for information. Once information flow exceeds the pilot's ability to mentally process the information, any additional information becomes unattended or displaces other tasks and information already being processed. For example, do not become distracted and fixate on an instrument light failure. This unnecessary focus displaces capability and prevents the pilot's ability to appreciate tasks of greater importance.

Aeronautical Decision-Making (ADM)

Flying safely requires the effective integration of three separate sets of skills: stick-and rudder skills needed to control the airplane; skills related to proficient operation of aircraft systems; and ADM skills. The ADM process addresses all aspects of decision-making in the flight deck and identifies the steps involved in good decision-making. While the ADM process does not eliminate errors, it helps the pilot recognize errors and enables the pilot to manage the error to minimize its effects. These steps are:

- Identifying personal attitudes hazardous to safe flight;

- Learning behavior modification techniques;

- Learning how to recognize and cope with stress;

- Developing risk assessment skills;

- Using all resources; and

- Evaluating the effectiveness of one's own personal ADM skills.

Ground Operations

The airport ramp can be a complex environment with airport personnel, passengers, trucks and other vehicles, airplanes, helicopters, and errant animals. The pilot is responsible for the operation of their airplane and must operate safely at all times. Ground operations provide unique hazards, and mitigating those hazards requires proper planning and situational awareness at all times in the ground environment. A fundamental ground operation mitigation tactic is for the pilot to always have reviewed the airport diagram prior to operating and have it readily available at all times. Whether departing to or from the ramp, the pilot must maintain a high level of awareness that requires preparation to maximize safety. This includes being familiar and competent with the following:

- Refueling operations
- Passenger and baggage security and loading
- Ramp and taxi operations
- Standard ramp signals

During refueling operations, it is advisable that the pilot remove all passengers from aircraft during fueling operations and witness the refueling to ensure that the correct fuel and quantity is dispensed into the airplane and that any caps and cowls are properly secured after refueling.

Passengers may have little experience with the open ramp of an airport. The pilot must ensure the safety of their passengers by only allowing them to undertake freedoms for which they have been given direction by the pilot. At no time should passengers be allowed to roam the ramp without an escort to ensure their safety and ramp security. Baggage loading and security should be directly supervised by the pilot. Unsecured baggage or improperly loaded baggage may adversely affect the center of gravity of the airplane.

Ramp traffic may vary from a deserted open space to a complex environment with heavy corporate or military aircraft. Powerful aircraft may produce an environment, from exhaust blast or rotor downwash, which could easily cause a light airplane to become uncontrollable. Mitigating these light airplane hazards is important to starting off on a safe flight.

Some ramps may be staffed by personnel to assist the pilot in managing a safe departure from the ramp to the taxiway. These personnel use standard hand signals and the pilot should be familiar with the meaning of those signals. *[Figure 2-12]*

Engine Starting

Airplane engines vary substantially and specific procedures for engine starting must be accomplished in reference to approved engine start checklist as detailed in the airplane's AFM/POH. However, some generally accepted hazard mitigation practices and procedures are outlined.

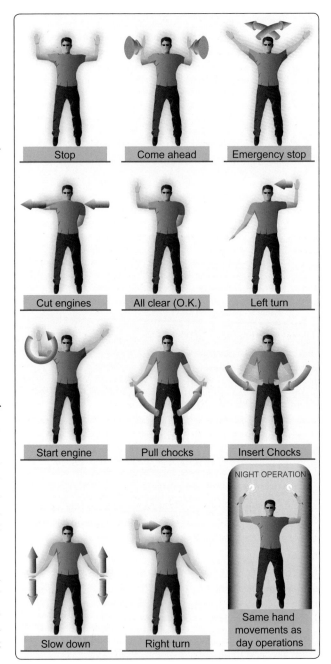

Figure 2-12. *Standard hand signals used to assist pilots in managing a safe departure from the ramp to the taxiway or runway.*

Prior to engine start, the pilot must ensure that the ramp area surrounding the airplane is clear of persons, equipment, and other hazards from coming into contact with the airplane or the propeller. Also, an awareness of what is behind the airplane prior to engine start is standard practice. A propeller or other engine thrust can produce substantial velocities, result in damage to property, and injure those on the ground. The hazard of debris being blown into persons or property must be mitigated by the pilot. At all times before engine start, the anti-collision lights should be turned on. For night operations, the position (navigation) lights should also be on. Finally, just prior to starter engagement, the pilot should always call "CLEAR" out of the side window and wait for a response from anyone who may be nearby before engaging the starter.

When activating the starter, the wheel brakes must be depressed and one hand is to be kept on the throttle to manage the initial starting engine speed. Ensuring that properly operating brakes are engaged prior to starter engagement prevents the airplane from rapidly lunging forward. After engine start, the pilot manipulates the throttle to set the engine revolutions per minute (rpm) at the AFM/POH prescribed setting. In general, 1,000 rpm is recommended following engine start to allow oil pressure to rise and minimize undue engine wear due to insufficient lubrication at high rpm. It is important in low temperatures that an airplane engine use the proper grade of oil for the operating temperature range and engine preheat when temperatures approach and descend below freezing.

The oil pressure must be monitored after engine start to ensure that pressure is increasing toward the AFM/POH specified value. The AFM/POH specifies an oil pressure range for the engine, if the limits are not reached and maintained, serious internal engine damage is likely. In most conditions, oil pressure should rise to at least the lower limit within 30 seconds. To prevent damage, the engine should be shut down immediately if the oil pressure does not rise to the AFM/POH values within the required time.

Engine starters are electric motors designed to produce rapid rotation of the engine crankshaft for starting. These electric motors are not designed for continuous duty and should the engine not start readily, avoid continuous starter operation for periods longer than 30 seconds without a cool down period of at least 30 seconds to 1 minute (some AFM/POH specify times greater than these given). Engine starter motors service life is drastically shortened from high heat through overuse.

Although quite rare, the starter motor may remain electrically and mechanically engaged after engine start. This can be detected by a continuous and very high current draw on the ammeter. Some airplanes also have a starter engaged warning light specifically for this purpose. The engine should be shut down immediately if this occurs.

The pilot should be attentive for sounds, vibrations, smell, or smoke that are not consistent with normal operational experience. Any concerns should lead to a shutdown and further investigation.

Hand Propping

A spinning propeller can be lethal should it strike someone. Historically, when aircraft lacked electrical systems, it was necessary to "hand prop" an aircraft for starting. Hand propping an aircraft is a hazardous procedure when done perfectly. The consequences of not mitigating the hazards associated with hand propping can lead to serious injury, fatalities, and runaway airplanes. All alternatives must be considered prior to hand propping an aircraft and, when a decision is made to do so, the procedure must be carried out only by competent persons who have been trained to accomplish the procedure, understand how to mitigate the hazards, and take all the necessary precautions.

Even though today most airplanes are equipped with electric starters, it is still helpful if a pilot is familiar with the procedures and dangers involved in starting an aircraft engine by turning the propeller by hand; however, a person unfamiliar with the controls must never be allowed to occupy the pilot's seat when hand propping.

It is critical that the procedure never be attempted alone. Hand propping should only be attempted when two properly trained people, both familiar and experienced with the airplane and hand propping techniques, are available to perform the procedure. The first person is responsible for directing the procedure including pulling the propeller blades through. The second person must be seated in the airplane to ensure that the brakes are set, and controls are properly exercised, and to follow direction of the person pulling the propeller.

When hand propping is necessary, the ground surface near the propeller should be stable and free of debris—loose gravel, wet grass, mud, oil, ice, or snow might cause the person pulling the propeller through to slip into the rotating blades as the engine starts. Unless a firm footing is available, relocate the airplane to mitigate this dire consequences hazard.

Both participants should discuss the procedure and agree on voice commands and expected action. To begin the procedure, the fuel system and engine controls (tank selector, primer, pump, throttle, and mixture) are set for a normal start. The ignition/magneto switch should be checked to be sure that it is OFF. Then the descending propeller blade should be rotated so that it assumes a position slightly above the horizontal. The person doing the hand propping should face the descending blade squarely and stand slightly less than one arm's length from the blade. If a stance too far away were assumed, it would be necessary to lean forward in an unbalanced condition to reach the blade, which may cause the person to fall forward into the rotating blades when the engine starts.

The procedure and commands for hand propping are:

- Person out front says, "GAS ON, SWITCH OFF, THROTTLE CLOSED, BRAKES SET."

- Pilot seat occupant, after making sure the fuel is ON, mixture is RICH, magneto switch is OFF, throttle is CLOSED, and brakes are SET, says, "GAS ON, SWITCH OFF, THROTTLE CLOSED, BRAKES SET."
- Person out front, after pulling the propeller through to prime the engine says, "BRAKES AND CONTACT."
- Pilot seat occupant checks the brakes SET and turns the magnetos switch ON, then says, "BRAKES AND CONTACT."

The propeller is swung by forcing the blade downward rapidly, pushing with the palms of both hands. If the blade is gripped tightly with the fingers, the person's body may be drawn into the propeller blades should the engine misfire and rotate momentarily in the opposite direction. As the blade is pushed down, the person should step backward, away from the propeller. If the engine does not start, the propeller should not be repositioned for another attempt until it is verified that the magneto switch is turned OFF.

The words CONTACT (magnetos ON) and SWITCH OFF (magnetos OFF) are used because they are significantly different from each other. Under noisy conditions or high winds, the words CONTACT and SWITCH OFF are less likely to be misunderstood than SWITCH ON and SWITCH OFF.

When removing the wheel chocks or untying the tail after the engine starts, it is critical that everyone involved remember that the propeller is nearly invisible. Serious injuries and fatalities have occurred when people who have just started an engine walk or reach into the propeller arc to remove the chocks, reach the cabin, or in an attempt to reach the tail of the airplane. Before the wheel chocks are removed, the throttle should be set to idle and the chocks approached only from the rear of the propeller. One should never approach the wheel chocks from the front or the side.

The procedures for hand propping should always be in accordance with the AFM/POH and only accomplished if no alternatives are available, and then only by persons who are competent with hand propping procedures. The consequences of the hazards associated with hand propping are serious to fatal.

Taxiing

Taxiing is the controlled movement of the airplane under its own power while on the surface. Since an airplane is moved under its own power between a parking area and the runway, the pilot must thoroughly understand and be proficient in taxi procedures.

An essential requirement in conducting safe taxi operations is where the pilot maintains situational awareness of the ramp, parking areas, taxiways, runway environment, and the persons, equipment and aircraft at all times. Without such awareness, safety may be compromised. Depending on the airport, parking, ramp, and taxiways may or may not be controlled. As such, it is important that the pilot completely understand the environment in which they are operating. At small, rural airports these areas may be desolate with few aircraft which limits the potential hazards; however, as the complexity of the airport increases so does the potential for hazards. Regardless of the complexity, some generally accepted procedures are appropriate.

- The pilot should make themselves familiar with the parking, ramp, and taxi environment. This can be done by having an airport diagram, if available, out and in view at all times. *[Figure 2-13]*
- The pilot must be vigilant of the entire area around the airplane to ensure that the airplane clears all obstructions. If, at any time, there is doubt about a safe clearance from an object, the pilot should stop the airplane and check the clearance. It may be necessary to have the airplane towed or physically moved by a ground crew.
- When taxiing, the pilot's eyes should be looking outside the airplane scanning from side to side while looking both near and far to assess routing and potential conflicts.
- A safe taxiing speed must be maintained. The primary requirements for safe taxiing are positive control, the ability to recognize any potential hazards in time to avoid them, and the ability to stop or turn where and when desired, without undue reliance on the brakes. Pilots should proceed at a cautious speed on congested or busy ramps. Normally, the speed should be at the rate where movement of the airplane is dependent on the throttle. That is, slow enough so when the throttle is closed, the airplane can be stopped promptly.
- The pilot should accurately place the aircraft centered on the taxiway at all times. Some taxiways have above ground taxi lights and signage that could impact the airplane or propellers if the pilot does not exercise accurate control. When yellow taxiway centerline stripes are marked, this is more easily accomplished by the pilot visually placing the centerline stripe so it is under the center of the airplane fuselage.
- When taxiing, the pilot must slow down before attempting a turn. Sharp high-speed turns place undesirable side loads on the landing gear and may

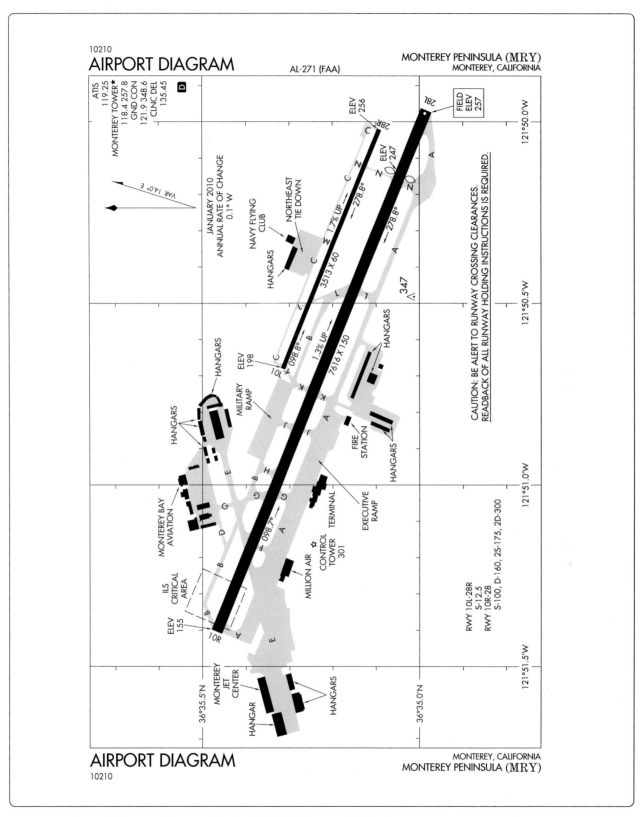

Figure 2-13. *Airport Diagram of Monterey Peninsula (MRY), Monterey, California.*

result in tire damage or an uncontrollable swerve or a ground loop. Swerves are most likely to occur when turning from a downwind heading toward an upwind heading. In moderate to high-wind conditions, the airplane may weathervane increasing the swerving tendency.

Steering is accomplished with rudder pedals and brakes. To turn the airplane on the ground, the pilot should apply the rudder in the desired direction of turn and use the appropriate power or brake to control the taxi speed. The rudder pedal should be held in the direction of the turn until just short of the point where the turn is to be stopped. Rudder pressure is then released or opposite pressure is applied as needed.

More engine power may be required to start the airplane moving forward, or to start a turn, than is required to keep it moving in any given direction. When using additional power, the throttle should immediately be retarded once the airplane begins moving to prevent excessive acceleration.

The brakes should be tested for proper operation as soon as the airplane is put in motion. Applying power to start the airplane moving forward slowly, then retarding the throttle and simultaneously applying just enough pressure to one side, then the other to confirm proper function and reaction of both brakes. This is best if the airplane has individual left/right brakes to stop the airplane. If braking performance is unsatisfactory, the engine should be shut down immediately.

When taxiing at appropriate speeds in no-wind conditions, the aileron and elevator control surfaces have little or no effect on directional control of the airplane. These controls should not be considered steering devices and should be held in a neutral position. *[Figure 2-14]*

The presence of moderate to strong headwinds and/or a strong propeller slipstream makes the use of the elevator necessary to maintain control of the pitch attitude while taxiing. This becomes apparent when considering the lifting action that may be created on the horizontal tail surfaces by either of those two factors. The elevator control in nosewheel-type airplanes should be held in the neutral position, while in tailwheel-type airplanes, it should be held in the full aft position to hold the tail down.

Downwind taxiing usually requires less engine power after the initial ground roll is begun, since the wind is pushing the airplane forward. To avoid overheating the brakes and controlling the airplane's speed when taxiing downwind, the pilot must keep engine power to a minimum. Rather than continuously riding the brakes to control speed, it is appropriate to apply brakes only occasionally. Other than

Figure 2-14. *Control positions of the nosewheel airplane.*

sharp turns at low speed, the throttle should always be at idle before the brakes are applied. It is a common error to taxi with a power setting that requires controlling taxi speed with the brakes.

When taxiing with a quartering headwind, the wing on the upwind side (the side that the wind is coming from) tends to be lifted by the wind unless the aileron control is held in that direction (upwind aileron UP). Moving the aileron into the UP position reduces the effect of the wind striking that wing, thus reducing the lifting action. This control movement also causes the downwind aileron to be placed in the DOWN position, thus a small amount of lift and drag on the downwind wing, further reducing the tendency of the upwind wing to rise.

When taxiing with a quartering tailwind, the elevator should be held in the DOWN position, and the upwind aileron, DOWN. Since the wind is striking the airplane from behind, these control positions reduce the tendency of the wind to get under the tail and the wing and to nose the airplane over. The application of these crosswind taxi corrections helps to minimize the weathervaning tendency and ultimately results in making the airplane easier to steer.

Normally, all turns should be started using the rudder pedal to steer the nosewheel. To tighten the turn after full pedal deflection is reached, the brake may be applied as needed. When stopping the airplane, it is advisable to always stop with the nosewheel straight ahead to relieve any side load on the nosewheel and to make it easier to start moving ahead.

During crosswind taxiing, even the nosewheel-type airplane has some tendency to weathervane. However, the weathervaning tendency is less than in tailwheel-type airplanes because the main wheels are located behind the airplane's center of gravity, and the nosewheel's ground friction helps to resist the tendency. The nosewheel linkage from the rudder pedals provides adequate steering control for safe and efficient ground handling, and normally, only rudder pressure is necessary to correct for a crosswind.

Taxiing checklists are sometimes specified by the AFM/POH, and the pilot must accomplish any items that are required. If there are no specific checklist items, taxiing still provides an opportunity to verify the operation and cross-check of the flight instruments. In general, the flight instruments should indicate properly with the airspeed at or near zero (depending on taxi speed, wind speed and direction, and lower limit sensitivity); the attitude indictor should indicate pitch and roll level (depending on airplane attitude) with no flags; the altimeter should indicate the proper elevation within prescribed limits; the turn indictor should show the correct direction of turn with the ball movement toward the outside of the turn with no flags; the directional gyro should be set and crossed checked to the magnetic compass and verified accurate to the direction of taxi; and the vertical speed indictor (VSI) should read zero. These checks can be accomplished on conventional mechanical instrumented aircraft or glass cockpits.

Before-Takeoff Check

The before-takeoff check is the systematic AFM/POH procedure for checking the engine, controls, systems, instruments, and avionics prior to flight. Normally, the before-takeoff checklist is performed after taxiing to a run-up position near the takeoff end of the runway. Many engines require that the oil temperature reach a minimum value as stated in the AFM/POH before takeoff power is applied. Taxiing to the run-up position usually allows sufficient time for the engine to warm up to at least minimum operating temperatures; however, the pilot verifies that temperatures are in their proper range prior to the application of high power.

A suitable location for run-up should be firm (a smooth, paved or turf surface if possible) and free of debris. Otherwise, the propeller may pick up pebbles, dirt, mud, sand, or other loose objects and hurl them backwards. This damages the propeller and may damage the tail of the airplane. Small chips in the leading edge of the propeller form stress risers or high stress concentrations. These are highly undesirable and may lead to cracks and possible propeller blade failure. The airplane should also be positioned clear of other aircraft and the taxiway. There should not be anything behind the airplane that might be damaged by the propeller airflow blasting rearward.

Before beginning the before-takeoff check, after the airplane is properly positioned for the run-up, it should be allowed to roll forward slightly to ensure that the nosewheel or tailwheel is in alignment with the longitudinal axis of the airplane.

While performing the before-takeoff checklist in accordance with the airplane's AFM/POH, the pilot must divide their attention between the inside and outside of the airplane. If the parking brake slips, or if application of the toe brakes is inadequate for the amount of power applied, the airplane could rapidly move forward and go unnoticed if pilot attention is fixed only inside the airplane. A good operational practice is to split attention from one item inside to a look outside.

Air-cooled engines generally are tightly cowled and equipped with baffles that direct the flow of air to the engine in sufficient volumes for cooling while in flight; however, on the ground, much less air is forced through the cowling and around the baffling. Prolonged ground operations may cause cylinder overheating long before there is an indication of rising oil temperature. To minimize overheating during engine run-up, it is recommended that the airplane be headed as nearly as possible into the wind and, if equipped, engine instruments that indicate cylinder head temperatures should be monitored. Cowl flaps, if available, should be set according to the AFM/POH.

Each airplane has different features and equipment and the before-takeoff checklist provided in airplane's AFM/POH must be used to perform the run-up. Many critical systems are checked and set during the before-takeoff checklist. Most airplanes have at least the following systems checked and set:

- Fuel System—set per the AFM/POH and verified ON and the proper and correct fuel tanks selected.

- Trim—set for takeoff position which includes the elevator and may also include rudder and aileron trim.

- Flight Controls—checked throughout their entire operating range. This includes full aileron, elevator, and rudder deflection in all directions. Often, pilots do not exercise a full range of movement of the flight controls, which is not acceptable.

- Engine Operation—checked to ensure that temperatures and pressures are in their normal ranges; magneto or Full Authority Digital Engine Control (FADEC) operation on single or dual ignition are acceptable and within limits; and, if equipped, carburetor heat is functioning. If the airplane is equipped with a constant speed or feathering propeller, that its operation is acceptable; and at minimum idle, the engine rpm continues to run smoothly.

- Electrical System—verified to ensure voltages are within operating range and that the system shows the battery system charging.

- Vacuum System—must show an acceptable level of vacuum, which is typically between 4.8 and 5.2 inches of mercury ("Hg) at 2,000 rpm. Refer to the AFM/POH for the manufacturer's values. It is important to ensure that mechanical gyroscopic instruments have adequate time to spool up to acceptable rpm in order for them to indicate properly. A hasty and quick taxi and run-up does not allow mechanical gyroscopic instruments to indicate properly and a departure into instrument meteorological conditions (IMC) is unadvisable.

- Flight Instruments—rechecked and set for the departure. Verify that the directional gyro and the magnetic compass are in agreement. If the directional gyro has a heading bug, it may be set to the runway heading that is in use or as assigned by air traffic control (ATC).

- Avionics—set with the appropriate frequencies, initial navigation sources and courses, autopilot preselects, transponder codes, and other settings and configurations based on the airplane's equipment and flight requirements.

- Takeoff Briefing—made out loud by the pilot even when no other person is there to listen. A sample takeoff briefing may be the following:

  "This will be normal takeoff (use normal, short, or soft as appropriate) from runway (use runway assigned), wind is from the (direction and speed), rotation speed is (use the specified or calculated manufacturer's takeoff or rotation speed (V_R)), an initial turn to (use planned heading) and climb to (use initial altitude in feet). The takeoff will be rejected for engine failure below V_R, applying appropriate braking, stopping ahead. Engine failure after V_R and with runway remaining, I will lower pitch to best glide speed, land, and apply appropriate braking, stopping straight ahead. Engine failure after V_R and with no runway remaining, I will lower pitch to best glide speed, no turns will be made prior to (insert appropriate altitude), land in the most suitable area, and apply appropriate braking, avoiding hazards on the ground as much possible. If time permits, fuel, ignition, and electrical systems will be switched off."

Takeoff Checks:

Runway numbers on paved runways agree with magnetic compass and heading indicators before beginning takeoff roll. The last check on engines as power is brought to full takeoff power includes:

- Is power correct?
- RPM normal?
- Engine smooth?

- Engine instruments normal and in green ranges?
- Doors latched and windows closed as required?
- Controls held so rudder is used to keep airplane parallel to centerline and ailerons are used to keep airplane on centerline?

After-Landing

During the after-landing roll, while maintaining airplane track over runway centerline with ailerons and heading down runway with rudder pedals, the airplane should be gradually slowed to normal taxi speed with normal brake pressure before turning off of the landing runway. Any significant degree of turn at faster speeds could result in subsequent damage to the landing gear, tires, brakes, or the airplane structure.

To give full attention to controlling the airplane during the landing roll, the after-landing checklist should be performed only after the airplane is brought to a complete stop beyond the runway holding position markings. There have been many cases where a pilot has mistakenly manipulated the wrong handle and retracted the landing gear, instead of the flaps, due to improper division of attention while the airplane was moving. However, this procedure may be modified if the manufacturer recommends that specific after-landing items be accomplished during landing rollout. For example, when performing a short-field landing, the manufacturer may recommend retracting the flaps on rollout to improve braking. In this situation, the pilot should make a positive identification of the flap control handle before retracting the flaps.

Clear of Runway and Stopped

Because of different configurations and equipment in various airplanes, the after-landing checklist within the AFM/POH must be used. Some of the items may include:

- Power—set to the AFM/POH values such as throttle 1,000 rpm, propeller full forward, mixture leaned.
- Fuel—may require switching tanks and fuel pumps switched off.
- Flaps—set to the retracted position.
- Cowl flaps—may be opened or closed depending on temperature conditions.
- Trim—reset to neutral or takeoff position.
- Lights—may be switched off if not needed, such as strobe lights.
- Avionics—may be switched off or to standby, such as the transponder and frequencies changed to contact ground control or Common Traffic Advisory Frequency (CTAF), as required.

- Install chocks and release parking brake in accordance with AFM/POH.

Parking

Unless parking in a designated, supervised area, the pilot should select a location and heading that prevents propeller or jet blast of other airplanes from striking the airplane unnecessarily. Whenever possible, the airplane should be parked headed into the existing or forecast wind. Often airports have airplane tie downs located on ramp areas which may or may not be aligned with the wind or provide a significant choice in parking location. After stopping in the desired direction, the airplane should be allowed to roll straight ahead enough to straighten the nosewheel or tailwheel.

Engine Shutdown

The pilot should always use the procedures in the airplane's AFM/POH shutdown checklist for shutting down the engine and securing the airplane. Important items may include:

- Parking Brake—set to ON.
- Throttle—set to IDLE or 1,000 rpm. If turbocharged, observe the manufacturer's spool down procedure.
- Magneto Switch Test—turn momentarily OFF then quickly ON again at idle rpm to check for proper operation of switch in the OFF position.
- Propeller—set to FULL INCREASE, if equipped.
- Avionics—turn OFF.
- Alternator—turn OFF.
- Mixture—set to IDLE CUTOFF.
- Magneto Switch—turn ignition switch to OFF when engine stops.
- Master Switch—turn to OFF.
- Secure—install control locks and anti-theft security locks.

Post-Flight

A flight is not complete until the engine is shut down and the airplane is secured. A pilot should consider this an essential part of any flight.

Securing and Servicing

After engine shutdown and deplaning passengers, the pilot should accomplish a post-flight inspection. This includes a walk around to inspect the general condition of the aircraft. Inspect near and around the cowling for signs of oil or fuel streaks and around the oil breather for excessive oil discharge. Inspect under wings and other fuel tank locations for fuel stains. Inspect landing gear and tires for damage and brakes for any leaking hydraulic fluid. Inspect cowling inlets for obstructions.

Oil levels should be checked and quantities brought to AFM/POH levels. Fuel should be added based on the immediate use of the airplane. If the airplane is going to be inactive, it is a good operating practice to fill the fuel tanks to prevent water condensation from forming inside the tank. If another flight is planned, the fuel tanks should be filled based on the flight planning requirements for that flight.

The aircraft should be hangared or tied down, flight controls secured, and security locks in place. The type of tie downs may vary significantly from chains to well-worn ropes. Chains are not flexible and as such should not be made taught as to allow the airplane some movement and prevent airframe structural damage. Tie down ropes are flexible and may be reasonably cinched to the airplane's tie down rings. Consider utilizing pitot tube covers, cowling inlet covers, rudder gust locks, window sunscreens, and propeller security locks to further enhance the safety and security of the airplane.

Hangaring is not without hazards to the airplane. The pilot should ensure that enough space is allocated to the airplane so it is free from any impact to the hangar, another aircraft, or vehicle. The airplane should be inspected after hangaring to ensure that no damage was imparted on the airplane.

Chapter Summary

In this chapter emphasis was placed on determining the airworthiness of the airplane, preflight visual inspection, managing risk and pilot-available resources, safe surface-based operations, and the adherence to and proper use of the AFM/POH and checklists. To maximize the safety of flight operations, a pilot must recognize that flight safety begins by properly preparing for flight and by managing the airplane, environment, resources, and themselves until the airplane is returned to its tie-down or hangar at the termination of flight. This is accomplished by the pilot ensuring that the airplane is in a safe condition for flight and it meets all the regulatory requirements of 14 CFR part 91 by an effective and continuous assessment of the risks and utilization of resources, and by the pilot honestly evaluating and determining their preparedness and continuation for acting as PIC.

Chapter 3
Basic Flight Maneuvers

Introduction

Airplanes operate in an environment that is unlike an automobile. Drivers tend to drive with a fairly narrow field of view and focus primarily on forward motion. Beginning pilots tend to practice the same. Flight instructors face the challenge of teaching beginning pilots about attitude awareness, which requires understanding the motions of flight. An airplane rotates in bank, pitch, and yaw while also moving horizontally, vertically, and laterally. The four fundamentals (straight-and-level flight, turns, climbs, and descents) are the principle maneuvers that control the airplane through the six motions of flight.

The Four Fundamentals

To master any subject, one must first master the fundamentals. An attempt to move on to advanced maneuvers prior to mastering the four fundamentals hinders the learning process. To be a competent pilot first requires that the pilot is skilled in the basics of fundamental airmanship. This requires mastery of the four basic flight maneuvers upon which all flying tasks are based: straight-and-level flight, turns, climbs, and descents.

Consider the following: a takeoff is a combination of straight-and-level and a climb, turning on course to the first navigation fix after departure is a climb and a turn, and the landing at the destination is a combination of airplane ground handling, acceleration, pitch and a climb.

The flight instructor must impart competent knowledge of these basic flight maneuvers so that the beginning pilot is able to combine them at a performance level that at least meets the Federal Aviation Administration (FAA) Practical Test Standards (PTS) or Airman Certification Standards (ACS), as appropriate. The importance of this phase of flight training cannot be overstated. As the beginning pilot progresses to more complex flight maneuvers, any deficiencies in the mastery of the four fundamentals are likely to become barriers to effective and efficient learning. Many beginning pilot difficulties in advanced maneuvers are likely caused by a lack of understanding, training, or practice in the four fundamentals.

Effect and Use of the Flight Controls

The airplane flies in an environment that allows it to travel up and down as well as left and right. That up or down can be relative to the flight conditions. If the airplane is right side up relative to the horizon, forward control stick or wheel (elevator control) movement will result in a loss of altitude. If the same airplane is upside down relative to the horizon that same forward control movement will result in a gain of altitude. In any regard, that forward movement of the elevator control will always move the airplane in the same direction relative to the pilot's perspective. Therefore, the airplane controls always function the same relative to the pilot. Depending on the airplane's orientation to the Earth, the same control actions may result in different movements of the airplane. *[Figure 3-1]* The pilot is always considered the referenced center of effect as the flight controls are used. *[Figure 3-2]* The following is always true, regardless of the airplane's attitude in relation to the Earth's horizon.

With the pilot's hand:

- When pulling the elevator pitch control toward the pilot, which is an aft movement of the aileron and elevator controls, control stick, or side stick controller (referred to as adding back pressure), the airplane's nose will rotate backwards relative to the pilot around the pitch (lateral) axis of the airplane. Think of this movement from the pilot's feet to the pilot's head.

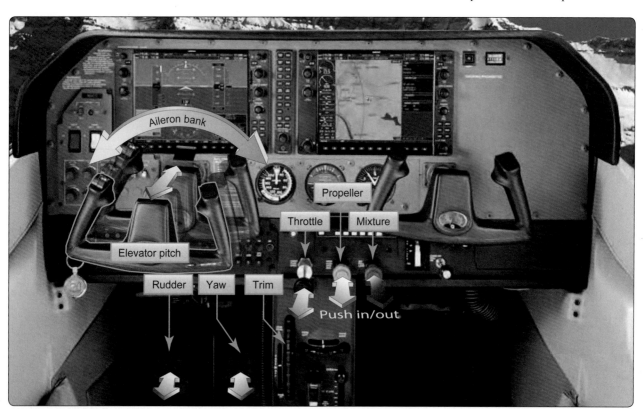

Figure 3-1. *Basic flight controls and instrument panel.*

Primary control surface	Airplane movement	Axes of rotation
Aileron	Roll	Longitudinal
Elevator/stabilator	Pitch	Lateral
Rudder	Yaw	Vertical

Figure 3-2. *The pilot is always considered the referenced center of effect as the flight controls are used.*

- When pushing the elevator pitch control toward the instrument panel, which is the forward movement of the aileron and elevator controls, control stick, or side stick controller (referred to as increasing forward pressure), the airplane rotates the nose forward relative to the pilot around the pitch axis of the airplane. Think of this movement from the pilot's head to the pilot's feet.

- When right pressure is applied to the aileron control, which is a clockwise rotation of aileron and elevator controls or the right deflection of the control stick or side stick controller, the airplane's right wing banks (rolls) lower in relation to the pilot. Think of this movement from the pilot's head to the pilot's right hip.

- When left pressure is applied to the aileron control, which is a counterclockwise rotation of aileron and elevator controls or the left deflection of the control stick or side stick controller, the airplane's left wing banks (rolls) lower in relation to the pilot. Think of this movement from the pilot's head to the pilot's left hip.

With the pilot's feet:

- When forward pressure is applied to the right rudder pedal, the airplane's nose moves (yaws) to the right in relation to the pilot. Think of this movement from the pilot's left shoulder to the pilot's right shoulder.

- When forward pressure is applied to the left rudder pedal, the airplane's nose moves (yaws) to the left in relation to the pilot. Think of this movement from the pilot's right shoulder to the pilot's left shoulder.

While in flight, the flight controls have a resistance to a pilot's movement due to the airflow over the airplane's control surfaces, and the control surfaces remain in a fixed position as long as all forces acting upon them remain balanced. The amount of force that the passing airflow exerts on a control surface is governed by the airspeed and the degree that the surface is moved out of its streamlined position. This resistance increases as airspeed increases and decreases as airspeed decreases. While the airflow over the control surfaces changes during various flight maneuvers, it is not the amount of control surface movement that is important. What is important, is that the pilot maneuvers the airplane by applying sufficient flight control pressures to obtain the desired result.

The pitch and roll flight controls (aileron and elevator controls, stick, or side-stick control) should be held lightly with the fingers and not grabbed or squeezed by the hand. When flight control pressure is applied to change a control surface position, pressure should only be exerted on the aileron and elevator controls with the fingers. This is an important concept and habit to learn which benefits the pilot as they progress to greater challenges such as instrument flying. A common error with beginning pilots is that they grab the aileron and elevator controls with a closed palm with such force that the sensitive feeling is lost. This must be avoided as it prevents the development of "feel," which is an important aspect of airplane control.

The pilot's feet should rest comfortably against the rudder pedals. Both heels should support the weight of the feet on the cockpit floor with the ball of each foot touching the individual rudder pedals. The legs and feet should be relaxed. When using the rudder pedals, pressure should be applied smoothly and evenly by pressing with the ball of one foot. Since the rudder pedals are interconnected through springs or a direct mechanical linkage and act in opposite directions, when pressure is applied to one rudder pedal, foot pressure on the opposite rudder pedal must be relaxed proportionately. Remember, the ball of each foot must rest comfortably on the rudder pedals so that even slight pressure changes can be felt.

In summary, during flight, it is pressure the pilot exerts on the aileron and elevator controls and rudder pedals that causes the airplane to move about the roll (longitudinal), pitch (lateral), and yaw (vertical) axes. When a control surface is moved out of its streamlined position (even slightly), the air flowing across the surface exerts a force against that surface and it tries to return it to its streamlined position. It is this force that the pilot feels as resistance on the aileron and elevator controls and the rudder pedals.

Feel of the Airplane

The ability to sense a flight condition, such as straight-and-level flight or a dive, without relying on cockpit instrumentation is often called "feeling the airplane." Examples of this "feel" may be sounds of the airflow across the airframe, vibrations felt through the controls, engine and propeller sounds and vibrations at various flight attitudes, and the sensations felt by the pilot through physical accelerations.

Humans sense "feel" through kinesthesis (the ability to sense movement through the body) and proprioception (unconscious perception of movement and spatial orientation). These stimuli are detected by nerves and by the semicircular canals of the inner ear. When properly developed, kinesthesis can provide the pilot with critical information about changes in the airplane's direction and speed of motion; however, there are limits in kinesthetic sense and when relied upon solely without visual information, as when flying in instrument meteorological conditions (IMC), ultimately leads to disorientation and loss of aircraft control.

Developing this "feel" takes time and exposure in a particular airplane and only comes with dedicated practice at the various flight conditions so that a pilot's senses are trained by the sounds, vibrations, and forces produced by the airplane. The following are some important examples:

- Rushing air past a cockpit creates a distinctive noise pattern and as the level of sound increases, it likely indicates that the airplane's airspeed is increasing and that the pitch attitude is decreasing. As the noise decreases, the airplane's pitch attitude is likely increasing and its airspeed decreasing.

- The sound of the engine in cruise flight is different from that in a climb and different again when in a dive. In fixed-pitch propeller airplanes, when the airplane's pitch attitude increases, the engine sound decreases and as pitch attitude decreases, the engine noise increases.

- In a banked turn, the pilot is forced downward into the seat due to the resultant load factor. The increased G force of a turn feels the same as the pull up from a dive, and the decreased G force from leveling out feels the same as lowering the nose out of a climb.

Sources of actual "feel" are very important to the pilot. This actual feel is the result of acceleration, which is simply how fast velocity is changing. Acceleration describes the rate of change in both the magnitude and the direction of velocity. These accelerations impart forces on the airplane and its occupants during flight. The pilot can sense these forces through pressures into or out of the seat; or shift the pilot from side to side in their seat as the airplane slips or skids. These forces need not be strong, only perceptible by the pilot to be useful. An accomplished pilot who has excellent "feel" for the airplane is able to detect even the smallest accelerations.

A flight instructor should direct the beginner pilot to be aware of these senses and teach an awareness of their meaning and their relationship to the various conditions of flight. To do this effectively, the flight instructor must fully understand the difference between perceiving and reacting to sound, vibrations, and forces versus merely noticing them. A pilot who develops a "feel" for the airplane early in flight training is likely to have less difficulty advancing in their flight training.

Attitude Flying

An airplane's attitude is determined by the angular difference between a specific airplane's axis and the natural horizon. A false horizon can occur when the natural horizon is obscured or not readily apparent. This is an important concept because it requires the pilot to develop a pictorial sense of this natural horizon. Pitch attitude is the angle formed between the airplane's longitudinal axis, which extends from the nose to the tail of the airplane, and the natural horizon. Bank attitude is the angle formed by the airplane's lateral axis, which extends from wingtip to wingtip, and the natural horizon. *[Figures 3-3A and 3-3B]* Angular difference about

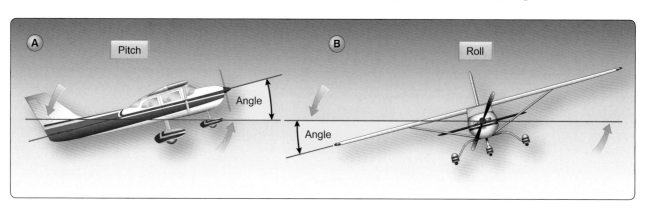

Figure 3-3. *(A) Pitch attitude is the angle formed between the airplane's longitudinal axis. (B) Bank attitude is the angle formed by the airplane's lateral axis.*

the airplane's vertical axis (yaw) is an attitude relative to the airplane's direction of flight but not relative to the natural horizon.

Controlling an airplane requires one of two methods to determine the airplane's attitude in reference to the horizon. When flying "visually" in visual meteorological conditions (VMC), a pilot uses their eyes and visually references the airplane's wings and cowling to establish the airplane's attitude to the natural horizon (a visible horizon). If no visible horizon can be seen due to whiteouts, haze over the ocean, night over a dark ocean, etc., it is IMC for practical and safety purposes. *[Figure 3-4]* When flying in IMC or when cross-checking the visual references, the airplane's attitude is controlled by the pilot referencing the airplane's mechanical or electronically generated instruments to determine the airplane's attitude in relationship to the natural horizon.

Airplane attitude control is composed of four components: pitch control, bank (roll) control, power control, and trim.

- Pitch control—controlling of the airplane's pitch attitude about the lateral axis by using the elevator to raise and lower the nose in relation to the natural horizon or to the airplane's flight instrumentation.

- Bank control—controlling of the airplane about the airplane's longitudinal axis by use of the ailerons to attain a desired bank angle in relation to the natural horizon or to the airplane's instrumentation.

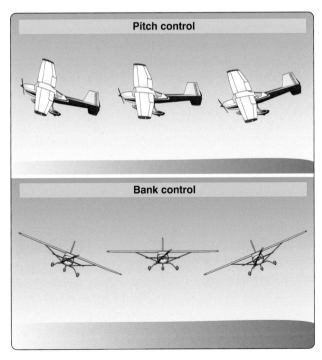

Figure 3-4. *Airplane attitude is based on relative positions of the nose and wings on the natural horizon.*

- Power control—in most general aviation (GA) airplanes is controlled by the throttle and is used when the flight situation requires a specific thrust setting or for a change in thrust to meet a specific objective.

- Trim control—used to relieve the control pressures held by the pilot on the flight controls after a desired attitude has been attained.

Note: Yaw control is used to cancel out the effects of yaw induced changes, such as adverse yaw and effects of the propeller.

Integrated Flight Instruction

When introducing basic flight maneuvers to a beginning pilot, it is recommended that the "Integrated" or "Composite" method of flight instruction be used. This means the use of outside references and flight instruments to establish and maintain desired flight attitudes and airplane performance. *[Figure 3-5]* When beginning pilots use this technique, they achieve a more precise and competent overall piloting ability. Although this method of airplane control may become second nature with experience, the beginning pilot must make a determined effort to master the technique.

As the beginner pilot develops a competent skill in visual reference flying, the flight instructor should further develop the beginner pilot's effectiveness through the use of integrated flight instruction; however, it is important that the beginner pilot's visual skills be sufficiently developed for long-term, safe, and effective aircraft control. *[Figure 3-5]*

The basic elements of integrated flight instruction are as follows:

- The pilot visually controls the airplane's attitude in reference outside to the natural horizon. At least 90 percent of the pilot's attention should be devoted to outside visual references and scanning for airborne traffic. The process of visually evaluating pitch and bank attitude is nearly an imperceptible continuous stream of attitude information. If the attitude is found to be other than desired, the pilot should make precise, smooth, and accurate flight control corrections to return the airplane to the desired attitude. Continuous visual checks of the outside references and immediate corrections made by the pilot minimize the chance for the airplane to deviate from the desired heading, altitude, and flightpath.

- The airplane's attitude is validated by referring to flight instruments and confirming performance. If the flight instruments display that the airplane's performance is in need of correction, the required correction must be determined and then precisely,

90 percent of the time, the pilot's attention should be outside the flight deck.
No more than 10 percent of the pilot's attention should be inside the flight deck.

Figure 3-5. *Integrated flight instruction teaches pilots to use both external and cockpit attitude references.*

smoothly, and accurately applied with reference to the natural horizon. The airplane's attitude and performance are then rechecked by referring to flight instruments. The pilot then maintains the corrected attitude by reference to the natural horizon.

- The pilot should monitor the airplane's performance by making quick snap-shots of the flight instruments. No more than 10 percent of the pilot's attention should be inside the cockpit. The pilot must develop the skill to quickly focus on the appropriate flight instruments and then immediately return to the visual outside references to control the airplane's attitude.

The pilot should become familiar with the relationship between outside visual references to the natural horizon and the corresponding flight instrument indications. For example, a pitch attitude adjustment may require a movement of the pilot's reference point of several inches in relation to the natural horizon but correspond to a seemingly insignificant movement of the reference bar on the airplane's attitude indicator. Similarly, a deviation from a desired bank angle, which is obvious when referencing the airplane's wingtips or cowling relative to the natural horizon, may be imperceptible on the airplane's attitude indicator to the beginner pilot.

The most common error made by the beginner pilot is to make pitch or bank corrections while still looking inside the cockpit. It is also common for beginner pilots to fixate on the flight instruments—a conscious effort is required by them to return to outside visual references. For the first several hours

of instruction, flight instructors may choose to use flight instrument covers to develop a beginning pilot's skill or to correct a pilot's poor habit of fixating on instruments by forcing them to use outside visual references for aircraft control.

The use of integrated flight instruction does not, and is not intended to prepare pilots for flight in instrument weather conditions. The most common error made by the beginning student is to make pitch or bank corrections while still looking inside the cockpit. Control pressure is applied, but the beginning pilot, not being familiar with the intricacies of flight by references to instruments, including such things as instrument lag and gyroscopic precession, will invariably make excessive attitude corrections and end up "chasing the instruments." Airplane attitude by reference to the natural horizon, however, is immediate in its indications, accurate, and presented many times larger than any instrument could be. Also, the beginning pilot must be made aware that anytime, for whatever reason, airplane attitude by reference to the natural horizon cannot be established and/or maintained, the situation should be considered a bona fide emergency.

Straight-and-Level Flight

Straight-and-level flight is flight in which heading and altitude are constantly maintained. The four fundamentals are in essence a derivation of straight-and-level flight. As such, the need to form proper and effective skills in flying straight and level should not be understated. Precise mastery of straight-and-level flight is the result of repetition and effective practice. Perfection in straight-and-level flight comes only as a result of the pilot understanding the effect and use of the flight controls, properly using the visual outside references, and the utilization of snap-shots from the flight instruments in a continuous loop of information gathering. A pilot must make effective, timely, and proportional corrections for deviations in the airplane's direction and altitude from unintentional slight turns, descents, and climbs to master straight-and-level flight.

Straight-and-level flight is a matter of consciously fixing the relationship of a reference point on the airplane in relation to the natural horizon. *[Figure 3-6]* The establishment of reference points should be initiated on the ground as the reference points depends on the pilot's seating position, height, and manner of sitting. It is important that the pilot sit in a normal manner with the seat position adjusted, which allows for the pilot to see adequately over the instrument panel while being able to fully depress the rudder pedals to their maximum forward position without straining or reaching.

With beginner pilots, a flight instructor will likely use a dry erase marker or removable tape to make reference lines on the windshield or cowling to help the beginner pilot establish visual reference points. Vertical reference lines are best established on the ground, such as when the airplane is placed on a marked centerline, with the beginner pilot seated in proper position. Horizontal reference lines are best established with the airplane in flight, such as during slow flight and cruise configurations. The horizon reference point is always being the same, no matter what altitude, since the point is always on the horizon, although the distance to the horizon will be further as altitude increases. There are multiple horizontal reference lines due to the pitch attitude requirements of the maneuver; however, these teaching aids are generally needed for only a short period of time until the beginning pilot understands where and when to look during the various maneuvers.

Straight Flight

Maintaining a constant direction or heading is accomplished by visually checking the lateral level relationship of the airplane's wingtips to the natural horizon. Depending on whether the airplane is a high wing or low wing, both wingtips should be level and equally above or below the natural horizon. Any necessary bank corrections are made with the pilot's coordinated use of ailerons and rudder. *[Figure 3-7]* The pilot should understand that anytime the wings are banked, the airplane turns. The objective of straight flight is to detect small deviations as soon as they occur, thereby necessitating only minor flight control corrections. The bank attitude information can also be obtained from a quick scan of the attitude indicator (which shows the position of the airplane's wings relative to the horizon) and the heading indicator (which indicates whether flight control pressure is necessary to change the bank attitude to return to straight flight).

Figure 3-6. *Nose reference for straight-and-level flight.*

Figure 3-7. *Wingtip reference for straight-and-level flight.*

It is possible to maintain straight flight by simply exerting the necessary pressure with the ailerons or rudder independently in the desired direction of correction. However, the practice of using the ailerons and rudder independently is not correct and makes precise control of the airplane difficult. The correct bank flight control movement requires the coordinated use of ailerons and rudder. Straight-and-level flight requires almost no application of flight control pressures if the airplane is properly trimmed and the air is smooth. For that reason, the pilot must not form the habit of unnecessarily moving the flight controls. The pilot must learn to recognize when corrections are necessary and then to make a measured flight control response precisely, smoothly, and accurately.

Pilots may tend to look out to one side continually, generally to the left due to the pilot's left seat position and consequently focus attention in that direction. This not only gives a restricted angle from which the pilot is to observe but also causes the pilot to exert unconscious pressure on the flight controls in that direction. It is also important that the pilot not fixate in any one direction and continually scan outside the airplane, not only to ensure that the airplane's attitude is correct, but also to ensure that the pilot is considering other factors for safe flight. Continually observing both wingtips has advantages other than being the only positive check for leveling the wings. This includes looking for aircraft traffic, terrain and weather influences, and maintaining overall situational awareness.

Level Flight

In learning to control the airplane in level flight, it is important that the pilot be taught to maintain a light touch on the flight controls using fingers rather than the common problem of a tight-fisted palm wrapped around the flight controls. The pilot should exert only enough pressure on the flight controls to produce the desired result. The pilot should learn to associate the apparent movement of the references with the control pressures which produce attitude movement. As a result, the pilot can develop the ability to adjust the change desired in the airplane's attitude by the amount and direction of pressures applied to the flight controls without the pilot excessively referring to instrument or outside references for each minor correction.

The pitch attitude for level flight is first obtained by the pilot being properly seated, selecting a point toward the airplane's nose as a reference, and then keeping that reference point in a fixed position relative to the natural horizon. *[Figure 3-8]* The principles of attitude flying require that the reference point to the natural horizon position should be cross-checked against the flight instruments to determine if the pitch attitude is correct. If not, such as trending away from the desired altitude, the pitch attitude should be readjusted in relation to the natural horizon and then the flight instruments cross-checked to determine if altitude is now being corrected or maintained. In level flight maneuvers, the terms "increase

Figure 3-8. *Nose reference for level flight.*

the back pressure" or "increase pitch attitude" implies raising the airplane's nose in relation to the natural horizon and the terms "decreasing the pitch attitude" or "decrease pitch attitude" means lowering the nose in relation to the natural horizon. The pilot's primary reference is the natural horizon.

For all practical purposes, the airplane's airspeed remains constant in straight-and-level flight if the power setting is also constant. Intentional airspeed changes, by increasing or decreasing the engine power, provide proficiency in maintaining straight-and-level flight as the airplane's airspeed is changing. Pitching moments may also be generated by extension and retraction of flaps, landing gear, and other drag producing devices, such as spoilers. Exposure to the effect of the various configurations should be covered in any specific airplane checkout.

A common error of a beginner pilot is attempting to hold the wings level by only observing the airplane's nose. Using this method, the nose's short horizontal reference line can cause slight deviations to go unnoticed; however, deviations from level flight are easily recognizable when the pilot references the wingtips and, as a result, the wingtips should be the pilot's primary reference for maintaining level bank attitude. This technique also helps eliminate the potential for flying the airplane with one wing low and correcting heading errors with the pilot holding opposite rudder. A pilot with a bad habit of dragging one wing low and compensating with opposite rudder pressure will have difficulty in mastering other flight maneuvers.

Common errors in the performance of straight-and-level flight are:

- Attempting to use improper pitch and bank reference points on the airplane to establish attitude.

- Forgetting the location of preselected reference points on subsequent flights.

- Attempting to establish or correct airplane attitude using flight instruments rather than the natural horizon.

- "Chasing" the flight instruments rather than adhering to the principles of attitude flying.

- Mechanically pushing or pulling on the flight controls rather than exerting accurate and smooth pressure to affect change.

- Not scanning outside the cockpit to look for other aircraft traffic, weather and terrain influences, and not maintaining situational awareness.

- A tight palm grip on the flight controls resulting in a desensitized feeling of the hand and fingers, which results in overcontrolling the airplane.

- Habitually flying with one wing low or maintaining directional control using only the rudder control.

- Failure to make timely and measured control inputs when deviations from straight-and-level flight are detected.

- Inadequate attention to sensory inputs in developing feel for the airplane.

Trim Control

Proper trim technique is an important and often overlooked basic flying skill. An improperly trimmed airplane requires constant flight control pressures from the pilot, produces tension and fatigue, distracts the pilot from outside visual scanning, and contributes to abrupt and erratic airplane attitude control inputs.

Trim control surfaces are required to offset any constant flight control pressure inputs provided by the pilot. For example, elevator trim is a typical trim in light GA airplanes and is used to null the pressure exerted by the pilot on the pitch flight control, which is being held to produce the tail down force required for a specific angle of attack (AOA). *[Figure 3-9]* This relieves the pilot from holding a constant pressure on the flight controls to maintain a particular pitch attitude and provides an opportunity for the pilot to divert attention to other tasks, such as evaluating the airplane's attitude in relation to the natural horizon, scanning for aircraft traffic, and maintaining situational awareness.

Because of their relatively low power, speed, and cost constraints, not all light airplanes have a complete set (elevator, rudder, and aileron) trim controls that are adjustable from inside the cockpit. Nearly all light airplanes are equipped with at least a cockpit adjustable elevator trim. As airplanes increase in power, weight, and complexity, cockpit adjustable trim systems for the rudder and aileron may be available.

In airplanes where multiple trim axes are available, the rudder should be trimmed first. Rudder, elevator and then aileron should be trimmed next in sequence; however, if the airspeed is varying, continuous attempts to trim the rudder and aileron produces unnecessary pilot workload and distraction. Attempts to trim the rudder at varying airspeeds are impractical in many propeller airplanes because of the built-in compensation for the effect of a propeller's left turning tendencies. The correct procedure is when the pilot has established a constant airspeed and pitch attitude, the pilot should then hold the wings level with aileron flight control pressure while rudder control pressure is trimmed out. Finally, aileron trim should then be adjusted to relieve any aileron flight control pressure.

Elevator trim wheel

Elevator trim indicator

Figure 3-9. *Elevator trim is used in airplanes to null the pressure exerted by the pilot on the pitch flight control.*

A properly trimmed airplane is an indication of good piloting skills. Any control forces that the pilot feels should be a result of deliberate flight control pressure inputs during a planned change in airplane attitude, not a result of forces being applied by the airplane. A common trim control error is the tendency for the pilot to overcontrol the airplane with trim adjustments. Attempting to fly the airplane with the trim is a common fault in basic flying technique even among experienced pilots. The airplane attitude must be established first and held with the appropriate flight control pressures, and then the flight control pressures trimmed out so that the airplane maintains the desired attitude without the pilot exerting flight control pressure.

Level Turns

A turn is initiated by banking the wings in the desired direction of the turn through the pilot's use of the aileron flight controls. Left aileron flight control pressure causes the left wing to lower in relation to the pilot. Right aileron flight control pressure causes the right wing to lower in relation to the pilot. In other words, to turn left, lower left wing with aileron by left stick. To turn right, lower right wing with right stick. Depending on bank angle and airplane engineering, at many bank angles, the airplane will continue to turn with ailerons neutralized. So the sequence should be like the following: (1) bank airplane, adding either enough power or pitching up to compensate for the loss of lift (change in vector angle of lift); (2) neutralize controls as necessary to stop bank from increasing and hold desired bank angle; (3) use the opposite stick (aileron) to return airplane to level; (4) then take that control out to again neutralize the ailerons (along with either power or pitch reduction) for level flight. *[Figure 3-10]*

A turn is the result of the following:

- The ailerons bank the wings and so determine the rate of turn for a given airspeed. Lift is divided into both vertical and horizontal lift components as a result of the bank. The horizontal component of lift moves the airplane toward the banked direction.

- The elevator pitches the nose of the airplane up or down in relation to the pilot and perpendicular to the wings. If the pilot does not add power, and there is sufficient airspeed margin, the pilot must slightly increase the pitch to increase wing lift enough to replace the wing lift being diverted into turning force so as to maintain the current altitude.

- The vertical fin on an airplane does not produce lift. Rather the vertical fin on an airplane is a stabilizing surface and produces no lift if the airplane is flying straight ahead. The vertical fin's purpose is to keep the aft end of the airplane behind the front end.

- The throttle provides thrust which may be used for airspeed to tighten the turn.

- The pilot uses the rudder to offset any adverse yaw developed by wing's differential lift and the engine/propeller. The rudder does not turn the airplane. The rudder is used to maintain coordinated flight.

For purposes of this discussion, turns are divided into three classes: shallow, medium, and steep.

- Shallow turns—bank angle is approximately 20° or less. This shallow bank is such that the inherent lateral stability of the airplane slowly levels the wings unless aileron pressure in the desired direction of bank is held by the pilot to maintain the bank angle.

- Medium turns—result from a degree of bank between approximately 20° to 45°. At medium bank angles, the airplane's inherent lateral stability does not return the wings to level flight. As a result, the airplane tends to remain at a constant bank angle without any flight control pressure held by the pilot. The pilot neutralizes the aileron flight control pressure to maintain the bank.

- Steep turns—result from a degree of bank of approximately 45° or more. The airplane continues in the direction of the bank even with neutral flight controls unless the pilot provides opposite flight control aileron pressure to prevent the airplane from overbanking. The amount of opposite flight control pressures is dependent on various factors, such as bank angle and airspeed. In general, a noticeable level of opposite aileron flight control pressure is required by the pilot to prevent overbanking.

When an airplane is flying straight and level, the total lift is acting perpendicular to the wings and to the Earth. As the airplane is banked into a turn, total lift is the resultant of two components: vertical and horizontal. *[Figure 3-11]* The vertical lift component continues to act perpendicular to the Earth and opposes gravity. The horizontal lift component acts parallel to the Earth's surface opposing centrifugal force. These two lift components act at right angles to each other, causing the resultant total lifting force to act perpendicular to the banked wing of the airplane. It is the horizontal lift component that begins to turn the airplane and not the rudder.

In constant altitude, constant airspeed turns, it is necessary to increase the AOA of the wing when rolling into the turn by increasing back pressure on the elevator, as well as the addition of power to counter the loss of speed due to increased drag. This is required because total lift has

Figure 3-10. *Level turn to the left.*

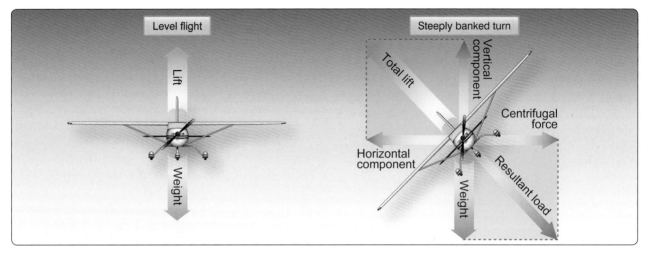

Figure 3-11. *When the airplane is banked into a turn, total lift is the resultant of two components: vertical and horizontal.*

divided into vertical and horizontal components of lift. In order to maintain altitude, the total lift (since total lift acts perpendicular to the wing) must be increased to meet the vertical component of lift requirements (to balance weight and load factor) for level flight.

The purpose of the rudder in a turn is to coordinate the turn. As lift increases, so does drag. When the pilot deflects the ailerons to bank the airplane, both lift and drag are increased on the rising wing and, simultaneously, lift and drag are decreased on the lowering wing. *[Figure 3-12]* This increased drag on the rising wing and decreased drag on the lowering wing results in the airplane yawing opposite to the direction of turn. To counteract this adverse yaw, rudder pressure is applied simultaneously with aileron in the desired direction of turn. This action is required to produce a coordinated turn. Coordinated flight is important to maintaining control of

Figure 3-12. *The rudder opposes adverse yaw to help coordinate the turn.*

the airplane. Situations can develop when a pilot is flying in uncoordinated flight and depending on the flight control deflections, may support pro-spin flight control inputs. This is especially hazardous when operating at low altitudes, such as when operating in the airport traffic pattern. Pilots must learn to fly with coordinated control inputs to prevent unintentional loss of control when maneuvering in certain situations.

During uncoordinated flight, the pilot may feel that they are being pushed sideways toward the outside or inside of the turn. *[Figure 3-13]* A skid is when the pilot may feel that they are being pressed toward the outside of the turn and toward the inside of the turn during a slip. The ability to sense a skid or slip is developed over time and as the "feel" of flying develops, a pilot should become highly sensitive to a slip or skid without undue reliance on the flight instruments.

Turn Radius

To understand the relationship between airspeed, bank, and radius of turn, it should be noted that the rate of turn at any given true airspeed depends on the horizontal lift component. The horizontal lift component varies in proportion to the amount of bank. Therefore, the rate of turn at a given airspeed increases as the angle of bank is increased. On the other hand, when a turn is made at a higher airspeed at a given bank angle, the inertia is greater and the horizontal lift component required for the turn is greater, causing the turning rate to become slower. *[Figure 3-14]* Therefore, at a given angle of bank, a higher airspeed makes the radius of turn larger because the airplane turns at a slower rate.

As the radius of the turn becomes smaller, a significant difference develops between the airspeed of the inside wing and the airspeed of the outside wing. The wing on the outside of the turn travels a longer path than the inside wing, yet both complete their respective paths in the same unit of time.

Figure 3-13. *Indications of a slip and skid.*

Therefore, the outside wing travels at a faster airspeed than the inside wing and, as a result, it develops more lift. This creates an overbanking tendency that must be controlled by the use of opposite aileron when the desired bank angle is reached. *[Figure 3-15]* Because the outboard wing is developing more lift, it also produces more drag. The drag causes a slight slip during steep turns that must be corrected by use of the rudder.

Establishing a Turn

On most light single-engine airplanes, the top surface of the engine cowling is fairly flat, and its horizontal surface to the natural horizon provides a reasonable indication for initially setting the degree of bank angle. *[Figure 3-16]* The pilot should then cross-check the flight instruments to verify that the correct bank angle has been achieved. Information

Figure 3-14. *Angle of bank and airspeed regulate rate and radius of turn.*

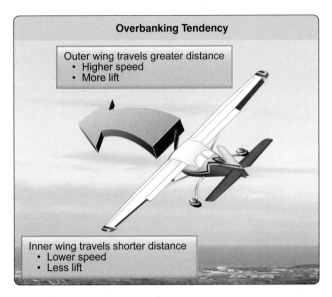

Figure 3-15. *Overbanking tendency.*

Beginning pilots should not use large aileron and rudder control inputs. This is because large control inputs produce rapid roll rates and allows little time for the pilot to evaluate and make corrections. Smaller flight control inputs result in slower roll rates and provide for more time to accurately complete the necessary pitch and bank corrections.

Some additional considerations for initiating turns are the following:

- If the airplane's nose starts to move before the bank starts, the rudder is being applied too soon.
- If the bank starts before the nose starts turning or the nose moves in the opposite direction, the rudder is being applied too late.
- If the nose moves up or down when entering a bank, excessive or insufficient elevator back pressure is being applied.

obtained from the attitude indicator shows the angle of the wing in relation to the horizon.

The pilot's seating position in the airplane is important as it affects the interpretation of outside visual references. A common problem is that a pilot may lean away from the turn in an attempt to remain in an upright position in relation to the horizon. This should be corrected immediately if the pilot is to properly learn to use visual references. *[Figure 3-17]*

Because most airplanes have side-by-side seating, a pilot does not sit on the airplane's longitudinal axis, which is where the airplane rotates in roll. The pilot sits slightly off to one side, typically the left, of the longitudinal axis. Due to parallax error, this makes the nose of the airplane appear to rise when making a left turn (due to pilot lowering in relation to the longitudinal axis) and the nose of the airplane appear to descend when making right turns (due to pilot elevating in relation to the longitudinal axis). *[Figure 3-18]*

After the bank has been established, all flight control pressures applied to the ailerons and rudder may be relaxed or adjusted, depending on the established bank angle, to compensate for the airplane's inherent stability or overbanking tendencies. The airplane should remain at the desired bank angle with the proper application of aileron pressures. If the desired bank angle is shallow, the pilot needs to maintain a small amount of aileron pressure into the direction of bank including rudder to compensate for yaw effects. For medium bank angles, the ailerons and rudder should be neutralized. Steep bank angles require opposite aileron and rudder to prevent the bank from steepening.

Back pressure on the elevator should not be relaxed as the vertical component of lift must be maintained if altitude is to be maintained. Throughout the turn, the pilot should reference the natural horizon, scan for aircraft traffic, and occasionally cross-check the flight instruments to verify performance. A reduction in airspeed is the result of increased drag but is generally not significant for shallow bank angles. In steeper turns, additional

Figure 3-16. *Visual reference for angle of bank.*

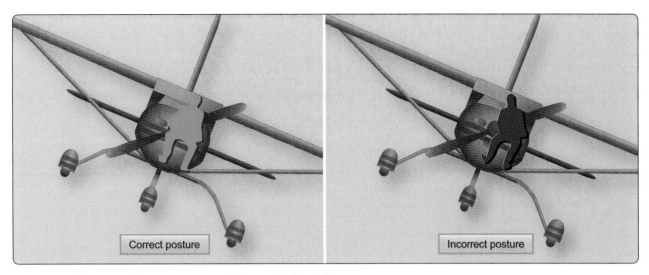

Figure 3-17. *Correct and incorrect posture while seated in the airplane.*

power may be required to maintain airspeed. If altitude is not being maintained during the turn, the pitch attitude should be corrected in relation to the natural horizon and cross-checked with the flight instruments to verify performance.

Steep turns require accurate, smooth, and timely flight control inputs. Minor corrections for pitch attitude are accomplished with proportional elevator back pressure while the bank angle is held constant with the ailerons. However, during steep turns, it is not uncommon for a pilot to allow the nose to get excessively low resulting in a significant loss in altitude in a very short period of time. The recovery sequence requires that the pilot first reduce the angle of bank with coordinated use of opposite aileron and rudder and then increase the pitch attitude by increasing elevator back pressure. If recovery from an excessively nose-low, steep bank condition is attempted by use of the elevator only, it only causes a steepening of the bank and unnecessary stress on the airplane. Steep turn

performance can be improved by an appropriate application of power to overcome the increase in drag and trimming additional elevator back pressure as the bank angle goes beyond 30°. This tends to reduce the demands for large control inputs from the pilot during the turn.

Since the airplane continues turning as long as there is any bank, the rollout from the turn must be started before reaching the desired heading. The amount of lead required to rollout on the desired heading depends on the degree of bank used in the turn. A rule of thumb is to lead by one-half the angle of bank. For example, if the bank is 30°, lead the rollout by 15°. The rollout from a turn is similar to the roll-in except the flight controls are applied in the opposite direction. Aileron and rudder are applied in the direction of the rollout or toward the high wing. As the angle of bank decreases, the elevator pressure should be relaxed as necessary to maintain altitude. As the wings become level, the flight control pressures should

Figure 3-18. *Parallax view.*

be smoothly relaxed so that the controls are neutralized as the airplane returns to straight-and-level flight. If trim was used, such as during a steep turn, forward elevator pressure may be required until the trim can be adjusted. As the rollout is being completed, attention should be given to outside visual references, as well as the flight instruments to determine that the wings are being leveled and the turn stopped.

For outside references, select the horizon and another point ahead. If those two points stay in alignment, the airplane is tracking to that point as long as there is not a crosswind requiring a crab angle. It would also be a good idea to include VFR references for heading as well and pitch. A pilot holds course in VFR by tracking to a point in front of the compass, with only glances at the compass to ensure he or she is still on course. This reliance on a surface point does not work when flying over water or flat snow covered surfaces. In these conditions, the pilot must rely on the compass or gyro-heading indicator.

Because the elevator and ailerons are on one control, practice is required to ensure that only the intended pressure is applied to the intended flight control. For example, a beginner pilot is likely to unintentionally add pressure to the pitch control when the only bank was intended. This cross-coupling may be diminished or enhanced by the design of the flight controls; however, practice is the appropriate measure for smooth, precise, and accurate flight control inputs. For example, diving when turning right and climbing when turning left in airplanes is common with stick controls, because the arm tends to rotate from the elbow joint, which induces a secondary arc control motion if the pilot is not extremely careful. Likewise, lowering the nose is likely to induce a right turn, and raising the nose to climb tends to induce a left turn. These actions would apply for a pilot using the right hand to move the stick. Airplanes with a control wheel may be less prone to these inadvertent actions, depending on control positions and pilot seating. In any case, the pilot must retain the proper sight picture of the nose following the horizon, whether up, down, left or right and isolate undesired motion. It is essential that flight control coordination be developed because it is the very basis of all fundamental flight maneuvers.

Common errors in level turns are:

- Failure to adequately clear in the direction of turn for aircraft traffic.
- Gaining or losing altitude during the turn.
- Not holding the desired bank angle constant.
- Attempting to execute the turn solely by instrument reference.
- Leaning away from the direction of the turn while seated.
- Insufficient feel for the airplane as evidenced by the inability to detect slips or skids without reference to flight instruments.
- Attempting to maintain a constant bank angle by referencing only the airplane's nose.
- Making skidding flat turns to avoid banking the airplane.
- Holding excessive rudder in the direction of turn.
- Gaining proficiency in turns in only one direction.
- Failure to coordinate the controls.

Climbs and Climbing Turns

When an airplane enters a climb, it changes its flightpath from level flight to a climb attitude. In a climb, weight no longer acts in a direction solely perpendicular to the flightpath. When an airplane enters a climb, excess lift must be developed to overcome the weight or gravity. This requirement to develop more lift results in more induced drag, which either results in decreased airspeed and/or an increased power setting to maintain a minimum airspeed in the climb. An airplane can only sustain a climb when there is sufficient thrust to offset increased drag; therefore, climb rate is limited by the excess thrust available.

The pilot should know the engine power settings, natural horizon pitch attitudes, and flight instrument indications that produce the following types of climb:

Normal climb—performed at an airspeed recommended by the airplane manufacturer. Normal climb speed is generally higher than the airplane's best rate of climb. The additional airspeed provides for better engine cooling, greater control authority, and better visibility over the nose of the airplane. Normal climb is sometimes referred to as cruise climb.

Best rate of climb (V_Y)—produces the most altitude gained over a given amount of time. This airspeed is typically used when initially departing a runway without obstructions until it is safe to transition to a normal or cruise climb configuration. Best angle of climb (V_X)—performed at an airspeed that produces the most altitude gain over a given horizontal distance. The best angle of climb results in a steeper climb, although the airplane takes more time to reach the same altitude than it would at best rate of climb airspeed. The best angle of climb is used to clear obstacles, such as a strand of trees, after takeoff. [Figure 3-19]

It should be noted that as altitude increases, the airspeed for best angle of climb increases and the airspeed for best rate of climb decreases. Performance charts contained in the Airplane Flight Manual or Pilot's Operating Handbook

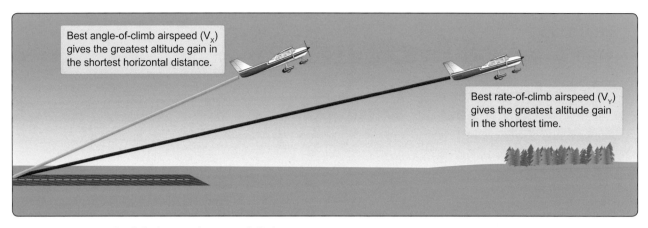

Best angle-of-climb airspeed (V_X) gives the greatest altitude gain in the shortest horizontal distance.

Best rate-of-climb airspeed (V_Y) gives the greatest altitude gain in the shortest time.

Figure 3-19. *Best angle of climb verses best rate of climb.*

(AFM/POH) must be consulted to ensure that the correct airspeed is used for the desired climb profile at the given environmental conditions. There is a point at which the best angle of climb airspeed and the best rate of climb airspeed intersect. This occurs at the absolute ceiling at which the airplane is incapable of climbing any higher. *[Figure 3-20]*

Establishing a Climb

A straight climb is entered by gently increasing back pressure on the elevator flight control to the pitch attitude referencing the airplane's nose to the natural horizon while simultaneously increasing engine power to the climb power setting. The wingtips should be referenced in maintaining the climb attitude while cross-checking the flight instruments to verify performance. In many airplanes, as power is increased, an increase in slipstream over the horizontal stabilizer causes the airplane's pitch attitude to increase greater than desired. The pilot should be prepared for slipstream effects but also for the effect of changing airspeed and changes in lift. The pilot should be prepared to use the required flight control pressures to achieve the desired pitch attitude.

If a climb is started from cruise flight, the airspeed gradually decreases as the airplane enters a stabilized climb attitude. The thrust required to maintain straight-and-level flight at a given airspeed is not sufficient to maintain the same airspeed in a climb. Increase drag in a climb stems from increased lift demands made upon the wing to increase altitude. Climbing requires an excess of lift over that necessary to maintain level flight. Increased lift will generate more induced drag. That increase in induced drag is why more power is needed and why a sustained climb requires an excess of thrust.

For practical purposes gravity or weight is a constant. Even using a vector diagram to show where more lift is necessary because the lift vector from the wings is no longer perpendicular to the wings, therefore more lift is needed from the wings which requires more thrust from the powerplant.

The power should be advanced to the recommended climb power. On airplanes equipped with an independently controllable-pitch propeller, this requires advancing the propeller control prior to increasing engine power. Some airplanes may be equipped with cowl flaps to facilitate effective engine cooling. The position of the cowl flaps should be set to ensure cylinder head temperatures remain within the manufacturer's specifications.

Engines that are normally aspirated experience a reduction of power as altitude is gained. As altitude increases, air density decreases, which results in a reduction of power. The indications show a reduction in revolutions per minute (rpm) for airplanes with fixed pitch propellers; airplanes that are equipped with controllable propellers show a decrease in manifold pressure. The pilot should reference the engine instruments to ensure that climb power is being

Figure 3-20. *Absolute ceiling.*

Figure 3-21. *Climb indications.*

maintained and that pressures and temperatures are within the manufacturer's limits. As power decreases in the climb, the pilot must continually advance the throttle or power lever to maintain specified climb settings.

The propeller effects during a climb and high power settings must be understood by the pilot. The propeller in most airplanes rotates clockwise when seen from the pilot's position. As pitch attitude is increased, the center of thrust from the propeller moves to the right and becomes asymmetrical. This asymmetric condition is often called "P-factor." This is the result of the increased AOA of the descending propeller blade, which is the right side of the propeller disc when seen from the cockpit. As the center of propeller thrust moves to the right, a left turning yawing moment moves the nose of the airplane to the left. This is compensated by the pilot through right rudder pressure. In addition, torque that acts opposite to the direction of propeller rotation causes the airplane to roll to the left. Under these conditions, torque and P-factor cause the airplane to roll and yaw to the left. To counteract this, right rudder and aileron flight control pressures must be used. During the initial practice of climbs, this may initially seem awkward; however, after some experience the correction for propeller effects becomes instinctive.

As the airspeed decreases during the climb's establishment, the airplane's pitch attitude tends to lower unless the pilot increases the elevator flight control pressure. Nose-up elevator trim should be used so that the pitch attitude can be maintained without the pilot holding back elevator pressure. Throughout the climb, since the power should be fixed at the climb power setting, airspeed is controlled by the use of elevator pressure. The pitch attitude to the natural horizon determines if the pitch attitude is correct and should be cross-checked to the flight instruments to verify climb performance. *[Figure 3-21]*

To return to straight-and-level flight from a climb, it is necessary to begin leveling-off prior to reaching the desired altitude. Level-off should begin at approximately 10 percent of the rate of climb. For example, if the airplane is climbing at 500 feet per minute (fpm), leveling off should begin 50 feet prior to reaching the desired altitude. The pitch attitude must be decreased smoothly and slowly to allow for the airspeed to increase; otherwise, a loss of altitude results if the pitch attitude is changed too rapidly without allowing the airspeed to increase proportionately.

After the airplane is established in level flight at a constant altitude, climb power should be retained temporarily so that the airplane accelerates to the cruise airspeed. When the airspeed reaches the desired cruise airspeed, the throttle setting and the propeller control, if equipped, should be set to the cruise power setting and the airplane re-trimmed.

Climbing Turns

In the performance of climbing turns, the following factors should be considered.

- With a constant power setting, the same pitch attitude and airspeed cannot be maintained in a bank as in a straight climb due to the increase in the total lift required.

- The degree of bank should not be too steep. A steep bank significantly decreases the rate of climb. The bank should always remain constant.

- It is necessary to maintain a constant airspeed and constant rate of turn in both right and left turns. The coordination of all flight controls is a primary factor.

- At a constant power setting, the airplane climbs at a slightly shallower climb angle because some of the lift is being used to turn the airplane.

All the factors that affect the airplane during level constant altitude turns affect the airplane during climbing turns. Compensation for the inherent stability of the airplane, overbanking tendencies, adverse yaw, propeller effects, reduction of the vertical component of lift, and increased drag must be managed by the pilot through the manipulation of the flight controls.

Climbing turns may be established by entering the climb first and then banking into the turn or climbing and turning simultaneously. During climbing turns, as in any turn, the loss of vertical lift must be compensated by an increase in pitch attitude. When a turn is coupled with a climb, the additional drag and reduction in the vertical component of lift must be further compensated for by an additional increase in elevator back pressure. When turns are simultaneous with a climb, it is most effective to limit the turns to shallow bank angles. This provides for an efficient rate of climb. If a medium or steep banked turn is used, climb performance is degraded or possibly non-existent.

Common errors in the performance of climbs and climbing turns are:

- Attempting to establish climb pitch attitude by primarily referencing the airspeed indicator resulting in the pilot chasing the airspeed.

- Applying elevator pressure too aggressively resulting in an excessive climb angle.

- Inadequate or inappropriate rudder pressure during climbing turns.

- Allowing the airplane to yaw during climbs usually due to inadequate right rudder pressure.

- Fixation on the airplane's nose during straight climbs, resulting in climbing with one wing low.

- Failure to properly initiate a climbing turn with a coordinated use of the flight controls, resulting in no turn but rather a climb with one wing low.

- Improper coordination resulting in a slip that counteracts the rate of climb, resulting in little or no altitude gain.

- Inability to keep pitch and bank attitude constant during climbing turns.

- Attempting to exceed the airplane's climb capability.

- Applying forward elevator pressure too aggressively during level-off resulting in a loss of altitude or G-force substantially less than one G.

Descents and Descending Turns

When an airplane enters a descent, it changes its flightpath from level flight to a descent attitude. *[Figure 3-22]* In a descent, weight no longer acts solely perpendicular to the flightpath. Since induced drag is decreased as lift is reduced in order to descend, excess thrust will provide higher airspeeds. The weight/gravity force is about the same. This causes an increase in total thrust and a power reduction is required to balance the forces if airspeed is to be maintained.

The pilot should know the engine power settings, natural horizon pitch attitudes, and flight instrument indications that produce the following types of descents:

Partial power descent—the normal method of losing altitude is to descend with partial power. This is often termed cruise or en route descent. The airspeed and power setting recommended by the AFM/POH for prolonged descent should be used. The target descent rate should be 500 fpm. The desired airspeed, pitch attitude, and power combination should be preselected and kept constant.

Descent at minimum safe airspeed—a nose-high, power-assisted descent condition principally used for clearing

Figure 3-22. *Descent indications.*

obstacles during a landing approach to a short runway. The airspeed used for this descent condition is recommended by the AFM/POH and is normally no greater than 1.3 V_{SO}. Some characteristics of the minimum safe airspeed descent are a steeper-than-normal descent angle, and the excessive power that may be required to produce acceleration at low airspeed should "mushing" and/or an excessive rate of descent be allowed to develop.

Emergency descent—some airplanes have a specific procedure for rapidly losing altitude. The AFM/POH specifies the procedure. In general, emergency descent procedures are high drag, high airspeed procedures requiring a specific airplane configuration (such as power to idle, propellers forward, landing gear extended, and flaps retracted) and a specific emergency descent airspeed. Emergency descent maneuvers often include turns.

Glides

A glide is a basic maneuver in which the airplane loses altitude in a controlled descent with little or no engine power; forward motion is maintained by gravity pulling the airplane along an inclined path and the descent rate is controlled by the pilot balancing the forces of gravity and lift. To level off from a partial power descent using a 1,000 feet per minute descent rate, use 10 percent (100 feet) as the lead point to begin raising the nose to stop descent and increasing power to maintain airspeed.

Although glides are directly related to the practice of power-off accuracy landings, they have a specific operational purpose in normal landing approaches, and forced landings after engine failure. Therefore, it is necessary that they be performed more subconsciously than other maneuvers because most of the time during their execution, the pilot will be giving full attention to details other than the mechanics of performing the maneuver. Since glides are usually performed relatively close to the ground, accuracy of their execution and the formation of proper technique and habits are of special importance.

The glide ratio of an airplane is the distance the airplane travels in relation to the altitude it loses. For example, if an airplane travels 10,000 feet forward while descending 1,000 feet, its glide ratio is 10 to 1.

The best glide airspeed is used to maximize the distance flown. This airspeed is important when a pilot is attempting to fly during an engine failure. The best airspeed for gliding is one at which the airplane travels the greatest forward distance for a given loss of altitude in still air. This best glide airspeed occurs at the highest lift-to-drag ratio (L/D). [Figure 3-23] When gliding at airspeed above or below the best glide airspeed, drag increases. Any change in the gliding

airspeed results in a proportional change in the distance flown. [Figure 3-24] As the glide airspeed is increased or decreased from best glide airspeed, the glide ratio is lessened.

Variations in weight do not affect the glide angle provided the pilot uses the proper airspeed. Since it is the L/D ratio that determines the distance the airplane can glide, weight does not affect the distance flown; however, a heavier airplane must fly at a higher airspeed to obtain the same glide ratio. For example, if two airplanes having the same L/D ratio but different weights start a glide from the same altitude, the heavier airplane gliding at a higher airspeed arrives at the same touchdown point in a shorter time. Both airplanes cover the same distance, only the lighter airplane takes a longer time.

Since the highest glide ratio occurs at maximum L/D, certain considerations must be given for drag producing components of the airplane, such as flaps, landing gear, and cowl flaps. When drag increases, a corresponding decrease in pitch attitude is required to maintain airspeed. As the pitch is lowered, the glide path steepens and reduces the distance traveled. To maximize the distance traveled during a glide, all drag producing components must be eliminated if possible.

Wind affects the gliding distance. With a tailwind, the airplane glides farther because of the higher groundspeed. Conversely, with a headwind, the airplane does not glide as far because of the slower groundspeed. This is important for a pilot to understand and manage when dealing with engine-related emergencies and any subsequent forced landing.

Certain considerations must be given to gliding flight. These considerations are caused by the absence of the propeller slipstream, compensation for p-factor in the airplane's design, and the effectiveness of airplane control surfaces

Figure 3-23. L/D_{MAX}.

Figure 3-24. *Best glide speed provides the greatest forward distance for a given loss of altitude.*

at slow speeds. With the absent propeller effects and the subsequent compensation for these effects, which is designed into many airplanes, it is likely that, during glides, slight left rudder pressure is required to maintain coordinated flight. In addition, the deflection of the flight controls to effect change is greater due to the relatively slow airflow over the control surfaces.

Minimum sink speed is used to maximize the time that the airplane remains in flight. It results in the airplane losing altitude at the lowest rate. Minimum sink speed occurs at an airspeed less than the best glide speed. It is important that pilots realize that flight at the minimum sink airspeed results in less distance traveled. Minimum sink speed is useful in flight situations where time in flight is more important than distance flown. An example is ditching an airplane at sea. Minimum sink speed is not an often published airspeed but generally is a few knots less than best glide speed.

In an emergency, such as an engine failure, attempting to apply elevator back pressure to stretch a glide back to the runway is likely to lead the airplane landing short and may even lead to loss of control if the airplane stalls. This leads to a cardinal rule of airplane flying that a student pilot must understand and appreciate: The pilot must never attempt to "stretch" a glide by applying back-elevator pressure and reducing the airspeed below the airplane's recommended best glide speed. The purpose of pitch control during the glide is to maintain the maximum L/D, which may require fore or aft flight control pressure to maintain best glide airspeed.

To enter a glide, the pilot should close the throttle and, if equipped, advance the propeller lever forward. With back pressure on the elevator flight control, the pilot should maintain altitude until the airspeed decreases to the recommended best glide speed. In most airplanes, as power is reduced, propeller slipstream decreases over the horizontal stabilizer, which decreases the tail-down force, and the airplane's nose tends to lower immediately. To keep pitch attitude constant after a power change, the pilot must counteract the pitch down with a simultaneous increase in elevator back pressure. If the pitch attitude is allowed to decrease during glide entry, excess airspeed is carried into the glide and retards the attainment of the correct glide angle

and airspeed. Speed should be allowed to dissipate before the pitch attitude is decreased. This point is particularly important for fast airplanes as they do not readily lose their airspeed—any slight deviation of the airplane's nose downwards results in an immediate increase in airspeed. Once the airspeed has dissipated to best glide speed, the pitch attitude should be set to maintain that airspeed. This should be done with reference to the natural horizon and with a quick reference to the flight instruments. When the airspeed has stabilized, the airplane should be trimmed to eliminate any flight control pressures held by the pilot. Precision is required in maintaining the best glide airspeed if the benefits are to be realized.

A stabilized, power-off descent at the best glide speed is often referred to as normal glide. The beginning pilot should memorize the airplane's attitude and speed with reference to the natural horizon and noting the sounds made by the air passing over the airplane's structure, forces on the flight controls, and the feel of the airplane. Initially, the beginner pilot may be unable to recognize slight variations in airspeed and angle of bank by vision or by the pressure required on the flight controls. The instructor should point out that an increase in sound levels denotes increasing speed, while a decrease in sound levels indicates decreasing speed. When a sound level change is perceived, a beginning pilot should cross-check the visual and pressure references. The beginning pilot must use all three airspeed references (sound, visual, and pressure) consciously until experience is gained, and then must remain alert to any variation in attitude, feel, or sound.

After a solid comprehension of the normal glide is attained, the beginning pilot should be instructed in the differences between normal and abnormal glides. Abnormal glides are those glides conducted at speeds other than the best glide speed. Glide airspeeds that are too slow or too fast may result in the airplane not being able to make the intended landing spot, flat approaches, hard touchdowns, floating, overruns, and possibly stalls and an accident.

Gliding Turns

The absence of the propeller slipstream, loss of effectiveness of the various flight control surfaces at lower airspeeds, and designed-in aerodynamic corrections complicates the task of flight control coordination in comparison to powered flight for the inexperienced pilot. These principles should be thoroughly explained by the flight instructor so that the beginner pilot may be aware of the necessary differences in coordination.

Three elements in gliding turns that tend to force the nose down and increase glide speed are:

- Decrease in lift due to the direction of the lifting force

- Excessive rudder inputs as a result of reduced flight control pressures

- The normal stability and inherent characteristics of the airplane to nose-down with the power off

These three factors make it necessary to use more back pressure on the elevator than is required for a straight glide or a level turn; and therefore, have a greater effect on control coordination. In rolling in or out of a gliding turn, the rudder is required to compensate for yawing tendencies; however, the required rudder pedal pressures are reduced as result of the reduced forces acting on the control surfaces. Because the rudder forces are reduced, the pilot may apply excessive rudder pedal pressures based on their experience with powered flight and overcontrol the aircraft causing slips and skids rather than coordinated flight. This may result in a much greater deflection of the rudder resulting in potentially hazardous flight control conditions.

Some examples of this hazard:

- A low-level gliding steep turn during an engine failure emergency. If the rudder is excessively deflected in the direction of the bank while the pilot is increasing elevator back pressure in an attempt to retain altitude, the situation can rapidly turn into an unrecoverable spin.

- During a power-off landing approach. The pilot depresses the rudder pedal with excessive pressure that leads to increased lift on the outside wing, banking the airplane in the direction of the rudder deflection. The pilot may improperly apply the opposite aileron to prevent the bank from increasing while applying elevator back pressure. If allowed to progress, this situation may result in a fully developed cross-control condition. A stall in this situation almost certainly results in a rapid and unrecoverable spin.

Level-off from a glide is really two different maneuvers depending on the type of glide:

1. In the event of a complete power failure, the best glide speed should be held until necessary to reconfigure for the landing, with planning for a steeper approach than usual when partial power is used for the approach to landing. A 10 percent lead (100 feet if the decent rate is 1,000 feet per minute) factor should be sufficient. That is what is given in the Instrument flying Handbook, so that should be the general rule of thumb for all publications.

2. In the case of a quicker descent or simulated power failure training, power should be applied as the 10% lead value appears on the altimeter to allow a slow but positive power application to maintain or increase airspeed while raising the nose to stop the descent. Retrim as necessary.

The level-off from a glide must be started before reaching the desired altitude because of the airplane's downward inertia. The amount of lead depends on the rate of descent and what airspeed is desired upon completion of the level off. For example, assume the aircraft is in a 500 fpm rate of descent, and the desired final airspeed is higher than the glide speed. The altitude lead should begin at approximately 100 feet above the target altitude and at the lead point, power should be increased to the appropriate level flight cruise power setting when the desired final airspeed is higher than the glide speed. At the lead point, power should be increased to the appropriate level flight cruise power setting. The airplane's nose tends to rise as airspeed and power increases and the pilot must smoothly control the pitch attitude so that the level-off is completed at the desired altitude and airspeed. When recovery is being made from a gliding turn, the back pressure on the elevator control, which was applied during the turn, must be decreased or the airplane's nose will pitch up excessively high resulting in a rapid loss of airspeed. This error requires considerable attention and conscious control adjustment before the normal glide can be resumed.

Common errors in the performance of descents and descending turns are:

- Failure to adequately clear for aircraft traffic in the turn direction or descent.

- Inadequate elevator back pressure during glide entry resulting in an overly steep glide.

- Failure to slow the airplane to approximate glide speed prior to lowering pitch attitude.

- Attempting to establish/maintain a normal glide solely by reference to flight instruments.

- Inability to sense changes in airspeed through sound and feel.

- Inability to stabilize the glide (chasing the airspeed indicator).

- Attempting to "stretch" the glide by applying back-elevator pressure.

- Skidding or slipping during gliding turns due to inadequate appreciation of the difference in rudder forces as compared to turns with power.

- Failure to lower pitch attitude during gliding turn entry resulting in a decrease in airspeed.

- Excessive rudder pressure during recovery from gliding turns.

- Inadequate pitch control during recovery from straight glide.

- Cross-controlling during gliding turns near the ground.

- Failure to maintain constant bank angle during gliding turns.

Chapter Summary

The four fundamental maneuvers of straight-and-level flight, turns, climbs, and descents are the foundation of basic airmanship. Effort and continued practice are required to master the fundamentals. It is important that a pilot consider the six motions of flight: bank, pitch, yaw and horizontal, vertical, and lateral displacement. In order for an airplane to fly from one location to another, it pitches, banks, and yaws while it moves over and above, in relationship to the ground, to reach its destination. The airplane must be treated as an aerodynamic vehicle that is subject to rigid aerodynamic laws. A pilot must understand and apply the principles of flight in order to control an airplane with the greatest margin of mastery and safety.

Chapter 4

Maintaining Aircraft Control: Upset Prevention and Recovery Training

Introduction

A pilot's fundamental responsibility is to prevent a loss of control (LOC). Loss of control in-flight (LOC-I) is the leading cause of fatal general aviation accidents in the U.S. and commercial aviation worldwide. LOC-I is defined as a significant deviation of an aircraft from the intended flightpath and it often results from an airplane upset. Maneuvering is the most common phase of flight for general aviation LOC-I accidents to occur; however, LOC-I accidents occur in all phases of flight.

To prevent LOC-I accidents, it is important for pilots to recognize and maintain a heightened awareness of situations that increase the risk of loss of control. Those situations include: uncoordinated flight, equipment malfunctions, pilot complacency, distraction, turbulence, and poor risk management – like attempting to fly in instrument meteorological conditions (IMC) when the pilot is not qualified or proficient. Sadly, there are also LOC-I accidents resulting from intentional disregard or recklessness.

Wing Washout
Wing root has greater angle of incidence than wing tip.

Power-On Stall and Recovery

Slow to lift-off speed, maintain altitude.

Set takeoff power, raise nose.

When stall occurs, reduce AOA, roll wings level, and add power as needed.

As flying speed returns, stop descent and establish a climb.

Maintain climb airspeed, raise landing gear and flaps, and trim.

Return to the desired flightpath.

To maintain aircraft control when faced with these or other contributing factors, the pilot must be aware of situations where LOC-I can occur, recognize when an airplane is approaching a stall, has stalled, or is in an upset condition, and understand and execute the correct procedures to recover the aircraft.

Defining an Airplane Upse

The term "upset" was formally introduced by an industry work group in 2004 in the "Pilot Guide to Airplane Upset Recovery," which is one part of the "Airplane Upset Recovery Training Aid." The working group was primarily focused on large transport airplanes and sought to come up with one term to describe an "unusual attitude" or "loss of control," for example, and to generally describe specific parameters as part of its definition. Consistent with the Guide, the FAA has defined an upset as an event that unintentionally exceeds the parameters normally experienced in flight or training. These parameters are:

- Pitch attitude greater than 25°, nose up

- Pitch attitude greater than 10°, nose down

- Bank angle greater than 45°

- Within the above parameters, but flying at airspeeds inappropriate for the conditions.

The reference to inappropriate airspeeds describes a number of undesired aircraft states, including stalls. However, stalls are directly related to angle of attack (AOA), not airspeed.

To develop the crucial skills to prevent LOC-I, a pilot must receive upset prevention and recovery training (UPRT), which should include: slow flight, stalls, spins, and unusual attitudes.

Upset training has placed more focus on prevention—understanding what can lead to an upset so a pilot does not find himself or herself in such a situation. If an upset does occur, however, upset training also reinforces proper recovery techniques. A more detailed discussion of UPRT to include its core concepts, what the training should include, and what airplanes or kinds of simulation can be used for the training can be found later in this chapter.

Coordinated Flight

Coordinated flight occurs whenever the pilot is proactively correcting for yaw effects associated with power (engine/propeller effects), aileron inputs, how an airplane reacts when turning, and airplane rigging. The airplane is in coordinated flight when the airplane's nose is yawed directly into the relative wind and the ball is centered in the slip/skid indicator. *[Figure 4-1]*

A pilot should develop a sensitivity to side loads that indicate the nose is not yawed into the relative wind, and the airplane

Figure 4-1. *Coordinated flight in a turn.*

is not slipping or skidding. A correction should be made by applying rudder pressure on the side toward which one feels a leaning sensation. This will be the same side to which the ball in the slip/skid indicator has slewed (i.e., the old saying "step on the ball").

Angle of Attack

The angle of attack (AOA) is the angle at which the chord of the wing meets the relative wind. The chord is a straight line from the leading edge to the trailing edge. At low angles of attack, the airflow over the top of the wing flows smoothly and produces lift with a relatively small amount of drag. As the AOA increases, lift as well as drag increases; however, above a wing's critical AOA, the flow of air separates from the upper surface and backfills, burbles and eddies, which reduces lift and increases drag. This condition is a stall, which can lead to loss of control if the AOA is not reduced.

It is important for the pilot to understand that a stall is the result of exceeding the critical AOA, not of insufficient airspeed. The term "stalling speed" can be misleading, as this speed is often discussed when assuming 1G flight at a particular weight and configuration. Increased load factor directly affects stall speed (as well as do other factors such as gross weight, center of gravity, and flap setting). Therefore, it is possible to stall the wing at any airspeed, at any flight attitude, and at any power setting. For example, if a pilot maintains airspeed and rolls into a coordinated, level 60° banked turn, the load factor is 2Gs, and the airplane will stall at a speed that is 40 percent higher than the straight-and-level stall speed. In that 2G level turn, the pilot has to increase AOA to increase the lift required to maintain altitude. At this condition, the pilot is closer to the critical AOA than during level flight and therefore closer to the higher speed that the airplane will stall at. Because "stalling speed" is not a constant number, pilots must understand the underlying factors that affect it in order to maintain aircraft control in all circumstances.

Slow Flight

Slow flight is when the airplane AOA is just under the AOA which will cause an aerodynamic buffet or a warning from a stall warning device if equipped with one. A small increase in AOA may result in an impending stall, which increases the risk of an actual stall. In most normal flight operations the airplane would not be flown close to the stall-warning AOA or critical AOA, but because the airplane is flown at higher AOAs, and thus reduced speeds in the takeoff/departure and approach/landing phases of flight, learning to fly at reduced airspeeds is essential. In these phases of flight, the airplane's close proximity to the ground would make loss of control catastrophic; therefore, the pilot must be proficient in slow flight.

The objective of maneuvering in slow flight is to understand the flight characteristics and how the airplane's flight controls feel near its aerodynamic buffet or stall-warning. It also helps to develop the pilot's recognition of how the airplane feels, sounds, and looks when a stall is impending. These characteristics include, degraded response to control inputs and difficulty maintaining altitude. Practicing slow flight will help pilots recognize an imminent stall not only from the feel of the controls, but also from visual cues, aural indications, and instrument indications.

For pilot training and testing purposes, slow flight includes two main elements:

1. Slowing to, maneuvering at, and recovering from an airspeed at which the airplane is still capable of maintaining controlled flight without activating the stall warning—5 to 10 knots above the 1G stall speed is a good target; and

2. Performing slow flight in configurations appropriate to takeoffs, climbs, descents, approaches to landing, and go-arounds.

Slow flight should be introduced with the airspeed sufficiently above the stall to permit safe maneuvering, but close enough to the stall warning for the pilot to experience the characteristics of flight at a very low airspeed. One way to determine the target airspeed is to slow the airplane to the stall warning when in the desired slow flight configuration, pitch the nose down slightly to eliminate the stall warning, add power to maintain altitude and note the airspeed.

When practicing slow flight, a pilot learns to divide attention between aircraft control and other demands. How the airplane feels at the slower airspeeds aids the pilot in learning that as airspeed decreases, control effectiveness decreases. For instance, reducing airspeed from 30 knots to 20 knots above the stalling speed will result in a certain loss of effectiveness of flight control inputs because of less airflow over the control surfaces. As airspeed is further reduced, the control effectiveness is further reduced and the reduced airflow over the control surfaces results in larger control movements being required to create the same response. Pilots sometimes refer to the feel of this reduced effectiveness as "sloppy" or "mushy" controls.

When flying above minimum drag speed (L/D$_{MAX}$), even a small increase in power will increase the speed of the airplane. When flying at speeds below L/D$_{MAX}$, also referred to as flying on the back side of the power curve, larger inputs in power or reducing the AOA will be required for the airplane to be able to accelerate. Since slow flight will be performed well below L/D$_{MAX}$, the pilot must be aware that large power inputs or a reduction in AOA will be required to prevent the aircraft from decelerating. It is important to note that when flying on the backside of the power curve, as the AOA increases toward the critical AOA and the airplane's speed continues to decrease, small changes in the pitch control result in disproportionally large changes in induced drag and therefore changes in airspeed. As a result, pitch becomes a more effective control of airspeed when flying below L/D$_{MAX}$ and power is an effective control of the altitude profile (i.e., climbs, descents, or level flight)

It is also important to note that an airplane flying below L/D$_{MAX}$, exhibits a characteristic known as "speed instability" and the airspeed will continue to decay without appropriate pilot action. For example, if the airplane is disturbed by turbulence and the airspeed decreases, the airspeed may continue to decrease without the appropriate pilot action of reducing the AOA or adding power. *[Figure 4-2]*

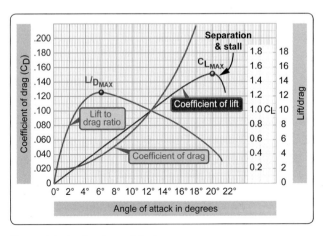

Figure 4-2. *Angle-of-attack in degrees.*

Performing the Slow Flight Maneuver

Slow flight should be practiced in straight-and-level flight, straight-ahead climbs and climbing medium-banked (approximately 20 degrees) turns, and straight-ahead power-off gliding descents and descending turns to represent the takeoff and landing phases of flight. Slow flight training should include slowing the airplane smoothly and promptly from cruising to approach speeds without changes in altitude or heading, and understanding the required power and trim settings to maintain slow flight. It should also include configuration changes, such as extending the landing gear and adding flaps, while maintaining heading and altitude. Slow flight in a single-engine airplane should be conducted so the maneuver can be completed no lower than 1,500 feet AGL, or higher, if recommended by the manufacturer. In all cases, practicing slow flight should be conducted at an adequate height above the ground for recovery should the airplane inadvertently stall.

To begin the slow flight maneuver, clear the area and gradually reduce thrust from cruise power and adjust the pitch to allow the airspeed to decrease while maintaining altitude. As the speed of the airplane decreases, note a change in the sound of the airflow around the airplane. As the speed approaches the target slow flight speed, which is an airspeed just above the stall warning in the desired configuration (i.e., approximately 5–10 knots above the stall speed for that flight condition), additional power will be required to maintain altitude. During these changing flight conditions, it is important to trim the airplane to compensate for changes in control pressures. If the airplane remains trimmed for cruising speed (a lower AOA), strong aft (back) control pressure is needed on the elevator, which makes precise control difficult unless the airplane is retrimmed.

Slow flight is typically performed and evaluated in the landing configuration. Therefore, both the landing gear and the flaps should be extended to the landing position. It is recommended the prescribed before-landing checks be completed to configure the airplane. The extension of gear and flaps typically occurs once cruise power has been reduced and at appropriate airspeeds to ensure limitations for extending those devices are not exceeded. Practicing this maneuver in other configurations, such as a clean or takeoff configuration, is also good training and may be evaluated on the practical test.

With an AOA just under the AOA which may cause an aerodynamic buffet or stall warning, the flight controls are less effective. *[Figure 4-3]* The elevator control is less responsive and larger control movements are necessary to retain control of the airplane. In propeller-driven airplanes, torque, slipstream effect, and P-factor may produce a strong

Figure 4-3. *Slow flight—low airspeed, high angle of attack, high power, and constant altitude.*

left yaw, which requires right rudder input to maintain coordinated flight. The closer the airplane is to the 1G stall, the greater the amount of right rudder pressure required.

Maneuvering in Slow Flight

When the desired pitch attitude and airspeed have been established in straight-and-level slow flight, the pilot must maintain awareness of outside references and continually cross-check the airplane's instruments to maintain control. The pilot should note the feel of the flight controls, especially the airspeed changes caused by small pitch adjustments, and the altitude changes caused by power changes. The pilot should practice turns to determine the airplane's controllability characteristics at this low speed. During the turns, it will be necessary to increase power to maintain altitude. Abrupt or rough control movements during slow flight may result in a stall. For instance, abruptly raising the flaps while in slow flight can cause the plane to stall.

The pilot should also practice climbs and descents by adjusting the power when stabilized in straight-and-level slow flight. The pilot should note the increased yawing tendency at high power settings and counter it with rudder input as needed.

To exit the slow flight maneuver, follow the same procedure as for recovery from a stall: apply forward control pressure to reduce the AOA, maintain coordinated flight and level the wings, and apply power as necessary to return to the desired flightpath. As airspeed increases, clean up the airplane by retracting flaps and landing gear if they were extended. A pilot should anticipate the changes to the AOA as the landing gear and flaps are retracted to avoid a stall.

Common errors in the performance of slow flight are:

- Failure to adequately clear the area

- Inadequate back-elevator pressure as power is reduced, resulting in altitude loss

- Excessive back-elevator pressure as power is reduced, resulting in a climb followed by a rapid reduction in airspeed

- Insufficient right rudder to compensate for left yaw

- Fixation on the flight instruments

- Failure to anticipate changes in AOA as flaps are extended or retracted

- Inadequate power management

- Inability to adequately divide attention between airplane control and orientation

- Failure to properly trim the airplane

- Failure to respond to a stall warning

Stalls

A stall is an aerodynamic condition which occurs when smooth airflow over the airplane's wings is disrupted, resulting in loss of lift. Specifically, a stall occurs when the AOA—the angle between the chord line of the wing and the relative wind—exceeds the wing's critical AOA. It is possible to exceed the critical AOA at any airspeed, at any attitude, and at any power setting. *[Figure 4-4]*

For these reasons, it is important to understand factors and situations that can lead to a stall, and develop proficiency in stall recognition and recovery. Performing intentional stalls will familiarize the pilot with the conditions that result in a stall, assist in recognition of an impending stall, and develop the proper corrective response if a stall occurs. Stalls are practiced to two different levels:

- Impending Stall—an impending stall occurs when the AOA causes a stall warning, but has not yet reached the critical AOA. Indications of an impending stall can include buffeting, stick shaker, or aural warning.

- Full Stall—a full stall occurs when the critical AOA is exceeded. Indications of a full stall are typically that an uncommanded nose-down pitch cannot be readily arrested, and this may be accompanied by an uncommanded rolling motion. For airplanes equipped with stick pushers, its activation is also a full stall indication.

Although it depends on the degree to which a stall has progressed, some loss of altitude is expected during recovery. The longer it takes for the pilot to recognize an impending stall, the more likely it is that a full stall will result. Intentional stalls should therefore be performed at an altitude that provides adequate height above the ground for recovery and return to normal level flight.

Stall Recognition

A pilot must recognize the flight conditions that are conducive to stalls and know how to apply the necessary corrective action. This level of proficiency requires learning to recognize an impending stall by sight, sound, and feel.

Stalls are usually accompanied by a continuous stall warning for airplanes equipped with stall warning devices. These devices may include an aural alert, lights, or a stick shaker all which alert the pilot when approaching the critical AOA. Certification standards permit manufacturers to provide the required stall warning either through the inherent aerodynamic qualities of the airplane or through a stall warning device that gives a clear indication of the impending stall. However, most vintage airplanes, and many types of light sport and experimental airplanes, do not have stall warning devices installed.

Other sensory cues for the pilot include:

- Feel—the pilot will feel control pressures change as speed is reduced. With progressively less resistance on the control surfaces, the pilot must use larger control movements to get the desired airplane response. The pilot will notice the airplane's reaction time to control movement increases. Just before the stall occurs, buffeting, uncommanded rolling, or vibrations may begin to occur.

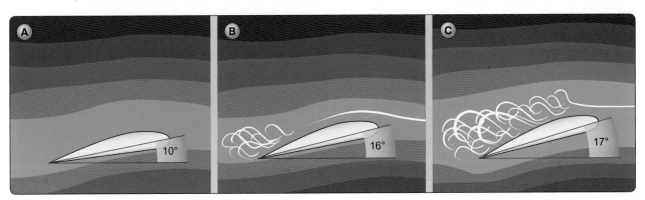

Figure 4-4. *Critical angle of attack and stall.*

- Vision—since the airplane can be stalled in any attitude, vision is not a foolproof indicator of an impending stall. However, maintaining pitch awareness is important.

- Hearing—as speed decreases, the pilot should notice a change in sound made by the air flowing along the airplane structure.

- Kinesthesia—the physical sensation (sometimes referred to as "seat of the pants" sensations) of changes in direction or speed is an important indicator to the trained and experienced pilot in visual flight. If this sensitivity is properly developed, it can warn the pilot of an impending stall.

Pilots in training must remember that a level-flight 1G stalling speed is valid only:

- In unaccelerated 1G flight

- In coordinated flight (slip-skid indicator centered)

- At one weight (typically maximum gross weight)

- At a particular center of gravity (CG) (typically maximum forward CG)

Angle of Attack Indicators

Learning to recognize stalls without relying on stall warning devices is important. However, airplanes can be equipped with AOA indicators that can provide a visual indication of the airplane's proximity to the critical AOA. There are several different kinds of AOA indicators with varying methods for calculating AOA, therefore proper installation and training on the use of these devices is important. AOA indicators measure several parameters simultaneously, determine the current AOA, and provide a visual image of the proximity to the critical AOA. *[Figure 4-5]* Some AOA indicators also provide aural indications, which can provide awareness to a change in AOA that is trending towards the critical AOA prior to installed stall warning systems. It's important to note that some indicators take flap position into consideration, but not all do.

Understanding what type of AOA indicator is installed on an airplane, how the particular device determines AOA, what the display is indicating and when the critical AOA is reached, and what the appropriate response is to those indications are all important components to AOA indicator training. It is also encouraged to conduct in-flight training to see the indications throughout various maneuvers, like slow flight, stalls, takeoffs, and landings, and to practice the appropriate responses to those indications. It is also important to note that some items may limit the effectiveness of an AOA indicator (e.g., calibration techniques, wing contamination, unheated probes/vanes). Pilots flying an airplane equipped with an AOA indicator should refer to the pilot handbook information

or contact the manufacturer for specific limitations applicable to that indicator type.

Stall Characteristics

Different airplane designs can result in different stall characteristics. The pilot should know the stall characteristics of the airplane being flown and the manufacturer's recommended recovery procedures. Factors that can affect the stall characteristics of an airplane include its geometry, CG, wing design, and high-lift devices. Engineering design variations make it impossible to specifically describe the stall characteristics for all airplanes; however, there are enough similarities in small general aviation training-type airplanes to offer broad guidelines.

Most training airplanes are designed so that the wings stall progressively outward from the wing roots (where the wing attaches to the fuselage) to the wingtips. Some wings are

Figure 4-5. *A conceptual representation of an AOA indicator. It is important to become familiar with the equipment installed in a specific airplane.*

manufactured with a certain amount of twist, known as washout, resulting in the outboard portion of the wings having a slightly lower AOA than the wing roots. This design feature causes the wingtips to have a smaller AOA during flight than the wing roots. Thus, the wing roots of an airplane exceed the critical AOA before the wingtips, meaning the wing roots stall first. Therefore, when the airplane is in a stalled condition, the ailerons should still have a degree of control effectiveness until/unless stalled airflow migrates outward along the wings. Although airflow may still be attached at the wingtips, a pilot should exercise caution using the ailerons prior to the reduction of the AOA because it can exacerbate the stalled condition. For example, if the airplane rolls left at the stall ("rolls-off"), and the pilot applies right aileron to try to level the wing, the downward-deflected aileron on the left wing produces a greater AOA (and more induced drag), and a more complete stall at the tip as the critical AOA is exceeded. This can cause the wing to roll even more to the left, which is why it is important to first reduce the AOA before attempting to roll the airplane.

The pilot must also understand how the factors that affect stalls are interrelated. In a power-off stall, for instance, the cues (buffeting, shaking) are less noticeable than in the power-on stall. In the power-off, 1G stall, the predominant cue may be the elevator control position (full up elevator against the stops) and a high descent rate.

Fundamentals of Stall Recovery

Depending on the complexity of the airplane, stall recovery could consist of as many as six steps. Even so, the pilot should remember the most important action to an impending stall or a full stall is to reduce the AOA. There have been numerous situations where pilots did not first reduce AOA, and instead prioritized power and maintaining altitude, which resulted in a loss of control. This section provides a generic stall recovery procedure for light general aviation aircraft adapted from a template developed by major airplane manufacturers and can be adjusted appropriately for the aircraft used.

[*Figure 4-6*] However, a pilot should always follow the aircraft-specific manufacturer's recommended procedures if published and current.

The recovery actions should be made in a procedural manner; they can be summarized in *Figure 4-6*. The following discussion explains each of the six steps:

1. Disconnect the wing leveler or autopilot (if equipped). Manual control is essential to recovery in all situations. Disconnecting this equipment should be done immediately and allow the pilot to move to the next crucial step quickly. Leaving the wing leveler or autopilot connected may result in inadvertent changes or adjustments to the flight controls or trim that may not be easily recognized or appropriate, especially during high workload situations.

2. a) Pitch nose-down control. Reducing the AOA is crucial for all stall recoveries. Push forward on the flight controls to reduce the AOA below the critical AOA until the impending stall indications are eliminated before proceeding to the next step.

    b) Trim nose-down pitch. If the elevator does not provide the needed response, pitch trim may be necessary. However, excessive use of pitch trim may aggravate the condition, or may result in loss of control or high structural loads.

3. Roll wings level. This orients the lift vector properly for an effective recovery. It is important not to be tempted to control the bank angle prior to reducing AOA. Both roll stability and roll control will improve considerably after getting the wings flying again. It is also imperative for the pilot to proactively cancel yaw with proper use of the rudder to prevent a stall from progressing into a spin.

4. Add thrust/power. Power should be added as needed, as stalls can occur at high power or low power settings, or at high airspeeds or low airspeeds. Advance the

Stall Recovery Template	
1. Wing leveler or autopilot	1. Disconnect
2. a) Pitch nose-down	2. a) Apply until impending stall indications are eliminated
b) Trim nose-down pitch	b) As needed
3. Bank	3. Wings Level
4. Thrust/Power	4. As needed
5. Speed brakes/spoilers	5. Retract
6. Return to the desired flight path	

Figure 4-6. *Stall recovery template.*

throttle promptly, but smoothly, as needed while using rudder and elevator controls to stop any yawing motion and prevent any undesirable pitching motion. Adding power typically reduces the loss of altitude during a stall recovery, but it does not eliminate a stall. The reduction in AOA is imperative. For propeller-driven airplanes, power application increases the airflow around the wing, assisting in stall recovery.

5. Retract speedbrakes/spoilers (if equipped). This will improve lift and the stall margin.

6. Return to the desired flightpath. Apply smooth and coordinated flight control movements to return the airplane to the desired flightpath being careful to avoid a secondary stall. The pilot should, however, be situationally aware of the proximity to terrain during the recovery and take the necessary flight control action to avoid contact with it.

The above procedure can be adapted for the type of aircraft flown. For example, a single-engine training airplane without an autopilot would likely only use four of the six steps. The first step is not needed therefore reduction of the AOA until the stall warning is eliminated is first. Use of pitch trim is less of a concern because most pilots can overpower the trim in these airplanes and any mistrim can be corrected when returning to the desired flightpath. The next step is rolling the wings level followed by the addition of power as needed all while maintaining coordinated flight. The airplane is not equipped with speedbrakes or spoilers therefore this step can be skipped and the recovery will conclude with returning to the desired flightpath.

Similarly, a glider pilot does not have an autopilot therefore the first step is the reduction of AOA until the stall warning is eliminated. The pilot would then roll wings level while maintaining coordinated flight. There is no power to add therefore this step would not apply. Retracting speedbrakes or spoilers would be the next step for a glider pilot followed by returning to the desired flightpath.

Stall Training

Practice in both power-on and power-off stalls is important because it simulates stall conditions that could occur during normal flight maneuvers. It is important for pilots to understand the possible flight scenarios in which a stall could occur. Stall accidents usually result from an inadvertent stall at a low altitude, with the recovery not completed prior to ground contact. For example, power-on stalls are practiced to develop the pilot's awareness of what could happen if the airplane is pitched to an excessively nose-high attitude immediately after takeoff, during a climbing turn, or when trying to clear an obstacle. Power-off turning stalls develop

the pilot's awareness of what could happen if the controls are improperly used during a turn from the base leg to the final approach. The power-off straight-ahead stall simulates the stall that could occur when trying to stretch a glide after the engine has failed, or if low on the approach to landing.

As in all maneuvers that involve significant changes in altitude or direction, the pilot must ensure that the area is clear of other air traffic at and below their altitude and that sufficient altitude is available for a recovery before executing the maneuver. It is recommended that stalls be practiced at an altitude that allows recovery no lower than 1,500 feet AGL for single-engine airplanes, or higher if recommended by the AFM/POH. Losing altitude during recovery from a stall is to be expected.

Approaches to Stalls (Impending Stalls), Power-On or Power-Off

An impending stall occurs when the airplane is approaching, but does not exceed the critical AOA. The purpose of practicing impending stalls is to learn to retain or regain full control of the airplane immediately upon recognizing that it is nearing a stall, or that a stall is likely to occur if the pilot does not take appropriate action. Pilot training should emphasize teaching the same recovery technique for impending stalls and full stalls.

The practice of impending stalls is of particular value in developing the pilot's sense of feel for executing maneuvers in which maximum airplane performance is required. These maneuvers require flight in which the airplane approaches a stall, but the pilot initiates recovery at the first indication, such as by a stall warning device activation.

Impending stalls may be entered and performed in the same attitudes and configurations as the full stalls or other maneuvers described in this chapter. However, instead of allowing the airplane to reach the critical AOA, the pilot must immediately reduce AOA once the stall warning device goes off, if installed, or recognizes other cues such as buffeting. Hold the nose down control input as required to eliminate the stall warning. Then level the wings maintain coordinated flight, and then apply whatever additional power is necessary to return to the desired flightpath. The pilot will have recovered once the airplane has returned to the desired flightpath with sufficient airspeed and adequate flight control effectiveness and no stall warning. Performance of the impending stall maneuver is unsatisfactory if a full stall occurs, if an excessively low pitch attitude is attained, or if the pilot fails to take timely action to avoid excessive airspeed, excessive loss of altitude, or a spin.

Full Stalls, Power-Off

The practice of power-off stalls is usually performed with normal landing approach conditions to simulate an accidental stall occurring during approach to landing. However, power-off stalls should be practiced at all flap settings to ensure familiarity with handling arising from mechanical failures, icing, or other abnormal situations. Airspeed in excess of the normal approach speed should not be carried into a stall entry since it could result in an abnormally nose-high attitude.

To set up the entry for a straight-ahead power-off stall, airplanes equipped with flaps or retractable landing gear should be in the landing configuration. After extending the landing gear, applying carburetor heat (if applicable), and retarding the throttle to idle (or normal approach power), hold the airplane at a constant altitude in level flight until the airspeed decelerates to normal approach speed. The airplane should then be smoothly pitched down to a normal approach attitude to maintain that airspeed. Wing flaps should be extended and pitch attitude adjusted to maintain the airspeed.

When the approach attitude and airspeed have stabilized, the pilot should smoothly raise the airplane's nose to an attitude that induces a stall. Directional control should be maintained and wings held level by coordinated use of the ailerons and rudder. Once the airplane reaches an attitude that will lead to a stall, the pitch attitude is maintained with the elevator until the stall occurs. The stall is recognized by the full-stall cues previously described.

Recovery from the stall is accomplished by reducing the AOA, applying as much nose-down control input as required to eliminate the stall warning, leveling the wings, maintaining coordinated flight, and then applying power as needed. Right rudder pressure may be necessary to overcome the engine torque effects as power is advanced and the nose is being lowered. *[Figure 4-7]* If simulating an inadvertent stall on approach to landing, the pilot should initiate a go-around by establishing a positive rate of climb. Once in a climb, the flaps and landing gear should be retracted as necessary.

Recovery from power-off stalls should also be practiced from shallow banked turns to simulate an inadvertent stall during a turn from base leg to final approach. During the practice of these stalls, take care to ensure that the airplane remains coordinated and the turn continues at a constant bank angle until the full stall occurs. If the airplane is allowed to develop a slip, the outer wing may stall first and move downward abruptly. The recovery procedure is the same, regardless of whether one wing rolls off first. The pilot must apply as much nose down control input as necessary to eliminate the stall warning, level the wings with ailerons, coordinate with rudder, and add power as needed. In the practice of turning stalls, no attempt should be made to stall or recover the airplane on a predetermined heading. However, to simulate a turn from base to final approach, the stall normally should be made to occur within a heading change of approximately 90°.

Full Stalls, Power-On
Power-on stall recoveries are practiced from straight climbs and climbing turns (15° to 20° bank) to help the pilot recognize the potential for an accidental stall during takeoff, go around, climb, or when trying to clear an obstacle. Airplanes equipped with flaps or retractable landing gear should normally be in the takeoff configuration; however, power-on stalls should also be practiced with the airplane in a clean configuration (flaps and gear retracted) to ensure practice with all possible takeoff and climb configurations. Power for practicing the takeoff stall recovery should be maximum power, although for some airplanes it may be reduced to a setting that will prevent an excessively high pitch attitude.

To set up the entry for power-on stalls, establish the airplane in the takeoff or climb configuration. Slow the airplane to normal lift-off speed while continuing to clear the area of other traffic. Upon reaching the desired speed, set takeoff power or the recommended climb power for the power-on stall (often referred to as a departure stall) while establishing a climb attitude. The purpose of reducing the airspeed to lift-off airspeed before the throttle is advanced to the recommended setting is to avoid an excessively steep nose-up attitude for a long period before the airplane stalls.

After establishing the climb attitude, smoothly raise the nose to increase the AOA, and hold that attitude until the full stall

Power-Off Stall and Recovery

| Establish normal approach. | Raise nose, maintain heading. | When stall occurs, reduce angle of attack, roll wings level, and add power as needed. | As flying speed returns, stop descent and establish a climb. | Maintain climb airspeed, raise landing gear and flaps, and trim. | Return to the desired flightpath. |

Figure 4-7. *Power-off stall and recovery.*

Figure 4-8. *Power-on stall.*

occurs. As described in connection with the stall characteristics discussion, continual adjustments must be made to aileron pressure, elevator pressure, and rudder pressure to maintain coordinated flight while holding the attitude until the full stall occurs. In most airplanes, as the airspeed decreases the pilot must move the elevator control progressively further back while simultaneously adding right rudder and maintaining the climb attitude until reaching the full stall.

The pilot must promptly recognize when the stall has occurred and take action to prevent a prolonged stalled condition. The pilot should recover from the stall by immediately reducing the AOA and applying as much nose-down control input as required to eliminate the stall warning, level the wings with ailerons, coordinate with rudder, and smoothly advance the power as needed. Since the throttle is already at the climb power setting, this step may simply mean confirming the proper power setting. *[Figure 4-8]*

The final step is to return the airplane to the desired flightpath (e.g., straight and level or departure/climb attitude). With sufficient airspeed and control effectiveness, return the throttle to the appropriate power setting.

Secondary Stall

A secondary stall is so named because it occurs after recovery from a preceding stall. It is typically caused by abrupt control inputs or attempting to return to the desired flightpath too

quickly and the critical AOA is exceeded a second time. It can also occur when the pilot does not sufficiently reduce the AOA by lowering the pitch attitude or attempts to break the stall by using power only. *[Figure 4-9]*

When a secondary stall occurs, the pilot should again perform the stall recovery procedures by applying nose-down elevator pressure as required to eliminate the stall warning, level the wings with ailerons, coordinate with rudder, and adjust power as needed. When the airplane is no longer in a stalled condition the pilot can return the airplane to the desired flightpath. For pilot certification, this is a demonstration-only maneuver; only flight instructor applicants may be required to perform it on a practical test.

Accelerated Stalls

The objectives of demonstrating an accelerated stall are to determine the stall characteristics of the airplane, experience stalls at speeds greater than the +1G stall speed, and develop the ability to instinctively recover at the onset of such stalls. This is a maneuver only commercial pilot and flight instructor applicants may be required to perform or demonstrate on a practical test. However, all pilots should be familiar with the situations that can cause an accelerated stall, how to recognize it, and the appropriate recovery action should one occur.

At the same gross weight, airplane configuration, CG location, power setting, and environmental conditions,

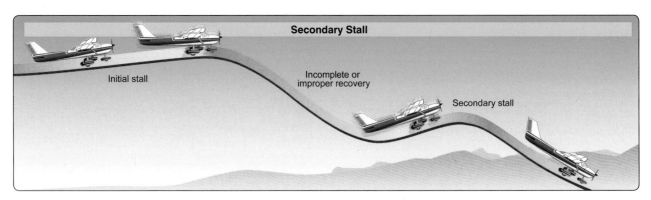

Figure 4-9. *Secondary stall.*

a given airplane consistently stalls at the same indicated airspeed provided the airplane is at +1G (i.e., steady-state unaccelerated flight). However, the airplane can also stall at a higher indicated airspeed when the airplane is subject to an acceleration greater than +1G, such as when turning, pulling up, or other abrupt changes in flightpath. Stalls encountered any time the G-load exceeds +1G are called "accelerated maneuver stalls". The accelerated stall would most frequently occur inadvertently during improperly executed turns, stall and spin recoveries, pullouts from steep dives, or when overshooting a base to final turn. An accelerated stall is typically demonstrated during steep turns.

A pilot should never practice accelerated stalls with wing flaps in the extended position due to the lower design G-load limitations in that configuration. Accelerated stalls should be performed with a bank of approximately 45°, and in no case at a speed greater than the airplane manufacturer's recommended airspeed or the specified design maneuvering speed (V_A).

It is important to be familiar with V_A, how it relates to accelerated stalls, and how it changes depending on the airplane's weight. V_A is the maximum speed at which the maximum positive design load limit can be imposed either by gusts or full one-sided deflection with one control surface without causing structural damage. Performing accelerated stalls at or below V_A allows the airplane to reach the critical AOA, which unloads the wing before it reaches the load limit. At speeds above V_A, the wing can reach the design load limit at an AOA less than the critical AOA. This means it is possible to damage the airplane before reaching the critical AOA and an accelerated stall. Knowing what V_A is for the weight of the airplane being flown is critical to prevent exceeding the load limit of the airplane during the maneuver.

There are two methods for performing an accelerated stall. The most common accelerated stall procedure starts from straight-and-level flight at an airspeed at or below V_A. Roll the airplane into a coordinated, level-flight 45° turn and then smoothly, firmly, and progressively increase the AOA through back elevator pressure until a stall occurs. Alternatively, roll the airplane into a coordinated, level-flight 45° turn at an airspeed above V_A. After the airspeed reaches V_A, or at an airspeed 5 to 10 percent faster than the unaccelerated stall speed, progressively increase the AOA through back elevator pressure until a stall occurs. The increased back elevator pressure increases the AOA, which increases the lift and thus the G load. The G load pushes the pilot's body down in the seat. The increased lift also increases drag, which may cause the airspeed to decrease. It is recommended that you know the published stall speed for 45° of bank, flaps up, before performing the maneuver. This speed is typically published in the AFM.

An airplane typically stalls during a level, coordinated turn similar to the way it does in wings level flight, except that the stall buffet can be sharper. If the turn is coordinated at the time of the stall, the airplane's nose pitches away from the pilot just as it does in a wings level stall since both wings will tend to stall nearly simultaneously. If the airplane is not properly coordinated at the time of stall, the stall behavior may include a change in bank angle until the AOA has been reduced. It is important to take recovery action at the first indication of a stall (if impending stall training/checking) or immediately after the stall has fully developed (if full stall training/checking) by applying forward elevator pressure as required to reduce the AOA and to eliminate the stall warning, level the wings using ailerons, coordinate with rudder, and adjust power as necessary. Stalls that result from abrupt maneuvers tend to be more aggressive than unaccelerated, +1G stalls. Because they occur at higher-than-normal airspeeds or may occur at lower-than-anticipated pitch attitudes, they can surprise an inexperienced pilot. A prolonged accelerated stall should never be allowed. Failure to take immediate steps toward recovery may result in a spin or other departure from controlled flight.

Cross-Control Stall

The objective of the cross-control stall demonstration is to show the effects of uncoordinated flight on stall behavior and to emphasize the importance of maintaining coordinated flight while making turns. This is a demonstration-only maneuver; only flight instructor applicants may be required to perform it on a practical test. However, all pilots should be familiar with the situations that can lead to a cross-control stall, how to recognize it, and the appropriate recovery action should one occur.

The aerodynamic effects of the uncoordinated, cross-control stall can surprise the unwary pilot because it can occur with very little warning and can be deadly if it occurs close to the ground. The nose may pitch down, the bank angle may suddenly change, and the airplane may continue to roll to an inverted position, which is usually the beginning of a spin. It is therefore essential for the pilot to follow the stall recovery procedure by reducing the AOA until the stall warning has been eliminated, then roll wings level using ailerons, and coordinate with rudder inputs before the airplane enters a spiral or spin.

A cross-control stall occurs when the critical AOA is exceeded with aileron pressure applied in one direction and rudder pressure in the opposite direction, causing uncoordinated flight. A skidding cross-control stall is most likely to occur in the traffic pattern during a poorly planned and executed base-to-final approach turn in which the airplane overshoots the runway centerline and the pilot attempts to correct back to centerline by increasing the bank angle, increasing back elevator pressure, and applying rudder in the direction of the turn (i.e., inside or bottom rudder pressure) to bring the nose around further to align it with the runway. The difference in lift between the inside and outside wing will increase, resulting in an unwanted increase in bank angle. At the same time, the nose of the airplane slices downward through the horizon. The natural reaction to this may be for the pilot to pull back on the elevator control, increasing the AOA toward critical. Should a stall be encountered with these inputs, the airplane may rapidly enter a spin. The safest action for an "overshoot" is to perform a go-around. At the relatively low altitude of a base-to-final approach turn, a pilot should be reluctant to use angles of bank beyond 30 degrees to correct back to runway centerline.

Before performing this stall, establish a safe altitude for entry and recovery in the event of a spin, and clear the area of other traffic while slowly retarding the throttle. The next step is to lower the landing gear (if equipped with retractable gear), close the throttle, and maintain altitude until the airspeed approaches the normal glide speed. To avoid the possibility of exceeding the airplane's limitations, do not extend the flaps. While the gliding attitude and airspeed are being established, the airplane should be retrimmed. Once the glide is stabilized, the airplane should be rolled into a medium-banked turn to simulate a final approach turn that overshoots the centerline of the runway.

During the turn, smoothly apply excessive rudder pressure in the direction of the turn but hold the bank constant by applying opposite aileron pressure. At the same time, increase back elevator pressure to keep the nose from lowering. All of these control pressures should be increased until the airplane stalls. When the stall occurs, recover by applying nose-down elevator pressure to reduce the AOA until the stall warning has been eliminated, remove the excessive rudder input and level the wings, and apply power as needed to return to the desired flightpath.

Elevator Trim Stall

The elevator trim stall demonstration shows what can happen when the pilot applies full power for a go-around without maintaining positive control of the airplane. *[Figure 4-10]* This is a demonstration-only maneuver; only flight instructor applicants may be required to perform it on a practical test. However, all pilots should be familiar with the situations that can cause an elevator trim stall, how to recognize it, and the appropriate recovery action should one occur.

This situation may occur during a go-around procedure from a normal landing approach or a simulated, forced-landing approach, or immediately after a takeoff, with the trim set for a normal landing approach glide at idle power. The objective of the demonstration is to show the importance of making smooth power applications, overcoming strong trim forces, maintaining positive control of the airplane to hold safe flight attitudes, and using proper and timely trim techniques. It also develops the pilot's ability to avoid actions that could result in this stall, to recognize when an elevator trim stall is approaching, and to take prompt and correct action to prevent a full stall condition. It is imperative to avoid the occurrence of an elevator trim stall during an actual go-around from an approach to landing.

At a safe altitude and after ensuring that the area is clear of other air traffic, the pilot should slowly retard the throttle and extend the landing gear (if the airplane is equipped with retractable gear). The next step is to extend the flaps to the one-half or full position, close the throttle, and maintain altitude until the airspeed approaches the normal glide speed.

When the normal glide is established, the pilot should trim the airplane nose-up for the normal landing approach glide. During this simulated final approach glide, the throttle is then advanced smoothly to maximum allowable power, just as it would be adjusted to perform a go-around.

The combined effects of increased propwash over the tail and elevator trim tend to make the nose rise sharply and turn to the

Figure 4-10. *Elevator trim stall.*

left. With the throttle fully advanced, the pitch attitude increases above the normal climbing attitude. When it is apparent the airplane is approaching a stall, the pilot must apply sufficient forward elevator pressure to reduce the AOA and eliminate the stall warning before returning the airplane to the normal climbing attitude. The pilot will need to adjust trim to relieve the heavy control pressures and then complete the normal go-around procedures and return to the desired flightpath. If taken to the full stall, recovery will require a significant nose-down attitude to reduce the AOA below its critical AOA, along with a corresponding significant loss of altitude.

Common Errors

Common errors in the performance of intentional stalls are:

- Failure to adequately clear the area

- Over-reliance on the airspeed indicator and slip-skid indicator while excluding other cues

- Inadvertent accelerated stall by pulling too fast on the controls during a power-off or power on stall entry

- Inability to recognize an impending stall condition

- Failure to take timely action to prevent a full stall during the conduct of impending stalls

- Failure to maintain a constant bank angle during turning stalls

- Failure to maintain proper coordination with the rudder throughout the stall and recovery

- Recovering before reaching the critical AOA when practicing the full stall maneuver

- Not disconnecting the wing leveler or autopilot, if equipped, prior to reducing AOA

- Recovery is attempted without recognizing the importance of pitch control and AOA

- Not maintaining a nose down control input until the stall warning is eliminated

- Pilot attempts to level the wings before reducing AOA

- Pilot attempts to recover with power before reducing AOA

- Failure to roll wings level after AOA reduction and stall warning is eliminated

- Inadvertent secondary stall during recovery

- Excessive forward-elevator pressure during recovery resulting in low or negative G load

- Excessive airspeed buildup during recovery

- Losing situational awareness and failing to return to desired flightpath or follow ATC instructions after recovery.

Spin Awareness

A spin is an aggravated stall that typically occurs from a full stall occurring with the airplane in a yawed state and results in the airplane following a downward corkscrew path. As the airplane rotates around a vertical axis, the outboard wing is less stalled than the inboard wing, which creates a rolling, yawing, and pitching motion. The airplane is basically descending due to gravity, rolling, yawing, and pitching in a spiral path. *[Figure 4-11]* The rotation results from an unequal AOA on the airplane's wings. The less-stalled rising wing has a decreasing AOA, where the relative lift increases and the drag decreases. Meanwhile, the descending wing has an increasing AOA, which results in decreasing relative lift and increasing drag.

A spin occurs when the airplane's wings exceed their critical AOA (stall) with a sideslip or yaw acting on the airplane at, or beyond, the actual stall. An airplane will yaw not only because of incorrect rudder application but because of adverse yaw created by aileron deflection; engine/prop effects, including p-factor, torque, spiraling slipstream, and gyroscopic precession; and wind shear, including wake turbulence. If the yaw had been created by the pilot because of incorrect

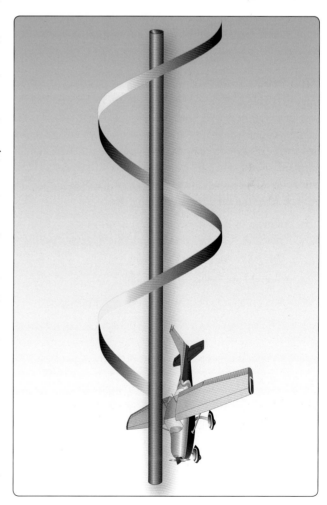

Figure 4-11. *Spin—an aggravated stall and autorotation.*

rudder use, the pilot may not be aware that a critical AOA has been exceeded until the airplane yaws out of control toward the lowering wing. A stall that occurs while the airplane is in a slipping or skidding turn can result in a spin entry and rotation in the direction of rudder application, regardless of which wingtip is raised. If the pilot does not immediately initiate stall recovery, the airplane may enter a spin.

Maintaining directional control and not allowing the nose to yaw before stall recovery is initiated is key to averting a spin. The pilot must apply the correct amount of rudder to keep the nose from yawing and the wings from banking.

Modern airplanes tend to be more reluctant to spin compared to older designs, however it is not impossible for them to spin. Mishandling the controls in turns, stalls, and flight at minimum controllable airspeeds can put even the most reluctant airplanes into an accidental spin. Proficiency in avoiding conditions that could lead to an accidental stall/spin situation, and in promptly taking the correct actions to recover to normal flight, is essential. An airplane must be stalled and yawed in order to enter a spin; therefore, continued practice in stall recognition and recovery helps the pilot develop a more instinctive and prompt reaction in recognizing an approaching spin. Upon recognition of a spin or approaching spin, the pilot should immediately execute spin recovery procedures.

Spin Procedures

The first rule for spin demonstration is to ensure that the airplane is approved for spins. Please note that this discussion addresses generic spin procedures; it does not cover special spin procedures or techniques required for a particular airplane. Safety dictates careful review of the AFM/POH and regulations before attempting spins in any airplane. The review should include the following items:

- The airplane's AFM/POH limitations section, placards, or type certification data to determine if the airplane is approved for spins

- Weight and balance limitations

- Recommended entry and recovery procedures

- The current 14 CFR Part 91 parachute requirements

Also essential is a thorough airplane preflight inspection, with special emphasis on excess or loose items that may affect the weight, center of gravity, and controllability of the airplane. It is also important to ensure that the airplane is within any CG limitations as determined by the manufacturer. Slack or loose control cables (particularly rudder and elevator) could prevent full anti-spin control deflections and delay or preclude recovery in some airplanes.

Prior to beginning spin training, clear the flight area above and below the airplane for other traffic. This task may be accomplished while slowing the airplane for the spin entry. In addition, all spin training should be initiated at an altitude high enough to complete recovery at or above 1,500 feet AGL.

It may be appropriate to introduce spin training by first practicing both power-on and power-off stalls in a clean configuration. This practice helps familiarize the pilot with the airplane's specific stall and recovery characteristics. In all phases of training, the pilot should take care with handling of the power (throttle), and apply carburetor heat, if equipped, according to the manufacturer's recommendations.

There are four phases of a spin: entry, incipient, developed, and recovery. *[Figure 4-12]*

Entry Phase

In the entry phase, the pilot intentionally or accidentally provides the necessary elements for the spin. The entry procedure for demonstrating a spin is similar to a power-off stall. During the entry, the pilot should slowly reduce power to idle, while simultaneously raising the nose to a pitch attitude that ensures a stall. As the airplane approaches a stall, smoothly apply full rudder in the direction of the desired spin rotation while applying full back (up) elevator to the limit of travel. Always maintain the ailerons in the neutral position during the spin procedure unless AFM/POH specifies otherwise.

Incipient Phase

The incipient phase occurs from the time the airplane stalls and starts rotating until the spin has fully developed. This phase may take two to four turns for most airplanes. In this phase, the aerodynamic and inertial forces have not achieved a balance. As the incipient phase develops, the indicated airspeed will generally stabilize at a low and constant airspeed and the symbolic airplane of the turn indicator should indicate the direction of the spin. The slip/skid ball is unreliable when spinning.

The pilot should initiate incipient spin recovery procedures prior to completing 360° of rotation. The pilot should apply full rudder opposite the direction of rotation. The turn indicator shows a deflection in the direction of rotation if disoriented.

Incipient spins that are not allowed to develop into a steady-state spin are the most commonly used maneuver in initial spin training and recovery techniques.

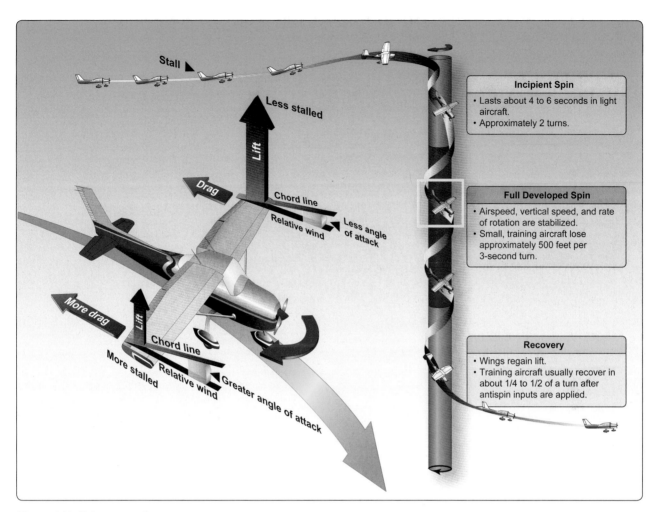

Figure 4-12. *Spin entry and recovery.*

The labels within the figure include:

Stall

Less stalled

Lift

Drag

Chord line

Relative wind

Less angle of attack

More drag

Lift

Chord line

More stalled

Relative wind

Greater angle of attack

Incipient Spin
- Lasts about 4 to 6 seconds in light aircraft.
- Approximately 2 turns.

Full Developed Spin
- Airspeed, vertical speed, and rate of rotation are stabilized.
- Small, training aircraft lose approximately 500 feet per 3-second turn.

Recovery
- Wings regain lift.
- Training aircraft usually recover in about 1/4 to 1/2 of a turn after antispin inputs are applied.

Developed Phase

The developed phase occurs when the airplane's angular rotation rate, airspeed, and vertical speed are stabilized in a flightpath that is nearly vertical. In the developed phase, aerodynamic forces and inertial forces are in balance, and the airplane's attitude, angles, and self-sustaining motions about the vertical axis are constant or repetitive, or nearly so. The spin is in equilibrium. It is important to note that some training airplanes will not enter into the developed phase but could transition unexpectedly from the incipient phase into a spiral dive. In a spiral dive the airplane will not be in equilibrium but instead will be accelerating and G load can rapidly increase as a result.

Recovery Phase

The recovery phase occurs when rotation ceases and the AOA of the wings is decreased below the critical AOA. This phase may last for as little as a quarter turn or up to several turns depending upon the airplane and the type of spin.

To recover, the pilot applies control inputs to disrupt the spin equilibrium by stopping the rotation and unstalling the wing. To accomplish spin recovery, always follow the manufacturer's recommended procedures. In the absence of the manufacturer's recommended spin recovery procedures and techniques, use the spin recovery procedures in *Figure 4-13*. If the flaps and/or retractable landing gear are extended prior to the spin, they should be retracted as soon as practicable after spin entry.

1. Reduce the Power (Throttle) to Idle

2. Position the Ailerons to Neutral

3. Apply Full Opposite Rudder against the Rotation

4. Apply Positive, Brisk, and Straight Forward Elevator (Forward of Neutral)

5. Neutralize the Rudder After Spin Rotation Stops

6. Apply Back Elevator Pressure to Return to Level Flight

Spin Recovery Template
1. Reduce the power (throttle) to idle
2. Position the ailerons to neutral
3. Apply full opposite rudder against the rotation
4. Apply positive, brisk, and straight forward elevator (forward of neutral)
5. Neutralize the rudder after spin rotation stops
6. Apply back elevator pressure to return to level flight

Figure 4-13. *Spin recovery template.*

The following discussion explains each of the six steps:

1. Reduce the Power (Throttle) to Idle. Power aggravates spin characteristics. It can result in a flatter spin attitude and usually increases the rate of rotation.

2. Position the Ailerons to Neutral. Ailerons may have an adverse effect on spin recovery. Aileron control in the direction of the spin may accelerate the rate of rotation, steepen the spin attitude and delay the recovery. Aileron control opposite the direction of the spin may cause flattening of the spin attitude and delayed recovery; or may even be responsible for causing an unrecoverable spin. The best procedure is to ensure that the ailerons are neutral.

3. Apply Full Opposite Rudder against the Rotation. Apply and hold full opposite rudder until rotation stops. Rudder tends to be the most important control for recovery in typical, single-engine airplanes, and its application should be brisk and full opposite to the direction of rotation. Avoid slow and overly cautious opposite rudder movement during spin recovery, which can allow the airplane to spin indefinitely, even with anti-spin inputs. A brisk and positive technique results in a more positive spin recovery.

4. Apply Positive, Brisk, and Straight Forward Elevator (Forward of Neutral). This step should be taken immediately after full rudder application. Do not wait for the rotation to stop before performing this step. The forceful movement of the elevator decreases the AOA and drives the airplane toward unstalled flight. In some cases, full forward elevator may be required for recovery. Hold the controls firmly in these positions until the spinning stops. (Note: If the airspeed is increasing, the airplane is no longer in a spin. In a spin, the airplane is stalled, and the indicated airspeed should therefore be relatively low and constant and not be accelerating.)

5. Neutralize the Rudder After Spin Rotation Stops. Failure to neutralize the rudder at this time, when airspeed is increasing, causes a yawing or sideslipping effect.

6. Apply Back Elevator Pressure to Return to Level Flight. Be careful not to apply excessive back elevator pressure after the rotation stops and the rudder has been neutralized. Excessive back elevator pressure can cause a secondary stall and may result in another spin. The pilot must also avoid exceeding the G-load limits and airspeed limitations during the pull out.

Again, it is important to remember that the spin recovery procedures and techniques described above are recommended for use only in the absence of the manufacturer's procedures. The pilot must always be familiar with the manufacturer's procedures for spin recovery.

Intentional Spins

If the manufacturer does not specifically approve an airplane for spins, intentional spins are not authorized by the CFRs or by this handbook. The official sources for determining whether the spin maneuver is approved are:

- Type Certificate Data Sheets or the Aircraft Specifications

- The limitation section of the FAA-approved AFM/POH. The limitation section may provide additional specific requirements for spin authorization, such as limiting gross weight, CG range, and amount of fuel.

- On a placard located in clear view of the pilot in the airplane (e.g., "NO ACROBATIC MANEUVERS INCLUDING SPINS APPROVED"). In airplanes placarded against spins, there is no assurance that recovery from a fully developed spin is possible.

Unfortunately, accident records show occurrences in which pilots intentionally ignored spin restrictions. Despite the installation of placards prohibiting intentional spins in these airplanes, some pilots and even some flight instructors attempt to justify the maneuver, rationalizing that the spin restriction results from a "technicality" in the airworthiness standards. They believe that if the airplane was spin tested during its certification process, no problem should result from demonstrating or practicing spins.

Such pilots overlook the fact that certification of a normal category airplane only requires the airplane to recover from a one-turn spin in not more than one additional turn or three seconds, whichever takes longer. In other words, the airplane may never be in a fully developed spin. Therefore, in airplanes placarded against spins, there is absolutely no assurance that recovery from a fully developed spin is possible under any circumstances. The pilot of an airplane placarded against intentional spins should assume that the airplane could become uncontrollable in a spin.

Weight and Balance Requirements Related to Spins

In airplanes that are approved for spins, compliance with weight and balance requirements is important for safe performance and recovery from the spin maneuver. Pilots must be aware that even minor weight or balance changes can affect the airplane's spin recovery characteristics. Such changes can either degrade or enhance the spin maneuver and/or recovery characteristics. For example, the addition of weight in the aft baggage compartment, or additional fuel, may still permit the airplane to be operated within CG, but could seriously affect the spin and recovery characteristics. An airplane that may be difficult to spin intentionally in the utility category (restricted aft CG and reduced weight) could have less resistance to spin entry in the normal category (less restricted aft CG and increased weight). This situation arises from the airplane's ability to generate a higher AOA. An airplane that is approved for spins in the utility category but loaded in accordance with the normal category may not recover from a spin that is allowed to progress beyond one turn.

Common Errors

Common errors in the performance of intentional spins are:

- Failure to apply full rudder pressure (to the stops) in the desired spin direction during spin entry

- Failure to apply and maintain full up-elevator pressure during spin entry, resulting in a spiral

- Failure to achieve a fully-stalled condition prior to spin entry

- Failure to apply full rudder (to the stops) briskly against the spin during recovery

- Failure to apply sufficient forward-elevator during recovery

- Waiting for rotation to stop before applying forward elevator

- Failure to neutralize the rudder after rotation stops, possibly resulting in a secondary spin

- Slow and overly cautious control movements during recovery

- Excessive back elevator pressure after rotation stops, possibly resulting in secondary stall

- Insufficient back elevator pressure during recovery resulting in excessive airspeed

Upset Prevention and Recovery

Unusual Attitudes Versus Upsets

An unusual attitude is commonly referenced as an unintended or unexpected attitude in instrument flight. These unusual attitudes are introduced to a pilot during student pilot training as part of basic attitude instrument flying and continue to be trained and tested as part of certification for an instrument rating, aircraft type rating, and an airline transport pilot certificate. A pilot is taught the conditions or situations that could cause an unusual attitude, with focus on how to recognize one, and how to recover from one.

As discussed at the beginning of this chapter, the term "upset" is inclusive of unusual attitudes. An upset is defined as an event that unintentionally exceeds the parameters normally experienced in flight or training. These parameters are:

- Pitch attitude greater than 25°, nose up

- Pitch attitude greater than 10°, nose down

- Bank angle greater than 45°

- Within the above parameters, but flying at airspeeds inappropriate for the conditions.

(Note: The reference to inappropriate airspeeds describes a number of undesired aircraft states, including stalls. However, stalls are directly related to AOA, not airspeed.)

Given the upset definition, there are a few key distinctions between an unusual attitude and an upset. First, an upset includes stall events where unusual attitude training typically does not. Second, an upset can include overspeeds or other inappropriate speeds for a given flight condition, which is also not considered part of unusual attitude training. Finally, an upset has defined parameters; an unusual attitude does not. For example, for training purposes an instructor could place the airplane in a 30° bank with a nose up pitch attitude of 15° and ask the student to recover and that would be considered an unusual attitude, but would not meet the upset parameters. While the information that follows in this section could apply to unusual attitudes, the focus will be on UPRT.

The top four causal and contributing factors that have led to an upset and resulted in LOC-I accidents are:

1. Environmental factors
2. Mechanical factors
3. Human factors
4. Stall-related factors

With the exception of stall-related factors, which were covered in the previous section, the remaining causal and contributing factors to LOC-I accidents will be discussed further below.

Environmental Factors

Turbulence, or a large variation in wind velocity over a short distance, can cause upset and LOC-I. Maintain awareness of conditions that can lead to various types of turbulence, such as clear air turbulence, mountain waves, wind shear, and thunderstorms or microbursts. In addition to environmentally-induced turbulence, wake turbulence from other aircraft can lead to upset and LOC-I.

Icing can destroy the smooth flow of air over the airfoil and increase drag while decreasing the ability of the airfoil to create lift. Therefore, it can significantly degrade airplane performance, resulting in a stall if not handled correctly.

Mechanical Factors

Modern airplanes and equipment are very reliable, but anomalies do occur. Some of these mechanical failures can directly cause a departure from normal flight, such as asymmetrical flaps, malfunctioning or binding flight controls, and runaway trim.

Upsets can also occur if there is a malfunction or misuse of the autoflight system. Advanced automation may tend to mask the cause of the anomaly. Disengaging the autopilot and the autothrottles allows the pilot to directly control the airplane and possibly eliminate the cause of the problem. For these reasons the pilot must maintain proficiency to manually fly the airplane in all flight conditions without the use of the autopilot/autothrottles.

Although these and other inflight anomalies may not be preventable, knowledge of systems and AFM/POH recommended procedures helps the pilot minimize their impact and prevent an upset. In the case of instrument failures, avoiding an upset and subsequent LOC-I may depend on the pilot's proficiency in the use of secondary instrumentation and partial panel operations.

Human Factors
VMC to IMC

Unfortunately, accident reports indicate that continued VFR flight from visual meteorological conditions (VMC) into marginal VMC and IMC is a factor contributing to LOC I. A loss of the natural horizon substantially increases the chances of encountering vertigo or spatial disorientation, which can lead to upset.

IMC

When operating in IMC, maintain awareness of conditions and use the fundamental instrument skills—cross-check, interpretation, and control—to prevent an upset.

Diversion of Attention

In addition to its direct impact, an inflight anomaly or malfunction can also lead to an upset if it diverts the pilot's attention from basic airplane control responsibilities. Failing to monitor the automated systems, over-reliance on those systems, or incomplete knowledge and experience with those systems can lead to an upset. Diversion of attention can also occur simply from the pilot's efforts to set avionics or navigation equipment while flying the airplane.

Task Saturation

The margin of safety is the difference between task requirements and pilot capabilities. An upset and eventual LOC-I can occur whenever requirements exceed capabilities. For example, an airplane upset event that requires rolling an airplane from a near-inverted to an upright attitude may demand piloting skills beyond those learned during primary training. In another example, a fatigued pilot who inadvertently encounters IMC at night coupled with a vacuum pump failure, or a pilot fails to engage pitot heat while flying in IMC, could become disoriented and lose control of the airplane due to the demands of extended—and unpracticed—partial panel flight. Additionally, unnecessary low-altitude flying and impromptu demonstrations for friends or others on the ground often lead pilots to exceed their capabilities, with fatal results.

Sensory Overload/Deprivation

A pilot's ability to adequately correlate warnings, annunciations, instrument indications, and other cues from the airplane during an upset can be limited. Pilots faced with upset situations can be rapidly confronted with multiple or simultaneous visual, auditory, and tactile warnings. Conversely, sometimes expected warnings are not provided when they should be; this situation can distract a pilot as much as multiple warnings can.

The ability to separate time-critical information from distractions takes practice, experience and knowledge of the airplane and its systems. Cross-checks are necessary not only to corroborate other information that has been presented, but also to determine if information might be missing or invalid. For example, a stall warning system may fail and therefore not warn a pilot of close proximity to a stall, other cues must be used to avert a stall and possible LOC-I. These cues include aerodynamic buffet, loss of roll authority, or inability to arrest a descent.

Spatial Disorientation

Spatial disorientation has been a significant factor in many airplane upset accidents. Accident data from 2008 to 2013 shows nearly 200 accidents associated with spatial disorientation with more than 70% of those being fatal. All pilots are susceptible to false sensory illusions while flying at night or in certain weather conditions. These illusions can lead to a conflict between actual attitude indications and what the pilot senses is the correct attitude. Disoriented pilots may not always be aware of their orientation error. Many airplane upsets occur while the pilot is engaged in some task that takes attention away from the flight instruments or outside references. Others perceive a conflict between bodily senses and the flight instruments, and allow the airplane to divert from the desired flightpath because they cannot resolve the conflict.

A pilot may experience spatial disorientation or perceive the situation in one of three ways:

1. Recognized spatial disorientation: the pilot recognizes the developing upset or the upset condition and is able to safely correct the situation.

2. Unrecognized spatial disorientation: the pilot is unaware that an upset event is developing, or has occurred, and fails to make essential decisions or take any corrective action to prevent LOC-I.

3. Incapacitating spatial disorientation: the pilot is unable to affect a recovery due to some combination of: (a) not understanding the events as they are unfolding, (b) lacking the skills required to alleviate or correct the situation, or (c) exceeding psychological or physiological ability to cope with what is happening.

For detailed information regarding causal factors of spatial disorientation, refer to Aerospace Medicine Spatial Disorientation and Aerospace Medicine Reference Collection, which provides spatial disorientation videos. This collection can be found online at: www.faa.gov/about/ office_org/ headquarters_offices/avs/offices/aam/cami/ library/online_libraries/aerospace_medicine/sd/videos/.

Startle Response

Startle is an uncontrollable, automatic muscle reflex, raised heart rate, blood pressure, etc., elicited by exposure to a sudden, intense event that violates a pilot's expectations.

Surprise Response

Surprise is an unexpected event that violates a pilot's expectations and can affect the mental processes used to respond to the event.

This human response to unexpected events has traditionally been underestimated or even ignored during flight training. The reality is that untrained pilots often experience a state of surprise or a startle response to an airplane upset event. Startle may or may not lead to surprise. Pilots can protect themselves against a debilitating surprise reaction or startle response through scenario-based training, and in such training, instructors can incorporate realistic distractions to help provoke startle or surprise. To be effective the controlled training scenarios must have a perception of risk or threat of consequences sufficient to elevate the pilot's stress levels. Such scenarios can help prepare a pilot to mitigate psychological/physiological reactions to an actual upset.

Upset Prevention and Recovery Training (UPRT)

Upsets are not intentional flight maneuvers, except in maneuver-based training; therefore, they are often unexpected. The reaction of an inexperienced or inadequately trained pilot to an unexpected abnormal flight attitude is usually instinctive rather than intelligent and deliberate. Such a pilot often reacts with abrupt muscular effort, which is without purpose and even hazardous in turbulent conditions, at excessive speeds, or at low altitudes.

Without proper upset recovery training on interpretation and airplane control, the pilot can quickly aggravate an abnormal flight attitude into a potentially fatal LOC-I accident. Consequently, UPRT is intended to focus education and training on the prevention of upsets, and on recovering from these events if they occur. *[Figure 4-14]*

- Upset prevention refers to pilot actions to avoid a divergence from the desired airplane state. Awareness and prevention training serve to avoid incidents; early recognition of an upset scenario coupled with appropriate preventive action often can mitigate a situation that could otherwise escalate into a LOC-I accident.

Figure 4-14. *Maneuvers that better prepare a pilot for understanding unusual attitudes and situations are representative of upset training.*

- Recovery refers to pilot actions that return an airplane that is diverging in altitude, airspeed, or attitude to a desired state from a developing or fully developed upset. Learn to initiate recovery to a normal flight mode immediately upon recognition of the developing upset condition. Ensure that control inputs and power adjustments applied to counter an upset are in direct proportion to the amount and rates of change of roll, yaw, pitch, or airspeed so as to avoid overstressing the airplane unless ground contact is imminent. Recovery training serves to reduce accidents as a result of an unavoidable or inadvertently encountered upset event.

UPRT Core Concepts

Airplane upsets are by nature time-critical events; they can also place pilots in unusual and unfamiliar attitudes that sometimes require counterintuitive control movements. Upsets have the potential to put a pilot into a life-threatening situation compounded by panic, diminished mental capacity, and potentially incapacitating spatial disorientation. Because real-world upset situations often provide very little time to react, exposure to such events during training is essential for pilots to reduce surprise and it mitigates confusion during unexpected upsets. The goal is to equip the pilot to promptly recognize an escalating threat pattern or sensory overload and quickly identify and correct an impending upset.

UPRT stresses that the first step is recognizing any time the airplane begins to diverge from the intended flightpath or airspeed. Pilots must identify and determine what, if any, action must be taken. As a general rule, any time visual cues or instrument indications differ from basic flight maneuver expectations, the pilot should assume an upset and cross-check to confirm the attitude, instrument error or instrument malfunction.

To achieve maximum effect, it is crucial for UPRT concepts to be conveyed accurately and in a non-threatening manner. Reinforcing concepts through positive experiences significantly improves a pilot's depth of understanding, retention of skills, and desire for continued training. Also, training in a carefully structured environment allows for exposure to these events and can help the pilot react more quickly, decisively, and calmly when the unexpected occurs during flight. However, like many other skills, the skills needed for upset prevention and recovery are perishable and thus require continuous reinforcement through training.

UPRT in the airplane and flight simulation training device (FSTD) should be conducted in both visual and simulated instrument conditions to allow pilots to practice recognition and recovery under both situations. UPRT should allow them to experience and recognize some of the physiological factors related to each, such as the confusion and disorientation that can result from visual cues in an upset event. Training that includes recovery from bank angles exceeding 90 degrees could further add to a pilot's overall knowledge and skills for upset recognition and recovery. For such training, additional measures should be taken to ensure the suitability of the airplane or FSTD and that instructors are appropriately qualified.

Upset prevention and recovery training is different from aerobatic training. *[Figure 4-15]* In aerobatic training, the pilot knows and expects the maneuver, so effects of startle or surprise are missing. The main goal of aerobatic training is to teach pilots how to intentionally and precisely maneuver an aerobatic-capable airplane in three dimensions. The primary goal of UPRT is to help pilots overcome sudden onsets of stress to avoid, prevent, and recover from unplanned excursions that could lead to LOC-I.

Aerobatics vs. UPRT Flight Training Methods		
ASPECT OF TRAINING	**AEROBATICS**	**UPSET PREVENTION AND RECOVERY TRAINING**
Primary Objective	Precision maneuvering capability	Safe, effective recovery from aircraft upsets
Secondary Outcome	Improved manual aircraft handling skills	Improved manual aircraft handling skills
Aerobatic Maneuvering	Primary mode of training	Supporting mode of training
Academics	Supporting role	Fundamental component
Training Resources Utilized	Aircraft (few exceptions)	Aircraft or a full-flight simulator

Figure 4-15. *Some differences between aerobatic training and upset prevention and recovery training.*

Comprehensive UPRT builds on three mutually supportive components: academics, airplane-based training and, typically at the transport category type-rating training level, use of FSTDs. Each has unique benefits and limitations but, when implemented cohesively and comprehensively throughout a pilot's career, the components can offer maximum preparation for upset awareness, prevention, recognition, and recovery.

Academic Material (Knowledge and Risk Management)

Academics establish the foundation for development of situational awareness, insight, knowledge, and skills. As in practical skill development, academic preparation should move from the general to specific while emphasizing the significance of each basic concept. Although academic preparation is crucial and does offer a level of mitigation of the LOC-I threat, long-term retention of knowledge is best achieved when applied and correlated with practical hands-on experience.

The academic material needs to build awareness in the pilot by providing the concepts, principles, techniques, and procedures for understanding upset hazards and mitigating strategies. Awareness of the relationship between AOA, G-load, lift, energy management, and the consequences of their mismanagement, is essential for assessing hazards, mitigating the risks, and acquiring and employing prevention skills. Training maneuvers should be designed to provide awareness of situations that could lead to an upset or LOC. With regard to the top four causal and contributing factors to LOC-I accidents presented earlier in this chapter, training should include scenarios that place the airplane and pilot in a simulated situation/environment that can lead to an upset.

The academics portion of UPRT should also address the prevention concepts surrounding Aeronautical Decision Making (ADM) and risk management (RM), and proportional counter response.

Prevention Through ADM and Risk Management

This element of prevention routinely occurs in a time-scale of minutes or hours, revolving around the concept of effective ADM and risk management through analysis, awareness, resource management, and interrupting the error chain through basic airmanship skills and sound judgment. For instance, imagine a situation in which a pilot assesses conditions at an airport prior to descent and recognizes those conditions as being too severe to safely land the airplane. Using situational awareness to avert a potentially threatening flight condition is an example of prevention of a LOC-I situation through effective risk management. Pilots should evaluate the circumstances for each flight (including the equipment and environment), looking specifically for scenarios that may require a higher level of risk management. These include situations which could result in low-altitude maneuvering, steep turns in the pattern, uncoordinated flight, or increased load factors.

Another part of ADM is crew resource management (CRM) or Single Pilot Resource Management (SRM). Both are relevant to the UPRT environment. When available, a coordinated crew response to potential and developing upsets can provide added benefits such as increased situational awareness, mutual support, and an improved margin of safety. Since an untrained crewmember can be the most unpredictable element in an upset scenario, initial UPRT for crew operations should be mastered individually before being integrated into a multi-crew, CRM environment. A crew must be able to accomplish the following:

- Communicate and confirm the situation clearly and concisely;

- Transfer control to the most situationally aware crewmember;

- Using standardized interactions, work as a team to enhance awareness, manage stress, and mitigate fear.

Prevention through Proportional Counter-Response

In simple terms, proportional counter response is the timely manipulation of flight controls and thrust, either as the sole pilot or crew as the situation dictates, to manage an airplane flight attitude or flight envelope excursion that was unintended or not commanded by the pilot.

The time-scale of this element of prevention typically occurs on the order of seconds or fractions of seconds, with the goal being able to recognize a developing upset and take proportionally appropriate avoidance actions to preclude the airplane entering a fully developed upset. Due to the sudden, surprising nature of this level of developing upset, there exists a high risk for panic and overreaction to ensue and aggravate the situation.

Recovery

Last but not least, the academics portion lays the foundation for development of UPRT skills by instilling the knowledge, procedures, and techniques required to accomplish a safe recovery. The airplane and FSTD-based training elements presented below serve to translate the academic material into structured practice. This can start with classroom visualization of recovery procedures and continue with repetitive skill practiced in an airplane, and then potentially further developed in the simulated environment.

In the event looking outside does not provide enough situational awareness of the airplane attitude, a pilot can use the flight instruments to recognize and recover from an upset. To recover from nose-high and nose-low attitudes, the pilot should follow the procedures recommended in the AFM/POH. In general, upset recovery procedures are summarized in *Figure 4-16*.

Upset Recovery Template
1. Disconnect the wing leveler or autopilot
2. Apply forward column or stick pressure to unload the airplane
3. Aggressively roll the wings to the nearest horizon
4. Adjust power as necessary by monitoring airspeed
5. Return to level flight

Figure 4-16. *Upset recovery template.*

Common Errors

Common errors associated with upset recoveries include the following:

- Incorrect assessment of what kind of upset the airplane is in

- Failure to disconnect the wing leveler or autopilot

- Failure to unload the airplane, if necessary

- Failure to roll in the correct direction

- Inappropriate management of the airspeed during the recovery

Roles of FSTDs and Airplanes in UPRT

Training devices range from aviation training devices (e.g., basic and advanced) to FSTDs (e.g., flight training devices (FTD) and full flight simulators (FFS)) and have a broad range of capabilities. While all of these devices have limitations relative to actual flight, only the higher fidelity devices (i.e., Level C and D FFS) are a satisfactory substitution for developing UPRT skills in the actual aircraft. Except for these higher fidelity devices, initial skill development should be accomplished in a suitable airplane, and the accompanying training device should be used to build upon these skills. *[Figure 4-17]*

Airplane-Based UPRT

Ultimately, the more realistic the training scenario, the more indelible the learning experience. Although creating a visual scene of a 110° banked attitude with the nose 30° below the horizon may not be technically difficult in a modern simulator, the learning achieved while viewing that scene from the security of the simulator is not as complete as when viewing the same scene in an airplane. Maximum learning is achieved when the pilot is placed in the controlled, yet adrenaline-enhanced, environment of upsets experienced

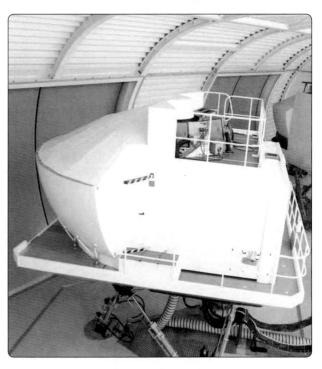

Figure 4-17. *A Level D full-flight simulator could be used for UPRT.*

while in flight. For these reasons, airplane-based UPRT improves a pilot's ability to overcome fear in an airplane upset event.

However, airplane-based UPRT does have limitations. The level of upset training possible may be limited by the maneuvers approved for the particular airplane, as well as by the flight instructor's own UPRT capabilities. For instance, UPRT conducted in the normal category by a typical CFI will necessarily be different from UPRT conducted in the aerobatic category by a CFI with expertise in aerobatics.

When considering upset training conducted in an aerobatic-capable airplane in particular, the importance of employing instructors with specialized UPRT experience in those airplanes cannot be overemphasized. Just as instrument or tailwheel instruction requires specific skill sets for those operations, UPRT demands that instructors possess the competence to oversee trainee progress, and the ability to intervene as necessary with consistency and professionalism. As in any area of training, the improper delivery of stall, spin and upset recovery training often results in negative learning, which could have severe consequences not only during the training itself, but in the skills and mindset pilots take with them into the cockpits of airplanes where the lives of others may be at stake.

All-Attitude/All-Envelope Flight Training Methods

Sound UPRT encompasses operation in a wide range of possible flight attitudes and covers the airplane's limit flight envelope. This training is essential to prepare pilots for unexpected upsets. As stated at the outset, the primary focus of a comprehensive UPRT program is the avoidance of, and safe recovery from, upsets. Much like basic instrument skills, which can be applied to flying a vast array of airplanes, the majority of skills and techniques required for upset recovery are not airplane specific. Just as basic instrument skills learned in lighter and lower performing airplanes are applied to more advanced airplanes, basic upset recovery techniques provide lessons that remain with pilots throughout their flying careers.

FSTD–based UPRT

UPRT can be effective in high fidelity devices (i.e. Level C and D FFS), however instructors and pilots must be mindful of the technical and physiological boundaries when using a particular FSTD for upset training. The FSTD must be qualified by the FAA National Simulator Program for UPRT; and, if the training is required for pilots by regulation, the course must also be FAA approved.

Spiral Dive

A spiral dive, a nose low upset, is a descending turn during which airspeed and G-load can increase rapidly and often results from a botched turn. In a spiral dive, the airplane is flying very tight circles, in a nearly vertical attitude and will be accelerating because it is no longer stalled. Pilots typically get into a spiral dive during an inadvertent IMC encounter, most often when the pilot relies on kinesthetic sensations rather than on the flight instruments. A pilot distracted by other sensations can easily enter a slightly nose low, wing low, descending turn and, at least initially, fail to recognize this error. Especially in IMC, it may be only the sound of increasing speed that makes the pilot aware of the rapidly developing situation. Upon recognizing the steep nose down attitude and steep bank, the startled pilot may react by pulling back rapidly on the yoke while simultaneously rolling to wings level. This response can create aerodynamic loads capable of causing airframe structural damage and /or failure.

1. Reduce Power (Throttle) to Idle
2. Apply Some Forward Elevator
3. Roll Wings Level
4. Gently Raise the Nose to Level Flight
5. Increase Power to Climb Power

The following discussion explains each of the five steps:

1. Reduce Power (Throttle) to Idle. Immediately reduce power to idle to slow the rate of acceleration.

2. Apply Some Forward Elevator. Prior to rolling the wings level, it is important to unload the G-load on the airplane ("unload the wing"). This is accomplished by applying some forward elevator pressure to return to about +1G. Apply just enough forward elevator to ensure that you are not aggravating the spiral with aft elevator. While generally a small input, this push has several benefits prior to rolling the wings level in the next step – the push reduces the AOA, reduces the G-load, and slows the turn rate while increasing the turn radius, and prevents a rolling pullout. The design limit of the airplane is lower during a rolling pullout, so failure to reduce the G-load prior to rolling the wings level could result in structural damage or failure.

3. Roll Wings Level. Roll to wings level using coordinated aileron and rudder inputs. Even though the airplane is in a nose-low attitude, continue the roll until the wings are completely level again before performing step four.

4. Gently Raise the Nose to Level Flight. It is possible that the airplane in a spiral dive might be at or even beyond V_{NE} (never exceed speed) speed. Therefore, the pilot must make all control inputs slowly and gently at this point to prevent structural failure. Raise the nose to a climb attitude only after speed decreases to safe levels.

Spiral Dive Recovery Template
1. Reduce power (throttle) to idle
2. Apply some forward elevator
3. Roll wings level
4. Gently raise the nose to level flight
5. Increase power to climb power

Figure 4-18. *Spiral dive recovery template.*

5. Increase Power to Climb Power. Once the airspeed has stabilized to V_Y, apply climb power and climb back to a safe altitude.

In general, spiral dive recovery procedures are summarized in *Figure 4-18*.

Common errors in the recovery from spiral dives are:

- Failure to reduce power first
- Mistakenly adding power
- Attempting to pull out of dive without rolling wings level
- Simultaneously pulling out of dive while rolling wings level
- Not unloading the Gs prior to rolling level
- Not adding power once climb is established

UPRT Summary
A significant point to note is that UPRT skills are both complex and perishable. Repetition is needed to establish the correct mental models, and recurrent practice/training is necessary as well. The context in which UPRT procedures are introduced and implemented is also an important consideration. The pilot must clearly understand, for example, whether a particular procedure has broad applicability, or is type-specific. To attain the highest levels of learning possible, the best approach starts with the broadest form of a given procedure, then narrows it down to type-specific requirements.

Chapter Summary
A pilot's most fundamental and important responsibility is to maintain aircraft control. Initial flight training thus provides skills to operate an airplane in a safe manner, generally within normal "expected" environments, with the addition of some instruction in upset and stall situations.

This chapter discussed the elements of basic aircraft control, with emphasis on AOA. It offered a discussion of circumstances and scenarios that can lead to LOC-I, including stalls and airplane upsets. It discussed the importance of developing proficiency in slow flight, stalls, and stall recoveries, spin awareness and recovery, upset prevention and recovery, and spiral dive recovery.

Pilots need to understand that primary training cannot cover all possible contingencies that an airplane or pilot may encounter, and therefore they should seek recurrent/additional training for their normal areas of operation, as well as to seek appropriate training that develops the aeronautical skill set beyond the requirements for initial certification.

For additional considerations on performing some of these maneuvers in multiengine airplanes and jet powered airplanes, refer to Chapters 12 and 15, respectively.

Additional advisory circular (AC) guidance is available at www.faa.gov:

- AC 61-67 (as revised), Stall and Spin Awareness Training;
- AC 120-109 (as revised), Stall Prevention and Recovery Training; and
- AC 120-111 (as revised), Upset Prevention and Recovery Training.

Chapter 5
Takeoffs and Departure Climbs

Introduction

A review of aircraft accident data shows that about twenty percent of all general aviation (GA) accidents occur during takeoff and departure climbs. Further breakdown of the data indicates that more than half of those accidents were the result of some sort of failure of the pilot, and twenty percent of the mishaps are the result of loss in control of the airplane. When compared to the entire profile of a normal flight, this phase of a flight is relatively short, but the pilot workload is greatest. This chapter discusses takeoffs and departure climbs in airplanes under normal conditions and under conditions that require maximum performance.

Though it may seem relatively simple, the takeoff often presents the most hazards of any part of a flight. The importance of thorough knowledge of procedures and techniques coupled with proficiency in performance cannot be overemphasized.

The discussion in this chapter is centered on airplanes with tricycle landing gear (nose-wheel). Procedures for conventional gear airplanes (tail-wheel) are discussed in Chapter 14. The manufacturer's recommended procedures pertaining to airplane configuration, airspeeds, and other information relevant to takeoffs and departure climbs in a specific make and model airplane are contained in the Federal Aviation Administration (FAA) approved Airplane Flight Manual and/or Pilot's Operating Handbook (AFM/POH) for that airplane. If any of the information in this chapter differs from the airplane manufacturer's recommendations as contained in the AFM/POH, the airplane manufacturer's recommendations take precedence.

Terms and Definitions

Although the takeoff and climb is one continuous maneuver, it will be divided into three separate steps for purposes of explanation: 1. takeoff roll, 2. lift-off, and 3. initial climb after becoming airborne. *[Figure 5-1]*

- Takeoff roll (ground roll) is the portion of the takeoff procedure during which the airplane is accelerated from a standstill to an airspeed that provides sufficient lift for it to become airborne.

- Lift-off is when the wings are lifting the weight of the airplane off the surface. In most airplanes, this is the result of the pilot rotating the nose up to increase the angle of attack (AOA).

- The initial climb begins when the airplane leaves the surface and a climb pitch attitude has been established. Normally, it is considered complete when the airplane has reached a safe maneuvering altitude or an en route climb has been established.

Prior to Takeoff

Before going to the airplane, the pilot should check the POH/AFM performance charts to determine the predicted performance and decide if the airplane is capable of a safe takeoff and climb for the conditions and location. *[Figure 5-2]* High density altitudes reduce engine and

Figure 5-1. *Takeoff and climb.*

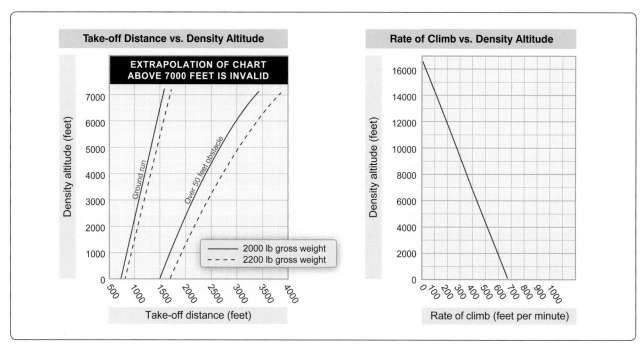

Figure 5-2. *Performance chart examples.*

propeller performance, increase takeoff rolls and decrease climb performance. A more detailed discussion of density altitude and how it affects airplane performance can be found in the Pilot's Handbook of Aeronautical Knowledge (FAA-H-8083-25, as revised).

All run up and pre-takeoff checklist items should be completed before taxiing onto the runway or takeoff area. As a minimum before every takeoff, all engine instruments should be checked for proper and usual indications, and all controls should be checked for full, free, and correct movement. In addition, the pilot must make certain that the approach and takeoff paths are clear of other aircraft. At nontowered airports, pilots should announce their intentions on the common traffic advisory frequency (CTAF) assigned to that airport. When operating from a towered airport, pilots must contact the tower operator and receive a takeoff clearance before taxiing onto the active runway.

It is not recommended to take off immediately behind another aircraft, particularly large, heavily loaded transport airplanes, because of the wake turbulence that is generated. If an immediate takeoff is necessary, plan to minimize the chances of flying through an aircraft's wake turbulence by avoiding the other aircraft's flightpath or rotate prior to the point at which the preceding aircraft rotated. While taxiing onto the runway, select ground reference points that are aligned with the runway direction to aid in maintaining directional control and alignment with the runway center line during the climb out. These may be runway centerline markings, runway lighting, distant trees, towers, buildings, or mountain peaks.

Normal Takeoff

A normal takeoff is one in which the airplane is headed into the wind; there are times that a takeoff with a tail wind is necessary. However, the pilot must consult the POH/AFM to ensure the aircraft is approved for a takeoff with a tail wind and that there is sufficient performance and runway length for the takeoff. Also, the takeoff surfaces are firm and of sufficient length to permit the airplane to gradually accelerate to normal lift-off and climb-out speed, and there are no obstructions along the takeoff path.

There are two reasons for making a takeoff as nearly into the wind as possible. First, since the airplane depends on airspeed, a headwind provides some of that airspeed even before the airplane begins to accelerate into the wind. Second, a headwind decreases the ground speed necessary to achieve flying speed. Slower ground speeds yield shorter ground roll distances and allow use of shorter runways while reducing wear and stress on the landing gear.

Takeoff Roll

For takeoff, use the rudder pedals in most general aviation airplanes to steer the airplane's nose wheel onto the runway centerline to align the airplane and nose wheel with the runway. After releasing the brakes, advance the throttle smoothly and continuously to takeoff power. An abrupt application of power may cause the airplane to yaw sharply to the left because of the torque effects of the engine and propeller. This is most apparent in high horsepower engines. As the airplane starts to roll forward, assure both feet are on the rudder pedals so that the toes or balls of the feet are on the rudder portions, not on the brake. At all times, monitor the engine instruments for indications of a malfunction during the takeoff roll.

In nose-wheel type airplanes, pressures on the elevator control are not necessary beyond those needed to steady it. Applying unnecessary pressure only aggravates the takeoff and prevents the pilot from recognizing when elevator control pressure is actually needed to establish the takeoff attitude.

As the airplane gains speed, the elevator control tends to assume a neutral position if the airplane is correctly trimmed. At the same time, the rudder pedals are used to keep the nose of the airplane pointed down the runway and parallel to the centerline. The effects of engine torque and P-factor at the initial speeds tend to pull the nose to the left (Torque and P-Factor will be discussed in greater detail in later chapter). The pilot must use whatever rudder pressure is needed to correct for these effects or winds. Use aileron controls into any crosswind to keep the airplane centered on the runway centerline. The pilot should avoid using the brakes for steering purposes as this will slow acceleration, lengthen the takeoff distance, and possibly result in severe swerving.

As the speed of the takeoff roll increases, more and more pressure will be felt on the flight controls, particularly the elevators and rudder. If the tail surfaces are affected by the propeller slipstream, they become effective first. As the speed continues to increase, all of the flight controls will gradually become effective enough to maneuver the airplane about its three axes. At this point, the airplane is being flown more than it is being taxied. As this occurs, progressively smaller rudder deflections are needed to maintain direction.

The feel of resistance to the movement of the controls and the airplane's reaction to such movements are the only real indicators of the degree of control attained. This feel of resistance is not a measure of the airplane's speed, but rather of its controllability. To determine the degree of controllability, the pilot must be conscious of the reaction of the airplane to the control pressures and immediately adjust the pressures as needed to control the airplane. The pilot must wait for the reaction of the airplane to the applied control pressures and attempt to sense the control resistance to pressure rather than attempt to control the airplane by movement of the controls.

A student pilot does not normally have a full appreciation of the variations of control pressures with the speed of the airplane. The student may tend to move the controls through wide ranges seeking the pressures that are familiar and expected and, as a consequence, overcontrol the airplane.

The situation may be aggravated by the sluggish reaction of the airplane to these movements. The flight instructor must help the student learn proper response to control actions and airplane reactions. The instructor should always stress using the proper outside reference to judge airplane motion. For takeoff, the student should always be looking far down the runway at two points aligned with the runway. The flight instructor should have the student pilot follow through lightly on the controls, feel for resistance, and point out the outside references that provide the clues for how much control movement is needed and how the pressure and response changes as airspeed increases. With practice, the student pilot should become familiar with the airplane's response to acceleration to lift off speed, corrective control movements needed, and the outside references necessary to accomplish the takeoff maneuver.

Lift-Off

Since a good takeoff depends on the proper takeoff attitude, it is important to know how this attitude appears and how it is attained. The ideal takeoff attitude requires only minimum pitch adjustments shortly after the airplane lifts off to attain the speed for the best rate of climb (V_Y). *[Figure 5-3]* The pitch attitude necessary for the airplane to accelerate to V_Y speed should be demonstrated by the instructor and memorized by the student. Flight instructors should be aware that initially, the student pilot may have a tendency to hold excessive back-elevator pressure just after lift-off, resulting in an abrupt pitch-up.

Figure 5-3. *Initial roll and takeoff attitude.*

Each type of airplane has a best pitch attitude for normal lift-off; however, varying conditions may make a difference in the required takeoff technique. A rough field, a smooth field, a hard surface runway, or a short or soft, muddy field all call for a slightly different technique, as will smooth air in contrast to a strong, gusty wind. The different techniques for those other-than-normal conditions are discussed later in this chapter.

When all the flight controls become effective during the takeoff roll in a nose-wheel type airplane, the pilot should gradually apply back-elevator pressure to raise the nose-wheel slightly off the runway, thus establishing the takeoff or lift-off attitude. This is the "rotation" for lift off and climb. As the airplane lifts off the surface, the pitch attitude to hold the climb airspeed should be held with elevator control and trimmed to maintain that pitch attitude without excessive control pressures. The wings should be leveled after lift-off and the rudder used to ensure coordinated flight.

After rotation, the slightly nose-high pitch attitude should be held until the airplane lifts off. Rudder control should be used to maintain the track of the airplane along the runway centerline until any required crab angle in level flight is established. Forcing it into the air by applying excessive back-elevator pressure would only result in an excessively high-pitch attitude and may delay the takeoff. As discussed earlier, excessive and rapid changes in pitch attitude result in proportionate changes in the effects of torque, thus making the airplane more difficult to control.

Although the airplane can be forced into the air, this is considered an unsafe practice and should be avoided under normal circumstances. If the airplane is forced to leave the ground by using too much back-elevator pressure before adequate flying speed is attained, the wing's AOA may become excessive, causing the airplane to settle back to the runway or even to stall. On the other hand, if sufficient back-elevator pressure is not held to maintain the correct takeoff attitude after becoming airborne, or the nose is allowed to lower excessively, the airplane may also settle back to the runway. This would occur because the AOA is decreased and lift diminished to the degree where it will not support the airplane. It is important, then, to hold the correct attitude constant after rotation or lift-off.

As the airplane leaves the ground, the pilot must keep the wings in a level attitude and hold the proper pitch attitude. Outside visual scans must be intensified at this critical point to attain/maintain proper airplane pitch and bank attitude. Due to the minimum airspeed, the flight controls are not as responsive, requiring more control movement to achieve an expected response. A novice pilot often has a tendency to fixate on the airplane's pitch attitude and/or the airspeed indicator and neglect bank control of the airplane. Torque from the engine tends to impart a rolling force that is most evident as the landing gear is leaving the surface.

During takeoffs in a strong, gusty wind, it is advisable that an extra margin of speed be obtained before the airplane is allowed to leave the ground. A takeoff at the normal takeoff speed may result in a lack of positive control, or a stall, when the airplane encounters a sudden lull in strong, gusty wind, or other turbulent air currents. In this case, the pilot should allow the airplane to stay on the ground longer to attain more speed; then make a smooth, positive rotation to leave the ground.

Initial Climb

Upon lift-off, the airplane should be flying at approximately the pitch attitude that allows it to accelerate to V_Y. This is the speed at which the airplane gains the most altitude in the shortest period of time.

If the airplane has been properly trimmed, some back-elevator pressure may be required to hold this attitude until the proper climb speed is established. Relaxation of any back-elevator pressure before this time may result in the airplane settling, even to the extent that it contacts the runway.

The airplane's speed will increase rapidly after it becomes airborne. Once a positive rate of climb is established, the pilot should retract the flaps and landing gear (if equipped). It is recommended that takeoff power be maintained until reaching an altitude of at least 500 feet above the surrounding terrain or obstacles. The combination of V_Y and takeoff power assures the maximum altitude gained in a minimum amount of time. This gives the pilot more altitude from which the airplane can be safely maneuvered in case of an engine failure or other emergency. A pilot should also consider flying at Vy versus a lower pitch for a cruise climb requires much quicker pilot response in the event of a powerplant failure to preclude a stall.

Since the power on the initial climb is set at the takeoff power setting, the airspeed must be controlled by making slight pitch adjustments using the elevators. However, the pilot should not fixate on the airspeed indicator when making these pitch changes, but should continue to scan outside to adjust the airplane's attitude in relation to the horizon. In accordance with the principles of attitude flying, the pilot should first make the necessary pitch change with reference to the natural horizon and hold the new attitude momentarily, and then glance at the airspeed indicator to verify if the new attitude is correct. Due to inertia, the airplane will not accelerate or decelerate immediately as the pitch is changed. It takes a little time for the airspeed to change. If the pitch attitude has been over or under corrected, the airspeed indicator will show a speed that is higher or lower than that desired. When this occurs, the cross-checking and appropriate pitch-changing process must be repeated until the desired climbing attitude is established. Pilots must remember the climb pitch will be lower when the airplane is heavily loaded, or power is limited by density altitude.

When the correct pitch attitude has been attained, the pilot should hold it constant while cross-checking it against the horizon and other outside visual references. The airspeed indicator should be used only as a check to determine if the attitude is correct.

After the recommended climb airspeed has been established and a safe maneuvering altitude has been reached, the pilot should adjust the power to the recommended climb setting and trim the airplane to relieve the control pressures. This makes it easier to hold a constant attitude and airspeed.

During initial climb, it is important that the takeoff path remain aligned with the runway to avoid drifting into obstructions or into the path of another aircraft that may be taking off from a parallel runway. A flight instructor should help the student identify two points inline ahead of the runway to use as a tracking reference. As long as those two points are inline, the airplane is remaining on the desired track. Proper scanning techniques are essential to a safe takeoff and climb, not only for maintaining attitude and direction, but also for avoiding collisions near the airport.

When the student pilot nears the solo stage of flight training, it should be explained that the airplane's takeoff performance will be much different when the instructor is not in the airplane. Due to decreased load, the airplane will become airborne earlier and climb more rapidly. The pitch attitude that the student has learned to associate with initial climb may also differ due to decreased weight, and the flight controls may seem more sensitive. If the situation is unexpected, it may result in increased tension that may remain until after the landing. Frequently, the existence of this tension and the uncertainty that develops due to the perception of an "abnormal" takeoff results in poor performance on the subsequent landing.

Common errors in the performance of normal takeoffs and departure climbs are:

- Failure to review AFM/POH and performance charts prior to takeoff.

- Failure to adequately clear the area prior to taxiing into position on the active runway.

- Abrupt use of the throttle.

- Failure to check engine instruments for signs of malfunction after applying takeoff power.

- Failure to anticipate the airplane's left turning tendency on initial acceleration.

- Overcorrecting for left turning tendency.

- Relying solely on the airspeed indicator rather than developing an understanding of visual references and tracking clues of airplane airspeed and controllability during acceleration and lift-off.

- Failure to attain proper lift-off attitude.

- Inadequate compensation for torque/P-factor during initial climb resulting in a sideslip.

- Overcontrol of elevators during initial climb-out and lack of elevator trimming.

- Limiting scan to areas directly ahead of the airplane (pitch attitude and direction), causing a wing (usually the left) to drop immediately after lift-off.

- Failure to attain/maintain best rate-of-climb airspeed (V_Y) or desired climb airspeed.

- Failure to employ the principles of attitude flying during climb-out, resulting in "chasing" the airspeed indicator.

Crosswind Takeoff

While it is usually preferable to take off directly into the wind whenever possible or practical, there are many instances when circumstances or judgment indicate otherwise. Therefore, the pilot must be familiar with the principles and techniques involved in crosswind takeoffs, as well as those for normal takeoffs. A crosswind affects the airplane during takeoff much as it does during taxiing. With this in mind, the pilot should be aware that the technique used for crosswind correction during takeoffs closely parallels the crosswind correction techniques used for taxiing.

Takeoff Roll

The technique used during the initial takeoff roll in a crosswind is generally the same as the technique used in a normal takeoff roll, except that the pilot must apply aileron pressure into the crosswind. This raises the aileron on the upwind wing, imposing a downward force on the wing to counteract the lifting force of the crosswind; and thus preventing the wing from rising. The pilot must remember that since the ailerons and rudder are deflected, drag will increase; therefore, less initial takeoff performance should be expected until the airplane is wings-level in coordinated flight in the climb.

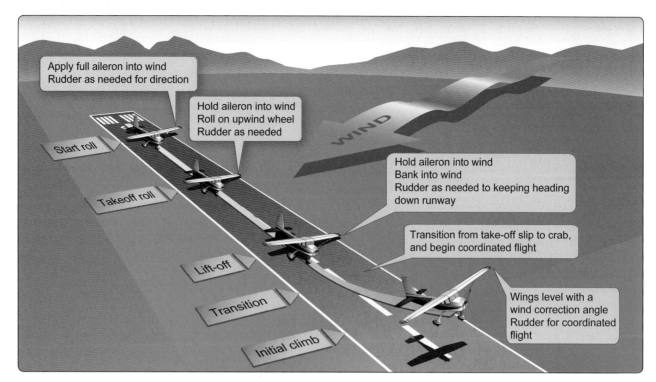

Figure 5-4. *Crosswind roll and takeoff climb.*

While taxiing into takeoff position, it is essential that the pilot check the windsock and other wind direction indicators for the presence of a crosswind. If a crosswind is present, the pilot should apply full aileron pressure into the wind while beginning the takeoff roll. The pilot should maintain this control position, as the airplane accelerates, until the ailerons become effective in maneuvering the airplane about its longitudinal axis. As the ailerons become effective, the pilot will feel an increase in pressure on the aileron control.

While holding aileron pressure into the wind, the pilot should use the rudder to maintain a straight takeoff path. *[Figure 5-4]* Since the airplane tends to weathervane into the wind while on the ground, the pilot will typically apply downwind rudder pressure. When the pilot increases power for takeoff, the resulting P-factor causes the airplane to yaw to the left. While this yaw may be sufficient to counteract the airplane's tendency to weathervane into the wind in a crosswind to the right, it may aggravate this tendency in a crosswind to the left. In any case, the pilot should apply rudder pressure in the appropriate direction to keep the airplane rolling straight down the runway.

As the forward speed of the airplane increases, the pilot should only apply enough aileron pressure to keep the airplane laterally aligned with the runway centerline. The rudders keep the airplane pointed parallel with the runway centerline, while the ailerons keep the airplane laterally aligned with the centerline. The crosswind component

effect will not completely vanish; therefore, the pilot must maintain some aileron pressure throughout the takeoff roll to keep the crosswind from raising the upwind wing. If the upwind wing rises, the amount of wing surface exposed to the crosswind will increase, which may cause the airplane to "skip." *[Figure 5-5]*

Figure 5-5. *Crosswind effect.*

This "skipping" is usually indicated by a series of very small bounces caused by the airplane attempting to fly and then settling back onto the runway. During these bounces, the crosswind also tends to move the airplane sideways, and these bounces develop into side-skipping. This side-skipping imposes severe side stresses on the landing gear and may result in structural failure.

During a crosswind takeoff roll, it is important that the pilot hold sufficient aileron pressure into the wind not only to keep the upwind wing from rising but to hold that wing down so that the airplane sideslips into the wind enough to counteract drift immediately after lift-off.

Lift-Off

As the nose-wheel raises off of the runway, the pilot should hold aileron pressure into the wind. This may cause the downwind wing to rise and the downwind main wheel to lift off the runway first, with the remainder of the takeoff roll being made on that one main wheel. This is acceptable and is preferable to side-skipping.

If a significant crosswind exists, the pilot should hold the main wheels on the ground slightly longer than in a normal takeoff so that a smooth but very definite lift-off can be made. This allows the airplane to leave the ground under more positive control and helps it remain airborne while the pilot establishes the proper amount of wind correction. More importantly, this procedure avoids imposing excessive side-loads on the landing gear and prevents possible damage that would result from the airplane settling back to the runway while drifting.

As both main wheels leave the runway, the airplane begins to drift sideways with the wind as ground friction is no longer a factor in preventing lateral movement. To minimize this lateral movement and to keep the upwind wing from rising, the pilot must establish and maintain the proper amount of crosswind correction prior to lift-off by applying aileron pressure into the wind. The pilot must also apply rudder pressure, as needed, to prevent weathervaning.

Initial Climb

If a proper crosswind correction is applied, the aircraft will maintain alignment with the runway while accelerating to takeoff speed and then maintain that alignment once airborne. As takeoff acceleration occurs, the efficiency of the up-aileron will increase with aircraft speed causing the upwind wing to produce greater downward force and, as a result, counteract the effect of the crosswind. The yoke, having been initially turned into the wind, can be relaxed to the extent necessary to keep the aircraft aligned with the runway. As the aircraft becomes flyable and airborne, the wing that is upwind will have a tendency to be lower relative the other wing requiring

simultaneous rudder input to maintain runway alignment. This will initially result in the aircraft to sideslip. However, as the aircraft establishes its climb, the nose should be turned into the wind to offset the crosswind, wings brought to level, and rudder input adjusted to maintain runway alignment (crabbing). *[Figure 5-6]* Firm and positive use of the rudder may be required to keep the airplane pointed down the runway or parallel to the centerline. Unlike landing, the runway alignment (staying over the runway and its extended centerline) is paramount to keeping the aircraft parallel to the centerline. The pilot must then apply rudder pressure firmly and aggressively to keep the airplane headed straight down the runway. However, because the force of a crosswind may vary markedly within a few hundred feet of the ground, the pilot should check the ground track frequently and adjust the wind correction angle, as necessary. The remainder of the climb technique is the same used for normal takeoffs and climbs.

The most common errors made while performing crosswind takeoffs include the following:

- Failure to review AFM/POH performance and charts prior to takeoff.
- Failure to adequately clear the area prior to taxiing onto the active runway.

Figure 5-6. *Crosswind climb flightpath.*

- Using less than full aileron pressure into the wind initially on the takeoff roll.

- Mechanical use of aileron control rather than judging lateral position of airplane on runway from visual clues and applying sufficient aileron to keep airplane centered laterally on runway.

- Side-skipping due to improper aileron application.

- Inadequate rudder control to maintain airplane parallel to centerline and pointed straight ahead in alignment with visual references.

- Excessive aileron input in the latter stage of the takeoff roll resulting in a steep bank into the wind at lift-off.

- Inadequate drift correction after lift-off.

Ground Effect on Takeoff

Ground effect is a condition of improved performance encountered when the airplane is operating very close to the ground. Ground effect can be detected and normally occurs up to an altitude equal to one wingspan above the surface. *[Figure 5-7]* Ground effect is most significant when the airplane maintains a constant attitude at low airspeed at low altitude (for example, during takeoff when the airplane lifts off and accelerates to climb speed, and during the landing flare before touchdown).

When the wing is under the influence of ground effect, there is a reduction in upwash, downwash, and wingtip vortices. As a result of the reduced wingtip vortices, induced drag is reduced. When the wing is at a height equal to ¼ the span, the reduction in induced drag is about 25 percent. When the wing is at a height equal to ¹/₁₀ the span, the reduction in induced drag is about 50 percent. At high speeds where parasite drag dominates, induced drag is a small part of the total drag. Consequently, ground effect is a greater concern during takeoff and landing.

At takeoff, the takeoff roll, lift-off, and the beginning of the initial climb are accomplished within the ground effect area.

The ground effect causes local increases in static pressure, which cause the airspeed indicator and altimeter to indicate slightly lower values than they should and usually cause the vertical speed indicator to indicate a descent. As the airplane lifts off and climbs out of the ground effect area, the following occurs:

- The airplane requires an increase in AOA to maintain lift coefficient.

- The airplane experiences an increase in induced drag and thrust required.

- The airplane experiences a pitch-up tendency and requires less elevator travel because of an increase in downwash at the horizontal tail.

- The airplane experiences a reduction in static source pressure and a corresponding increase in indicated airspeed.

Due to the reduced drag in ground effect, the airplane may seem to be able to take off below the recommended airspeed. However, as the airplane climbs out of ground effect below the recommended climb speed, initial climb performance will be much less than at Vy or even Vx. Under conditions of high-density altitude, high temperature, and/or maximum gross weight, the airplane may be able to lift off but will be unable to climb out of ground effect. Consequently, the airplane may not be able to clear obstructions. Lift off before attaining recommended flight airspeed incurs more drag, which requires more power to overcome. Since the initial takeoff and climb is based on maximum power, reducing drag is the only option. To reduce drag, pitch must be reduced which means losing altitude. Pilots must remember that many airplanes cannot safely takeoff at maximum gross weight at certain altitudes and temperatures, due to lack of performance. Therefore, under marginal conditions, it is important that the airplane takes off at the speed recommended for adequate initial climb performance.

Ground effect is important to normal flight operations. If the runway is long enough or if no obstacles exist, ground effect

Figure 5-7. *Takeoff in-ground effect area.*

can be used to the pilot's advantage by using the reduced drag to improve initial acceleration.

When taking off from an unsatisfactory surface, the pilot should apply as much weight to the wings as possible during the ground run and lift off, using ground effect as an aid, prior to attaining true flying speed. The pilot should reduce AOA to attain normal airspeed before attempting to fly out of the ground effect areas.

Short-Field Takeoff and Maximum Performance Climb

When performing takeoffs and climbs from fields where the takeoff area is short or the available takeoff area is restricted by obstructions, the pilot should operate the airplane at the maximum limit of its takeoff performance capabilities. To depart from such an area safely, the pilot must exercise positive and precise control of airplane attitude and airspeed so that takeoff and climb performance result in the shortest ground roll and the steepest angle of climb. *[Figure 5-8]* The pilot should consult and follow the performance section of the AFM/POH to obtain the power setting, flap setting, airspeed, and procedures prescribed by the airplane's manufacturer.

The pilot must have adequate knowledge in the use and effectiveness of the best angle-of-climb speed (V_X) and the best rate-of-climb speed (V_Y) for the specific make and model of airplane being flown in order to safely accomplish a takeoff at maximum performance.

V_X is the speed at which the airplane achieves the greatest gain in altitude for a given distance over the ground. It is usually slightly less than V_Y, which is the greatest gain in altitude per unit of time. The specific speeds to be used for a given airplane are stated in the FAA-approved AFM/POH. The pilot should be aware that, in some airplanes, a deviation of 5 knots from the recommended speed may result in a significant reduction in climb performance; therefore, the pilot must maintain precise control of the airspeed to ensure the maneuver is executed safely and successfully.

Takeoff Roll

Taking off from a short field requires the takeoff to be started from the very beginning of the takeoff area. At this point, the airplane is aligned with the intended takeoff path. If the airplane manufacturer recommends the use of flaps, they are extended the proper amount before beginning the takeoff roll. This allows the pilot to devote full attention to the proper technique and the airplane's performance throughout the takeoff.

The pilot should apply takeoff power smoothly and continuously, without hesitation, to accelerate the airplane as rapidly as possible. Some pilots prefer to hold the brakes until the maximum obtainable engine revolutions per minute (rpm) are achieved before allowing the airplane to begin its takeoff run. However, it has not been established that this procedure results in a shorter takeoff run in all light, single-engine airplanes. The airplane is allowed to roll with its full weight on the main wheels and accelerate to the lift-off speed. As the takeoff roll progresses, the pilot must adjust the airplane's pitch attitude and AOA to attain minimum drag and maximum acceleration. In nose-wheel type airplanes, this involves little use of the elevator control since the airplane is already in a low drag attitude.

Lift-Off

As V_X approaches, the pilot should apply back-elevator pressure until reaching the appropriate V_X attitude to ensure a smooth and firm lift-off, or rotation. Since the airplane accelerates more rapidly after lift-off, the pilot must apply additional back-elevator pressure to hold a constant airspeed. After becoming airborne, the pilot will maintain a wings-level climb at V_X until all obstacles have been cleared or; if no obstacles are present, until reaching an altitude of at least 50 feet above the takeoff surface. Thereafter, the pilot may lower the pitch attitude slightly and continue the climb at V_Y until reaching a safe maneuvering altitude. The pilot must always remember that an attempt to pull the airplane off the ground prematurely, or to climb too steeply, may cause the airplane to settle back to the runway or make contact with obstacles. Even if the airplane remains airborne, until the pilot reaches V_X, the initial climb will remain flat, which

Figure 5-8. *Short-field takeoff.*

diminishes the pilot's ability to successfully perform the climb and/or clear obstacles. *[Figure 5-9]*

The objective is to rotate to the appropriate pitch attitude at (or near) V_X. The pilot should be aware that some airplanes have a natural tendency to lift off well before reaching V_X. In these airplanes, it may be necessary to allow the airplane to lift-off in ground effect and then reduce pitch attitude to level until the airplane accelerates to V_X with the wheels just clear of the runway surface. This method is preferable to forcing the airplane to remain on the ground with forward-elevator pressure until V_X is attained. Holding the airplane on the ground unnecessarily puts excessive pressure on the nose-wheel and may result in "wheel barrowing." It also hinders both acceleration and overall airplane performance.

Initial Climb

On short-field takeoffs, the landing gear and flaps should remain in takeoff position until the airplane is clear of obstacles (or as recommended by the manufacturer) and V_Y has been established. Until all obstacles have been cleared, the pilot must maintain focus outside the airplane instead of reaching for landing gear or flap controls or looking inside the airplane for any reason. When the airplane is stabilized at V_Y, the landing gear (if retractable) and flaps should be retracted. It is usually advisable to raise the flaps in increments to avoid sudden loss of lift and settling of the airplane. Next, reduce the power to the normal climb setting or as recommended by the airplane manufacturer.

Common errors in the performance of short-field takeoffs and maximum performance climbs are:

- Failure to review AFM/POH and performance charts prior to takeoff.
- Failure to adequately clear the area.
- Failure to utilize all available runway/takeoff area.
- Failure to have the airplane properly trimmed prior to takeoff.
- Premature lift-off resulting in high drag.

- Holding the airplane on the ground unnecessarily with excessive forward-elevator pressure.
- Inadequate rotation resulting in excessive speed after lift-off.
- Inability to attain/maintain V_X.
- Fixation on the airspeed indicator during initial climb.
- Premature retraction of landing gear and/or wing flaps.

Soft/Rough-Field Takeoff and Climb

Takeoffs and climbs from soft fields require the use of operational techniques for getting the airplane airborne as quickly as possible to eliminate the drag caused by tall grass, soft sand, mud, and snow and may require climbing over an obstacle. The technique makes judicious use of ground effect to reduce landing gear drag and requires an understanding of the airplane's slow speed characteristics and responses. These same techniques are also useful on a rough field where the pilot should get the airplane off the ground as soon as possible to avoid damaging the landing gear.

Taking off from a soft surface or through soft surfaces or long, wet grass reduces the airplane's ability to accelerate during the takeoff roll and may prevent the airplane from reaching adequate takeoff speed if the pilot applies normal takeoff techniques. The pilot must be aware that the correct takeoff procedure for soft fields is quite different from the takeoff procedures used for short fields with firm, smooth surfaces. To minimize the hazards associated with takeoffs from soft or rough fields, the pilot should transfer the support of the airplane's weight as rapidly as possible from the wheels to the wings as the takeoff roll proceeds by establishing and maintaining a relatively high AOA or nose-high pitch attitude as early as possible. The pilot should lower the wing flaps prior to starting the takeoff (if recommended by the manufacturer) to provide additional lift and to transfer the airplane's weight from the wheels to the wings as early as possible. The pilot should maintain a continuous motion with sufficient power while lining up for the takeoff roll as stopping on a soft surface, such as mud or snow, might bog the airplane down.

Figure 5-9. *Effect of premature lift-off.*

Takeoff Roll

As the airplane is aligned with the takeoff path, the pilot should apply takeoff power smoothly and as rapidly as the powerplant can accept without faltering. As the airplane accelerates, the pilot should apply enough back-elevator pressure to establish a positive AOA and to reduce the weight supported by the nose-wheel.

When the airplane is held at a nose-high attitude throughout the takeoff run, the wings increasingly relieve the wheels of the airplane's weight as speed increases and lift develops, thereby minimizing the drag caused by surface irregularities or adhesion. If this attitude is accurately maintained, the airplane virtually flies itself off the ground, becoming airborne but at an airspeed slower than a safe climb speed because of ground effect. [*Figure 5-10*]

Lift-Off

After the airplane becomes airborne, the pilot should gently lower the nose with the wheels clear of the surface to allow the airplane to accelerate to V_Y, or V_X if obstacles must be cleared. Immediately after the airplane becomes airborne and while it accelerates, the pilot should be aware that, while transitioning out of the ground effect area, the airplane will have a tendency to settle back onto the surface. An attempt to climb prematurely or too steeply may cause the airplane to settle back to the surface as a result of the loss of ground effect. During the transition out of the ground effect area, the pilot should not attempt to climb out of ground effect before reaching the sufficient climb airspeed, as this may result in the airplane being unable to climb further, even with full power applied. Therefore, it is essential that the airplane remain in ground effect until at least V_X is reached. This requires a good understanding of the control pressures, aircraft responses, visual clues, and acceleration characteristics of that particular airplane.

Initial Climb

After a positive rate of climb is established, and the airplane has accelerated to V_Y, the pilot should retract the landing gear and flaps, if equipped. If departing from an airstrip with wet snow or slush on the takeoff surface, the gear should not be retracted immediately so that any wet snow or slush to be air-dried. In the event an obstacle must be cleared after a soft-field takeoff, the pilot should perform the climb-out at V_X until the obstacle has been cleared. The pilot should then adjust the pitch attitude to V_Y and retract the gear and flaps. The power can then be reduced to the normal climb setting. The pilot may then reduce power to normal climb setting.

Common errors in the performance of soft/rough field takeoff and climbs are:

- Failure to review AFM/POH and performance charts prior to takeoff.

- Failure to adequately clear the area.

- Insufficient back-elevator pressure during initial takeoff roll resulting in inadequate AOA.

- Failure to cross-check engine instruments for indications of proper operation after applying power.

- Poor directional control.

- Climbing too high after lift-off and not leveling off low enough to maintain ground effect altitude.

- Abrupt and/or excessive elevator control while attempting to level off and accelerate after liftoff.

- Allowing the airplane to "mush" or settle resulting in an inadvertent touchdown after lift-off.

- Attempting to climb out of ground effect area before attaining sufficient climb speed.

- Failure to anticipate an increase in pitch attitude as the airplane climbs out of ground effect.

Rejected Takeoff/Engine Failure

Emergency or abnormal situations can occur during a takeoff that require a pilot to reject the takeoff while still on the runway. Circumstances such as a malfunctioning powerplant, inadequate acceleration, runway incursion, or air traffic conflict may be reasons for a rejected takeoff.

Prior to takeoff, the pilot should identify a point along the runway at which the airplane should be airborne. If that

Figure 5-10. *Soft-field takeoff.*

point is reached and the airplane is not airborne, immediate action should be taken to discontinue the takeoff. Properly planned and executed, the airplane can be stopped on the remaining runway without using extraordinary measures, such as excessive braking that may result in loss of directional control, airplane damage, and/or personal injury.

In the event a takeoff is rejected, the power is reduced to idle and maximum braking applied while maintaining directional control. If it is necessary to shut down the engine due to a fire, the mixture control should be brought to the idle cutoff position and the magnetos turned off. In all cases, the manufacturer's emergency procedure should be followed.

Urgency characterizes all power loss or engine failure occurrences after lift-off. In most instances, the pilot has only a few seconds after an engine failure to decide what course of action to take and to execute it.

In the event of an engine failure on initial climb-out, the pilot's first responsibility is to maintain aircraft control. At a climb pitch attitude without power, the airplane is at or near a stalling AOA. At the same time, the pilot may still be holding right rudder. The pilot must immediately lower the nose to prevent a stall while moving the rudder to ensure coordinated flight. Attempting to turn back to the takeoff runway should not be attempted. The pilot should establish a controlled glide toward a plausible landing area, preferably straight ahead.

Noise Abatement

Aircraft noise problems are a major concern at many airports throughout the country. Many local communities have pressured airports into developing specific operational procedures that help limit aircraft noise while operating over nearby areas. As a result, noise abatement procedures have been developed for many of these airports that include standardized profiles and procedures to achieve these lower noise goals.

Airports that have noise abatement procedures provide information to pilots, operators, air carriers, air traffic facilities, and other special groups that are applicable to their airport. These procedures are available to the aviation community by various means. Most of this information comes from the Chart Supplements, local and regional publications, printed handouts, operator bulletin boards, safety briefings, and local air traffic facilities.

At airports that use noise abatement procedures, reminder signs may be installed at the taxiway hold positions for applicable runways to remind pilots to use and comply with noise abatement procedures on departure. Pilots who are not familiar with these procedures should ask the tower or air traffic facility for the recommended procedures. In any case, pilots should be considerate of the surrounding community while operating their airplane to and from such an airport. This includes operating as quietly, and safely as possible.

Chapter Summary

The takeoff and initial climb are relatively short phases required for every flight and are often taken for granted, yet 1 out of 5 accidents occur during this phase and half the mishaps are the result of pilot error. Becoming proficient in and applying the techniques and principles discussed in this chapter help pilots reduce their susceptibility to becoming a mishap statistic. The POH/AFM ground roll distances for take-off and landing added together provide a good estimate of the total runway needed to accelerate and then stop.

Chapter 6
Ground Reference Maneuvers

Introduction

Initial pilot training requires that a pilot understand the relationship of the various flight controls pressure inputs to the resulting attitudes of the airplane. This allows a pilot to develop a sense of feel and understand the various indications of airplane performance, such as pitch, roll, and yaw attitudes. With sufficient competency in this environment, the pilot is ready to apply these skills and place the airplane, not only in the correct attitude and power configuration, but also in orientation to specific ground-based references. These skills are the basis for traffic patterns, survey, photographic, sight-seeing, aerial application (crop dusting) and various other flight profiles requiring specific flightpaths referenced to points on the surface.

Too high

Pivotal alti

Too low

60°   45°

Wind

Steeper bank   Steeper bank

Shallowest bank   Shallowest bank

Shallower bank   Steepest bank   Steepest bank   Shallower ban

WIND

Steeper bank   Shallowest bank

Steepest bank

est bank

est bank   Shallower bank

A pilot must develop the proper coordination, timing, and attention to accurately and safely maneuver the airplane with regard to the required attitudes and ground references. Ground reference maneuvers are the principle flight maneuvers that combine the four fundamentals (straight-and-level, turns, climbs, and descents) into a set of integrated skills that the pilot uses in their everyday flight activity. A pilot must develop the skills necessary to accurately control, through the effect and use of the flight controls, the flightpath of the airplane in relationship to the ground. From every takeoff to every landing, a pilot exercises these skills in controlling the airplane.

The pilot should be introduced by their instructor to ground reference maneuvers as soon as the pilot shows proficiency in the four fundamentals. Accomplishing the ground reference maneuvers requires that the pilot competently manipulate the flight controls without any undue attention to mechanical flight control inputs—the pilot applies the necessary flight control pressures to affect the airplane's attitude and position by using the outside natural horizon and ground-based references with brief periods of scanning the flight instruments.

Maneuvering by Reference to Ground Objects

The purpose of ground reference maneuvers is to train pilots to accurately place the airplane in relationship to specific references and maintain a desired ground track. Such precision requires that a pilot simultaneously evaluate the airplane's attitude, reference points along the desired path, and the natural horizon. Vision is the most utilized sense in maneuvering in orientation to ground-based references; however, all senses are actively involved at different levels. For example, touch provides tactile feedback as to the required flight control pressures to overcome flight control surface forces that indirectly indicate the airplane's airspeed and aerodynamic load.

It is a common error for beginning pilots to fixate on a specific reference, such as a single location on the ground or the natural horizon. To be effective, the pilot must scan between several visual references to determine relative motion and to determine if the airplane is maintaining, or drifting to or from, the desired ground track. A pilot fixating on any one reference eliminates the ability to determine rate, which significantly degrades a pilot's performance. Visual scanning across several references allows the pilot to develop the important skill of determining the rate of closure to a specific point. Consider a skilled automobile driver in a simple intersection turn; the driver does not merely turn the steering wheel some degree and hope that it will work out. The skilled driver picks out several references, such as an island to their side, a painted lane line, or the opposing curb, and they use those

references to make almost imperceptible adjustments to the amount of deflection on the steering wheel, as well as the pressure on the accelerator pedal to smoothly join the lane into which they are turning. In the same manner, multiple references are required to precisely control the airplane in reference to the ground.

Not all ground-based references are visually equal and some understanding of those differences is important for their selection and use. For example, larger objects or references may appear closer than they actually are when compared to smaller objects or references. Also, prevailing visibility has a significant effect on the pilot's perception of the distance to a reference. Excellent visibilities with clear skies tend to make an object or reference appear closer than when compared to a hazy day with poor visibility. Another example is that rain can alter the visual image in a manner that an illusion of being at a higher altitude may be perceived, and brighter objects or references may appear closer than dimmer objects. Being aware of typical visual illusions helps a pilot select the best references for ground reference maneuvers. It is best, however sometimes impracticable, to find ground-based references that are similar in size and proportion.

Ground-based references can be numerous. Excellent examples are breakwaters, canals, fence lines, field boundaries, highways, railroad tracks, roads, pipe lines, power lines, water-tanks, and others; however, choices can be limited by geography, population density, infrastructure, or structures. Selecting a ground-based reference requires prior consideration, such as the type of maneuver being performed, altitude at which the maneuver will be performed, emergency landing requirements, density of structures, wind direction, visibility, and the type of airspace.

Division of attention is an important skill that a pilot must develop. A pilot must be able to fly the airplane affecting the flight controls in a manner they will place the airplane in the needed attitude while tracking a specific path over the ground. In addition, the pilot must be able to scan for hazards such as other aircraft, be immediately prepared for an emergency landing should the need arise, and scan the flight and engine instruments at regular intervals to ensure that a pending situation, such as decreasing oil pressure, does not turn into an unexpected incident.

Safety is paramount in all aspects of flying. Awareness and practice of safety-enhancing procedures must be constantly exercised. Ground reference maneuvers place the airplane in an environment where heightened awareness is needed. Pilots should be looking for other aircraft, including helicopters, radio towers, and assessing locations for emergency landings. Pilots should always clear the area with two 90°

clearing turns looking to the left and the right, as well as above and below the airplane. The maneuver area should not cause disturbances and be well away from groups of people, livestock, or communities. Before performing any maneuver, the pilot should complete the required checklist items, make any radio announcements (such as on a practice area frequency), and safety clearing turns. As a general note, a ground reference maneuver should not exceed a bank angle of 45° or an airspeed greater than maneuvering speed. As part of preflight planning, the pilot should determine the predicted (POH/AFM) stall speed at 50° or the highest bank angle planned plus some margin for error in maneuvering

Drift and Ground Track Control

Wind direction and velocity variations are the primary effects requiring corrections of the flightpath during ground reference maneuvers. Unlike an automobile, but similar to a boat or ship, wind directly influences the path that the airplane travels in reference to the ground. Whenever the airplane is in flight, the movement of the air directly affects the actual ground track of the airplane.

For example, an airplane is traveling at 90 knots (90 nautical miles per hour) and the wind is blowing from right to left at 10 knots. The airplane continues forward at 90 knots but also travels left 10 nautical miles for every hour of flight time. If the airplane, in this example doubles its speed to 180 knots, it still drifts laterally to the left 10 nautical miles every hour. The airplane travels within an often moving body of air, so traveling to a point on the surface requires compensation for the movement of the air mass.

Ground reference maneuvers are generally flown at altitudes between 600 and 1,000 feet above ground level (AGL). The pilot must consider the following when selecting the maneuvering altitude:

- The lower the maneuvering altitude, the faster the airplane appears to travel in relation to the ground.

- Drift should be easily recognizable from both sides of the airplane.

- The altitude should provide obstruction clearance of no less than 500 feet vertically above the obstruction and 2,000 feet horizontally.

- In case of an engine failure, the pilot must plan, consider, and be alert for forced landing areas while understanding that the lower the airplane's altitude, the less time there is to configure the airplane for an emergency landing and the shorter the glide distance.

- Any specific altitude required by test standards.

Correcting Drift During Straight-and-Level Flight

When flying straight and level and following a selected straight-line direct ground track, the preferred method of correcting for wind drift is to angle the airplane sufficiently into the wind to cancel the effect of the sideways drift caused by the wind. The wind's speed, the angle between the wind direction and the airplane's longitudinal axis, and the airspeed of the airplane determines the required wind correction angle. For example, an airplane with an airspeed of 100 knots, a 20 knot wind at 90° to the airplane's longitudinal axis, and a 12° angle into the wind is required to cancel the airplane's drift. If the wind in the above example is only 10 knots, the wind correction angle required to cancel the drift is six degrees. When the drift has been neutralized by heading the airplane into the wind, the airplane will fly the direct straight ground track.

To further illustrate this point, if a boat is crossing a river and the river's current is completely still, the boat could head directly to a point on the opposite shore on a straight course to that opposite point without any drift; however, rivers tend to have a downstream current that must be considered if the captain wants the boat to arrive at the opposite shore using a direct straight path. Any downstream current pushes the boat sideways and downstream at the speed of the current. To counteract this downstream movement, the boat must move upstream at the same speed as the river is moving the boat downstream. This is accomplished by angling the boat upstream sufficiently to counteract the downstream flow. If this is done, the boat follows a direct straight track across the river to the intended destination point. The amount of angle required is dependent on the forward speed of the boat and the speed of the current. The slower the forward speed of the boat and/or the faster speed of the current, the greater the angle must be to counteract the drift. The converse is also true. *[Figure 6-1]*

As soon as an airplane lifts off the surface and levels the wings, if there is any crosswind, the airplane will begin tracking sideways with the wind. Any wind not directly on the nose or tail of the airplane will drift the airplane sideways at a speed up to the speed of the wind. A wind that is directly to the right or the left (at a 90° angle) drifts the airplane sideways at the speed of the wind; when the wind is halfway between the side and the nose of the airplane (at a 45° angle), it drifts the airplane sideways at just over 70 percent of the speed of the wind. It should be understood that pilots do not calculate the required drift correction angles for ground reference maneuvers; they merely use the references and adjust the airplane's relationship to those references to cancel any drift. The groundspeed of the airplane is also affected

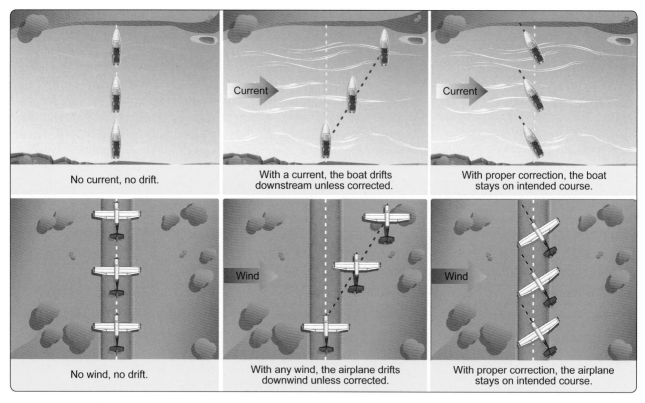

Figure 6-1. *Wind drift.*

by the wind. As the wind direction becomes parallel to the airplane's longitudinal axis, the magnitude of the wind's effect on the groundspeed is greater; as the wind becomes perpendicular to the longitudinal axis, the magnitude of the wind's effect on the groundspeed is less. In general, When the wind is blowing straight into the nose of the airplane, the groundspeed will be less than the airspeed. When the wind is blowing from directly behind the airplane, the groundspeed will be faster than the airspeed. In other words, when the airplane is headed upwind, the groundspeed is decreased; when headed downwind, the groundspeed is increased.

Constant Radius During Turning Flight

In a no-wind condition, the pilot can perform a ground-based constant radius turn by accurately maintaining a constant bank angle throughout the turn; however, with any wind the complexities of maintaining a ground-based constant radius turn increase. When wind is present, during ground reference maneuvers involving turns, the pilot must correct for wind drift. *[Figure 6-2]* Throughout the turn, the wind is acting on the airplane from a constantly changing angle—increasing or decreasing the groundspeed in a manner similar to straight flight. To follow a circular, constant radius ground track, the bank angle must vary to compensate for wind drift throughout the turn. The airplane's ground-based turn radius is affected by the airplane's groundspeed: the faster the groundspeed, the steeper the airplane must be banked to maintain a ground-

based constant radius turn. The converse is also true: the slower the groundspeed, the shallower the airplane needs to be banked to maintain a ground-based constant radius turn.

For a given true airspeed, the radius of turn in the air varies proportionally with the bank angle. To maintain the constant radius over the ground, the bank angle is proportional to ground speed. For example, an airplane is in the downwind position at 100 knots groundspeed. In this example, the wind is 10 knots, meaning that the airplane is at an airspeed of 90 knots (for this discussion, we ignore true, calibrated, and indicate airspeed and assume that they are all the same). If the pilot starts a downwind turn with a 45° "steepest" bank angle, the turn radius is approximately 890 feet. Let's assume the airplane is now upwind with a groundspeed of 80 knots. In order to maintain the 890-foot radius, the pilot must reduce the bank angle to a shallowest bank of approximately 33°. In another example, if the downwind is flown at an airspeed of 90 knots in a 10 knot tailwind with a desired turn radius of 2,000 feet, the "steepest" bank angle needs to be at approximately 24° and the upwind "shallowest" bank angle at approximately 16°.

To demonstrate the effect that wind has on turns, the pilot should select a straight-line ground reference, such as a road or railroad track. *[Figure 6-3]* Choosing a straight-line ground reference that is parallel to the wind, the airplane would be flown into the wind and directly over the selected

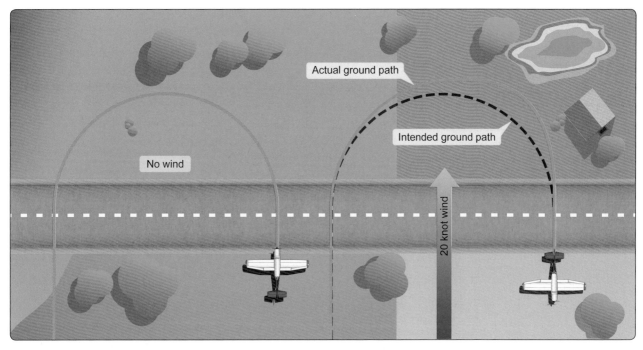

Figure 6-2. *Effect of wind during a turn.*

straight-line ground reference. Once a straight-line ground reference is established, the pilot makes a 360° constant medium banked turn. As the airplane completes the 360° turn, it should return directly over the straight-line ground reference but downwind from the starting point. Choosing a straight-line ground reference that has a crosswind, and using the same 360° constant medium-banked turn, demonstrates how the airplane drifts away from the reference even as the

pilot holds a constant bank angle. In both examples, the path over the ground is an elongated circle, although in reference to the air, the airplane flew a perfect continuous radius.

In order to compensate for the elongated, somewhat circular path over the ground, the pilot must adjust the bank angle as the groundspeed changes throughout the turn. Where groundspeed is the fastest, such as when the airplane is

Figure 6-3. *Effect of wind during turn.*

headed downwind, the turn bank angle must be steepest; where groundspeed is the slowest, such as when the airplane is headed upwind, the turn bank angle must be shallow. It is necessary to increase or decrease the angle of bank, which increases or decreases the rate of turn, to achieve the desired constant radius track over the ground.

Ground reference maneuvers should always be entered from a downwind position. This allows the pilot to establish the steepest bank angle required to maintain a constant radius ground track. If the bank is too steep, the pilot should immediately exit the maneuver and re-establish a lateral position that is further from the ground reference. The pilot should avoid bank angles in excess of 45° due to the increased stalling speed.

Tracking Over and Parallel to a Straight Line

The pilot should first be introduced to ground reference maneuvers by correcting for the effects of a crosswind over a straight-line ground reference, such as road or railroad tracks. If a straight road or railroad track is unavailable, the pilot will choose multiple references (three minimum) which, when an imaginary visual reference line is extended, represents a straight line. The reference should be suitably long so the pilot has sufficient time to understand the concepts of wind correction and practice the maneuver. Initially, the maneuver should be flown directly over the ground reference with the pilot angling the airplane's longitudinal axis into the wind sufficiently such as to cancel the effect of drift. The pilot should scan between far ahead and close to the airplane to practice tracking multiple references.

When proficiency has been demonstrated by flying directly over the ground reference line, the pilot should then practice flying a straight parallel path that is offset from the ground reference. The offset parallel path should not be more than three-fourths of a mile from the reference line. The maneuver should be flown offset from the ground references with the pilot angling the airplane's longitudinal axis into the wind sufficiently to cancel the effect of drift while maintaining a parallel track.

Rectangular Course

A principle ground reference maneuver is the rectangular course. *[Figure 6-4]* The rectangular course is a training maneuver in which the airplane maintains an equal distance from all sides of the selected rectangular references. The maneuver is accomplished to replicate the airport traffic pattern that an airplane typically maneuvers while landing. While performing the rectangular course maneuver, the pilot should maintain a constant altitude, airspeed, and distance from the ground references. The maneuver assists the pilot in practicing the following:

- Maintaining a specific relationship between the airplane and the ground.

- Dividing attention between the flightpath, ground-based references, manipulating the flight controls, and scanning for outside hazards and instrument indications.

- Adjusting the bank angle during turns to correct for groundspeed changes in order to maintain constant radius turns.

- Rolling out from a turn with the required wind correction angle to compensate for any drift cause by the wind.

- Establishing and correcting the wind correction angle in order to maintain the track over the ground.

- Preparing the pilot for the airport traffic pattern and subsequent landing pattern practice.

First, a square, rectangular field, or an area with suitable ground references on all four sides, as previously mentioned should be selected consistent with safe practices. The airplane should be flown parallel to and at an equal distance between one-half to three-fourths of a mile away from the field boundaries or selected ground references. The flightpath should be positioned outside the field boundaries or selected ground references so that the references may be easily observed from either pilot seat. It is not practicable to fly directly above the field boundaries or selected ground references. The pilot should avoid flying close to the references, as this will require the pilot to turn using very steep bank angles, thereby increasing aerodynamic load factor and the airplane's stall speed, especially in the downwind to crosswind turn.

The entry into the maneuver should be accomplished downwind. This places the wind on the tail of the airplane and results in an increased groundspeed. There should be no wind correction angle if the wind is directly on the tail of the airplane; however, a real-world situation results in some drift correction. The turn from the downwind leg onto the base leg is entered with a relatively steep bank angle. The pilot should roll the airplane into a steep bank with rapid, but not excessive, coordinated aileron and rudder pressures. As the airplane turns onto the following base leg, the tailwind lessens and becomes a crosswind; the bank angle is reduced gradually with coordinated aileron and rudder pressures. The pilot should be prepared for the lateral drift and compensate by turning more than 90° angling toward the inside of the rectangular course.

The next leg is where the airplane turns from a base leg position to the upwind leg. Ideally, the wind is directly on

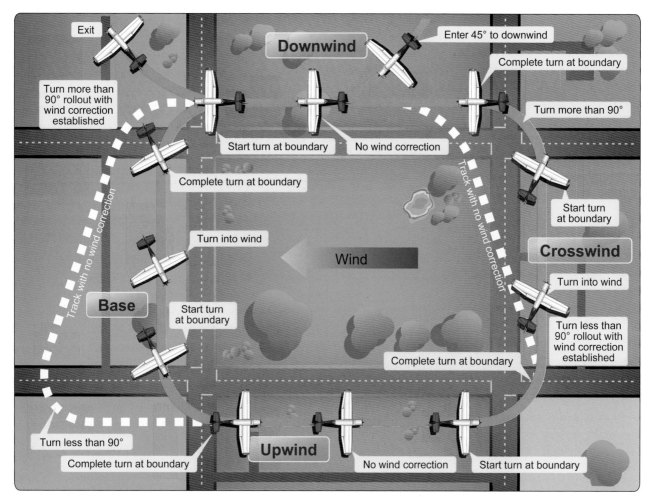

Figure 6-4. *Rectangular course.*

the nose of the airplane resulting in a direct headwind and decreased groundspeed; however, a real-world situation results in some drift correction. The pilot should roll the airplane into a medium banked turn with coordinated aileron and rudder pressures. As the airplane turns onto the upwind leg, the crosswind lessens and becomes a headwind, and the bank angle is gradually reduced with coordinated aileron and rudder pressures. Because the pilot was angled into the wind on the base leg, the turn to the upwind leg is less than 90°.

The next leg is where the airplane turns from an upwind leg position to the crosswind leg. The pilot should slowly roll the airplane into a shallow-banked turn, as the developing crosswind drifts the airplane into the inside of the rectangular course with coordinated aileron and rudder pressures. As the airplane turns onto the crosswind leg, the headwind lessens and becomes a crosswind. As the turn nears completion, the bank angle is reduced with coordinated aileron and rudder pressures. To compensate for the crosswind, the pilot must angle into the wind, toward the outside of the rectangular course, which requires the turn to be less than 90°.

The final turn is back to the downwind leg, which requires a medium-banked angle and a turn greater than 90°. The groundspeed will be increasing as the turn progresses and the bank should be held and then rolled out in a rapid, but not excessive, manner using coordinated aileron and rudder pressures.

For the maneuver to be executed properly, the pilot must visually utilize the ground-based, nose, and wingtip references to properly position the airplane in attitude and in orientation to the rectangular course. Each turn, in order to maintain a constant ground-based radius, requires the bank angle to be adjusted to compensate for the changing groundspeed—the higher the groundspeed, the steeper the bank. If the groundspeed is initially higher and then decreases throughout the turn, the bank angle should progressively decrease throughout the turn. The converse is also true, if the groundspeed is initially slower and then increases throughout the turn, the bank angle should progressively increase throughout the turn until rollout is started. Also, the rate for rolling in and out of the turn should be adjusted to prevent drifting in or out of the course. When the wind is from a direction that could drift the airplane into the course, the banking roll rate should be slow. When the wind is from a direction that could drift the airplane to the outside of the course, the banking roll rate should be quick.

The following are the most common errors made while performing rectangular courses:

- Failure to adequately clear the area above, below, and on either side of the airplane for safety hazards, initially and throughout the maneuver.

- Failure to establish a constant, level altitude prior to entering the maneuver.

- Failure to maintain altitude during the maneuver.

- Failure to properly assess wind direction.

- Failure to establish the appropriate wind correction angle.

- Failure to apply coordinated aileron and rudder pressure, resulting in slips and skids.

- Failure to manipulate the flight controls in a smooth and continuous manner.

- Failure to properly divide attention between controlling the airplane and maintaining proper orientation with the ground references.

- Failure to execute turns with accurate timing.

Turns Around a Point

Turns around a point are a logical extension of both the rectangular course and S-turns across a road. The maneuver is a 360° constant radius turn around a single ground-based reference point. [Figure 6-5] The principles are the same in any turning ground reference maneuver—higher groundspeeds require steeper banks and slower ground speeds require shallower banks. The objectives of turns around a point are as follows:

- Maintaining a specific relationship between the airplane and the ground.

- Dividing attention between the flightpath, ground-based references, manipulating of the flight controls, and scanning for outside hazards and instrument indications.

- Adjusting the bank angle during turns to correct for groundspeed changes in order to maintain a constant radius turn; steeper bank angles for higher ground speeds, shallow bank angles for slower groundspeeds.

- Improving competency in managing the quickly changing bank angles.

- Establishing and adjusting the wind correction angle in order to maintain the track over the ground.

- Developing the ability to compensate for drift in quickly changing orientations.

- Developing further awareness that the radius of a turn is correlated to the bank angle.

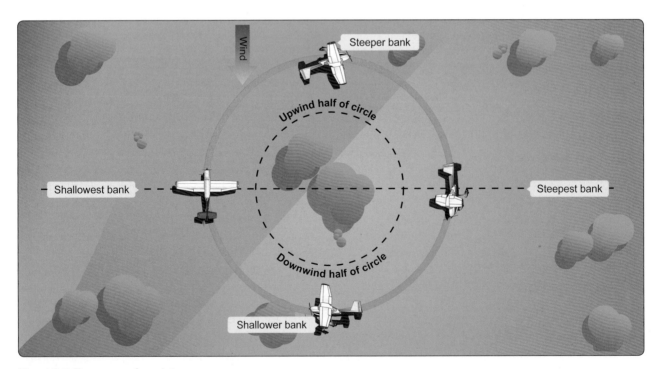

Figure 6-5. *Turns around a point.*

To perform a turn around a point, the pilot must complete at least one 360° turn; however, to properly assess wind direction, velocity, bank required, and other factors related to turns in wind, the pilot should complete two or more turns. As in other ground reference maneuvers, when wind is present, the pilot must a constantly adjust the airplane's bank and wind correction angle to maintain a constant radius turn around a point. In contrast to the ground reference maneuvers discussed previously in which turns were approximately limited to either 90° or 180°, turns around a point are consecutive 360° turns where, throughout the maneuver, the pilot must constantly adjust the bank angle and the resulting rate of turn in proportion to the groundspeed as the airplane sequences through the various wind directions. The pilot should make these adjustments by applying coordinated aileron and rudder pressure throughout the turn.

When performing a turn around a point, the pilot should select a prominent, ground-based reference that is easily distinguishable yet small enough to present a precise reference. *[Figure 6-6]* The pilot should enter the maneuver downwind, where the groundspeed is at its fastest, at the appropriate radius of turn and distance from the selected ground-based reference point. In a high-wing airplane, the lowered wing may block the view of the ground reference point, especially in airplanes with side-by-side seating during a left turn (assuming that the pilot is flying from the left seat). To prevent this, the pilot may need to change the maneuvering altitude or the desired turn radius. The pilot should ensure that the reference point is visible at all times throughout the maneuver, even with the wing lowered in a bank.

Upon entering the maneuver, depending on the wind's speed, it may be necessary to roll into the initial bank at a rapid rate so that the steepest bank is set quickly to prevent the airplane from drifting outside of the desired turn radius. This is best accomplished by repeated practice and assessing the required roll in rate. Thereafter, the pilot should gradually decrease the angle of bank until the airplane is headed directly upwind. As the upwind becomes a crosswind and then a downwind, the pilot should gradually steepen the bank to the steepest angle upon reaching the initial point of entry.

During the downwind half of the turn, the pilot should progressively adjust the airplane's heading toward the inside of the turn. During the upwind half, the pilot should progressively adjust the airplane's heading toward the outside of the turn. Recall from the previous discussion on wind correction angle that the airplane's heading should be ahead of its position over the ground during the downwind half of the turn behind its position during the upwind half. Remember that the goal is to make a constant radius turn over the ground and, because the airplane is flying through a moving air mass, the pilot must constantly adjust the bank angle to achieve this goal.

The following are the most common errors in the performance of turns around a point:

- Failure to adequately clear the area above, below, and on either side of the airplane for safety hazards, initially and throughout the maneuver.

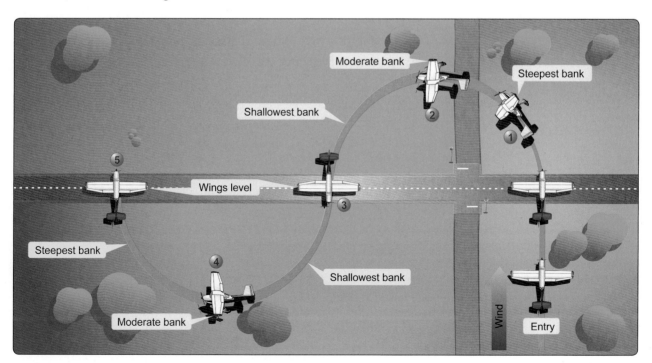

Figure 6-6. *S-turns.*

- Failure to establish a constant, level altitude prior to entering the maneuver.

- Failure to maintain altitude during the maneuver.

- Failure to properly assess wind direction.

- Failure to properly execute constant radius turns.

- Failure to manipulate the flight controls in a smooth and continuous manner.

- Failure to establish the appropriate wind correction angle.

- Failure to apply coordinated aileron and rudder pressure, resulting in slips or skids.

S-Turns

S-turns is a ground reference maneuver in which the airplane's ground track resembles two opposite but equal half-circles on each side of a selected ground-based straight-line reference. *[Figure 6-6]* This ground reference maneuver presents a practical application for the correction of wind during a turn. The objectives of S-turns across a road are as follows:

- Maintaining a specific relationship between the airplane and the ground.

- Dividing attention between the flightpath, ground-based references, manipulating the flight controls, and scanning for outside hazards and instrument indications.

- Adjusting the bank angle during turns to correct for groundspeed changes in order to maintain a constant radius turn—steeper bank angles for higher ground speeds, shallow bank angles for slower groundspeeds.

- Rolling out from a turn with the required wind correction angle to compensate for any drift cause by the wind.

- Establishing and correcting the wind correction angle in order to maintain the track over the ground.

- Developing the ability to compensate for drift in quickly changing orientations.

- Arriving at specific points on required headings.

With the airplane in the downwind position, the maneuver consists of crossing a straight-line ground reference at a 90° angle and immediately beginning a 180° constant radius turn. The pilot will then adjust the roll rate and bank angle for drift effects and changes in groundspeed, and re-cross the straight-line ground reference in the opposite direction just as the first 180° constant radius turn is completed. The pilot will then immediately begin a second 180° constant radius turn in the opposite direction, adjusting the roll rate and bank angle for drift effects and changes in groundspeed, again re-crossing the straight-line ground reference as the second 180° constant radius turn is completed. If the straight-line ground reference is of sufficient length, the pilot may complete as many as can be safely accomplished.

In the same manner as the rectangular course, it is standard practice to enter ground-based maneuvers downwind where groundspeed is greatest. As such, the roll into the turn must be rapid, but not aggressive, and the angle of bank must be steepest when initiating the turn. As the turn progresses, the bank angle and the rate of rollout must be decreased as the groundspeed decreases to ensure that the turn's radius is constant. During the first turn, when the airplane is at the 90° point, it will be directly crosswind. In addition to the rate of rollout and bank angle, the pilot must control the wind correction angle throughout the turn.

Controlling the wind correction angle during a turn can be complex to understand. The concept is best understood by comprehending the difference between the number of degrees that the airplane has turned over the ground verses the number of degrees that the airplane has turned in the air. For example, if the airplane is exactly crosswind, meaning directly at a point that is 90° to the straight-lined ground reference. If the wind, in this example, requires a 10° wind correction angle (for this example, this is a left turn with the crosswind from the left) the airplane would be at a heading that is 10° ahead when directly over the 90° ground reference point. In other words, the first 90° track over the ground would result in a heading change of 100° and the last 90° track over the ground would result in 80° of heading change.

As the turn progresses from a downwind position to an upwind position, the pilot must gradually decrease the bank angle with coordinated aileron and rudder pressure. The pilot should reference the airplane's nose, wingtips, and the ground references and adjust the rollout timing so that the airplane crosses the straight-line ground reference with the wings level, and at the proper heading, altitude, and airspeed. As the airplane re-crosses the straight-lined ground reference, the pilot should immediately begin the opposite turn—there should be no delay in rolling out from one turn and rolling into the next turn. Because the airplane is now upwind, the roll in should be smooth and gentle and the initial bank angle should be shallow. As the turn progresses, the wind changes from upwind, to crosswind, to downwind. In a similar manner described above, the pilot should adjust the bank angle to correct for changes in groundspeed. As the groundspeed increases, the pilot should increase the bank angle to maintain a constant radius turn. At the 90° crosswind position, the airplane should also have the correct wind correction angle. As the airplane turns downwind, the groundspeed increases;

the bank angle should be increased so that the rate of turn is used to maintain a constant radius turn.

The following are the most common errors made while performing S-turns across a road:

- Failure to adequately clear the area above, below, and on either side of the airplane for safety hazards, initially and throughout the maneuver.

- Failure to establish a constant, level altitude prior to entering the maneuver.

- Failure to maintain altitude during the maneuver.

- Failure to properly assess wind direction.

- Failure to properly execute constant radius turns.

- Failure to manipulate the flight controls in a smooth and continuous manner when transitioning into turns.

- Failure to establish the appropriate wind correction angle.

- Failure to apply coordinated aileron and rudder pressure, resulting in slips or skids.

Elementary Eights

Elementary eights are a family of maneuvers in which each individual maneuver is one that the airplane tracks a path over the ground similar to the shape of a figure eight. There are various types of eights, progressing from the elementary types to very difficult types in the advanced maneuvers. Each eight is intended to develop a pilot's flight control coordination skills, strengthen their awareness relative to the selected ground references, and enhance division of attention so that flying becomes more instinctive than mechanical. Eights require a greater degree of focused attention to the selected ground references; however, the real significance of eights is that pilot must strive for flight precision.

Elementary eights include eights along a road, eights across a road, and eights around pylons. Each of these maneuvers is a variation of a turn around a point. Each eight uses two ground reference points about which the airplane turns first in one direction and then the opposite direction—like a figure eight.

Eights maneuvers are designed for the following purposes:

- Further development of the pilot's skill in maintaining a specific relationship between the airplane and the ground references.

- Improving the pilot's ability to divide attention between the flightpath and ground-based references, manipulation of the flight controls, and scanning for outside hazards and instrument indications during both turning and straight-line flight.

- Developing the pilot's skills to visualize each specific segment of the maneuver and the maneuver as a whole, prior to execution.

- Developing a pilot's ability to intuitively manipulate flight controls to adjust the bank angle during turns to correct for groundspeed changes in order to maintain constant radius turns and proper ground track between ground references.

Eights Along a Road

Eights along a road is a ground reference maneuver in which the ground track consists of two opposite 360° adjacent turns with the center of each 360° turn and the adjacent turn point perpendicular or parallel to the straight-line ground reference (road, railroad tracks, fence line, pipeline right-of-way, etc.). *[Figure 6-7]* Like the other ground reference maneuvers, its objective is to further develop division of attention while compensating for drift, maintaining orientation with ground references, and maintaining a constant altitude.

Although eights along a road may be performed with the wind blowing parallel or perpendicular to the straight-line ground reference, only the perpendicular wind situation is explained since the principles involved are common to each. The pilot should select a straight-line ground reference that is perpendicular to the wind and position the airplane parallel to and directly above the straight-line ground reference. Since this places the airplane in a crosswind position, the pilot must compensate for the wind drift with an appropriate wind correction angle.

The following description is illustrated in *Figure 6-7*. The airplane is initially in a crosswind position, perpendicular to the wind, and over the ground-based reference. The first turn should be a left turn toward a downwind position starting with a steeping bank. When the entry is made into the turn, it requires that the turn begin with a medium bank and gradually steepen to its maximum bank angle when the airplane is directly downwind. As the airplane turns from downwind to crosswind, the bank angle needs to be gradually reduced since groundspeed is decreasing; however, the groundspeed only decreases by ½ of its velocity during the first ⅔ of the turn from downwind to crosswind.

The pilot must control the bank angle as well as the rate at which the bank angle is reduced so that the wind correction angle is correct. Assuming that the wind is coming from the right side of the airplane, the airplane heading should be slightly ahead of its position over the ground. When the airplane completes the first 180° of ground track, it is directly crosswind, and the airplane should be at the maximum wind correction angle.

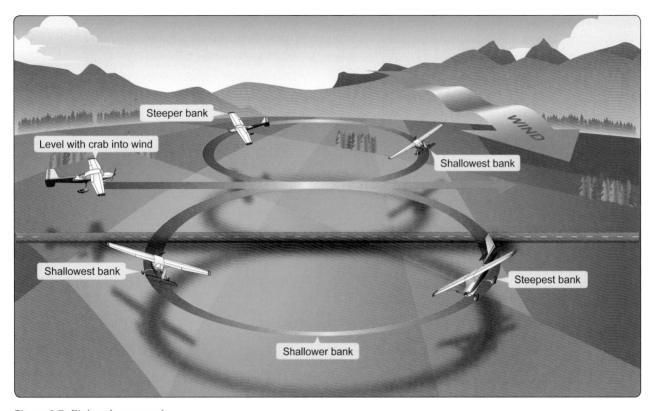

Figure 6-7. *Eights along a road.*

As the turn is continued toward the upwind, the airplane's groundspeed is decreasing, which requires the pilot to reduce the bank angle to slow the rate of turn. If the pilot does not reduce the bank angle, the continued high rate of turn would cause the turn to be completed prematurely. Another way to explain this effect is—the wind is drifting the airplane downwind at the same time its groundspeed is slowing; if the airplane has a steeper than required bank angle, its rate of turn will be too fast and the airplane will complete the turn before it has had time to return to the ground reference.

When the airplane is directly upwind, which is at 270° into the first turn, the bank angle should be shallow with no wind correction. As the airplane turns crosswind again, the airplane's groundspeed begins increasing; therefore, the pilot should adjust the bank angle and corresponding rate of turn proportionately in order to reach the ground reference at the completion of the 360° ground track. The pilot may vary the bank angle to correct for any previous errors made in judging the returning rate and closure rate. The pilot should time the rollout so that the airplane is straight-and-level over the starting point with enough drift correction to hold it over the straight-line ground reference. Assuming that the wind is now from the left, the airplane should be banked at a left wind correction angle.

After momentarily flying straight-and-level with the established wind correction, along the ground reference, the pilot should roll the airplane into a medium bank turn in the opposite direction to begin the 360° turn on the upwind side of the ground reference. The wind will decrease the airplane's groundspeed and drift the airplane back toward the ground reference; therefore, the pilot must decrease the bank slowly during the first 90° of the upwind turn in order to establish a constant radius. During the next 90° of turn, the pilot should increase the bank angle, since the groundspeed is increasing, to maintain a constant radius and establish the proper wind correction angle before reaching the 180° upwind position.

As the remaining 180° of turn continues, the wind becomes a tailwind and then a crosswind. Consistent with previous downwind and crosswind descriptions, the pilot must increase the bank angle as the airplane reaches the downwind position and decrease the bank angle as the airplane reaches the crosswind position. Further, the rate of roll in and roll out should be consistent with how fast the groundspeed changes during the turn. Remember, when turning from an upwind or downwind position to a crosswind position, the groundspeed changes by ½ during the first ⅔ of the 90° turn. The final ½ of the groundspeed changes in the last ⅓ of the turn. In contrast, when turning from a crosswind position to an upwind or downwind position, the groundspeed changes by ½ during the first ⅓ of the 90° turn. The final ½ of the groundspeed changes in the last ⅔ of the turn.

6-12

To successfully perform eights along a ground reference, the pilot must be able to smoothly and accurately coordinate changes in bank angle to maintain a constant radius turn and counteract drift. The speed in which the pilot can anticipate these corrections directly affects the accuracy of the overall maneuver and the amount of attention that can be directed toward scanning for outside hazards and instrument indications.

Eights Across A Road

This maneuver is a variation of eights along a road and involves the same principles and techniques. The primary difference is that at the completion of each loop of the figure eight, the airplane should cross an intersection of a specific ground reference point. *[Figure 6-8]*

The loops should be across the road and the wind should be perpendicular to the loops. Each time the reference is crossed, the crossing angle should be the same, and the wings of the airplane should be level. The eights may also be performed by rolling from one bank immediately to the other, directly over the reference.

Eights Around Pylons

Eights around pylons is a ground-reference maneuver with the same principles and techniques of correcting for wind drift as used in turns around a point and the same objectives as other ground track maneuvers. Eights around pylons utilizes two ground reference points called "pylons." Turns around each pylon are made in opposite directions to follow a ground track in the form of a figure 8. *[Figure 6-9]*

The pattern involves flying downwind between the pylons and upwind outside of the pylons. It may include a short period of straight-and-level flight while proceeding diagonally from one pylon to the other. The pylons should be on a line perpendicular to the wind. The maneuver should be started with the airplane on a downwind heading when passing equally between the pylons. The distance between the pylons and the wind velocity determines the initial angle of bank required to maintain a constant turn radius from the pylons during each turn. The steepest banks are necessary just after each turn entry and just before the rollout from each turn where the airplane is headed downwind and the groundspeed is highest; the shallowest banks are when the airplane is headed directly upwind and the groundspeed is lowest.

As in other ground reference maneuvers, the rate at which the bank angle must change depends on the wind velocity. If the airplane proceeds diagonally from one turn to the other, the rollout from each turn must be completed on the proper heading with sufficient wind correction angle to ensure that after brief straight-and-level flight, the airplane arrives at the point where a turn of the same radius can be made around the other pylon. The straight-and-level flight segments must be tangent to both circular patterns.

Common errors in the performance of elementary eights are:

- Failure to adequately clear the area above, below, and on either side of the airplane for safety hazards, initially and throughout the maneuver.

- Poor selection of ground references.

Figure 6-8. *Eights across a road.*

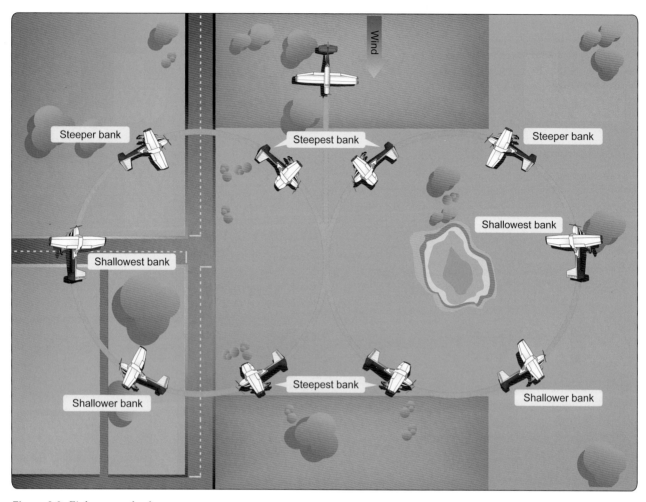

Figure 6-9. *Eights around pylons.*

- Failure to establish a constant, level altitude prior to entering the maneuver.

- Failure to maintain adequate altitude control during the maneuver.

- Failure to properly assess wind direction.

- Failure to properly execute constant radius turns.

- Failure to manipulate the flight controls in a smooth and continuous manner.

- Failure to establish the appropriate wind correction angles.

- Failure to apply coordinated aileron and rudder pressure, resulting in slips or skids.

- Failure to maintain orientation as the maneuver progresses.

Eights-on-Pylons

The eights-on-pylons is the most advanced and difficult of the ground reference maneuvers. Because of the techniques involved, the eights-on-pylons are unmatched for developing intuitive control of the airplane. Similar to eights around pylons except altitude is varied to maintain a specific visual reference to the pivot points.

The goal of the eights-on-pylons is to have an imaginary line that extends from the pilot's eyes to the pylon. This line must be imagined to always be parallel to the airplane's lateral axis. Along this line, the airplane appears to pivot as it turns around the pylon. In other words, if a taut string extended from the airplane to the pylon, the string would remain parallel to lateral axis as the airplane turned around the pylon. At no time should the string be at an angle to the lateral axis. *[Figure 6-10]* In explaining the performance of eights-on-pylons, the term "wingtip" is frequently considered as being synonymous with the proper visual reference line or pivot point on the airplane. This interpretation is not always correct. High-wing, low-wing, sweptwing, and tapered wing airplanes, as well as those with tandem or side-by-side seating, all present different angles from the pilot's eye to the wingtip. *[Figure 6-11]*

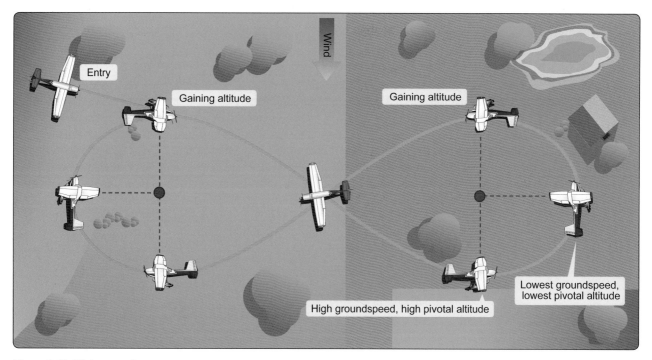

Figure 6-10. *Eights on pylons.*

The visual reference line, while not necessarily on the wingtip itself, may be positioned in relation to the wingtip (ahead, behind, above, or below), and differs for each pilot and from each seat in the airplane. This is especially true in tandem (fore and aft) seat airplanes. In side-by-side type airplanes, there is very little variation in the visual reference lines for different persons, if those persons are seated with their eyes at approximately the same level. Therefore, in the correct performance of eights-on-pylons, as in other maneuvers requiring a lateral reference, the pilot should use a visual reference line that, from eye level, parallels the lateral axis of the airplane.

The altitude that is appropriate for eights-on-pylons is called the "pivotal altitude" and is determined by the airplane's groundspeed. In previous ground-track maneuvers, the

Figure 6-11. *Line of sight.*

airplane flies a prescribed path over the ground and the pilot attempts to maintain the track by correcting for the wind. With eights-on-pylons, the pilot maintains lateral orientation to a specific spot on the ground. This develops the pilot's ability to maneuver the airplane accurately while dividing attention between the flightpath and the selected pylons on the ground.

An explanation of the pivotal altitude is also essential. First, a good rule of thumb for estimating the pivotal altitude is to square the groundspeed, then divide by 15 (if the groundspeed is in miles per hour) or divide by 11.3 (if the groundspeed is in knots), and then add the mean sea level (MSL) altitude of the ground reference. The pivotal altitude is the altitude at which, for a given groundspeed, the projection of the visual reference line to the pylon appears to pivot. *[Figure 6-12]* The pivotal altitude does not vary with the angle of bank unless the bank is steep enough to affect the groundspeed.

Distance from the pylon affects the angle of bank. At any altitude above that pivotal altitude, the projected reference line appears to move rearward in a circular path in relation to the pylon. Conversely, when the airplane is below the pivotal altitude, the projected reference line appears to move forward in a circular path. *[Figure 6-13]* To demonstrate this, the pilot will fly at maneuvering speed and at an altitude below the pivotal altitude, and then placed in a medium-banked turn. The projected visual reference line appears to move forward along the ground (pylon moves back) as the airplane turns. The pilot then executes a climb to an altitude well above the pivotal altitude. When the airplane is again at maneuvering speed, it is placed in a medium-banked turn. At the higher altitude, the projected visual reference line appears to move backward across the ground (pylon moves forward).

After demonstrating the maneuver at a high altitude, the pilot should reduce power and begin a descent at maneuvering speed in a continuing medium bank turn around the pylon. The apparent backward movement of the projected visual reference line with respect to the pylon will slow down as altitude is lost and will eventually stop for an instant. If the

pilot continues the descent below the pivotal altitude, the projected visual reference line with respect to the pylon will begin to move forward.

The altitude at which the visual reference line ceases to move across the ground is the pivotal altitude. If the airplane descends below the pivotal altitude, the pilot should increase power to maintain airspeed while regaining altitude to the point at which the projected reference line moves neither backward nor forward but actually pivots on the pylon. In this way, the pilot can determine the pivotal altitude of the airplane.

The pivotal altitude is critical and changes with variations in groundspeed. Since the headings throughout turns continuously vary from downwind to upwind, the groundspeed constantly changes. This results in the proper pivotal altitude varying slightly throughout the turn. The pilot should adjust for this by climbing or descending, as necessary, to hold the visual reference line on the pylons. This change in altitude is dependent on the groundspeed.

Selecting proper pylon is an important factor of successfully performing eights-on-pylons. They should be sufficiently prominent so the pilot can view them when completing the turn around one pylon and heading for the next. They should also be adequately spaced to provide time for planning the turns but not spaced so far apart that they cause unnecessary straight-and-level flight between the pylons. The selected pylons should also be at the same elevation, since differences of over few feet necessitate climbing or descending between each turn. The pilot should select two pylons along a line that lies perpendicular to the direction of the wind. The distance between the pylons should allow for the straight-and-level flight segment to last from 3 to 5 seconds.

The pilot should estimate the pivotal altitude during preflight planning. Weather reports and consultation with other pilots flying in the area may provide both the wind direction and velocity. If the references are previously known (many flight instructors already have these ground-based reference selected), the sectional chart will provide the MSL of the references, the Pilot's Operating Handbook (POH) provides the range of maneuvering airspeeds (based on weight), and the wind direction and velocity can be estimated to calculate the appropriate pivotal altitudes. The pilot should calculate the pivotal altitude for each position: upwind, downwind, and crosswind.

The pilot should begin the eight-on-pylons maneuver by flying diagonally crosswind between the pylons to a point downwind from the first pylon so that the first turn can be made into the wind. As the airplane approaches a position where the pylon appears to be just ahead of the wingtip, the

Groundspeed		Approximate Pivotal Altitude
Knots	MPH	
87	100	670
91	105	735
96	110	810
100	115	885
104	120	960
109	125	1050
113	130	1130

Figure 6-12. *Speed versus pivotal altitude.*

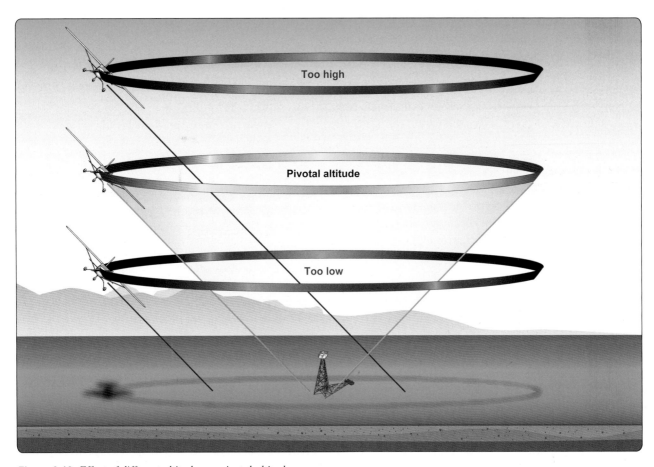

Figure 6-13. *Effect of different altitudes on pivotal altitude.*

pilot should begin the turn by lowering the upwind wing to the point where the visual reference line aligns with the pylon. The reference line should appear to pivot on the pylon. As the airplane heads upwind, the groundspeed decreases, which lowers the pivotal altitude. As a result, the pilot must descend to hold the visual reference line on the pylon. As the turn progresses on the upwind side of the pylon, the wind becomes more of a crosswind. Since this maneuver does not require the turn to be completed at a constant radius, the pilot does not need to apply drift correction to complete the turn.

If the visual reference line appears to move ahead of the pylon, the pilot should increase altitude. If the visual reference line appears to move behind the pylon, the pilot should decrease altitude. Deflecting the rudder to yaw the airplane and force the wing and reference line forward or backward to the pylon places the airplane in uncoordinated flight, at low altitude, with steep bank angles and must not be attempted.

As the airplane turns toward a downwind heading, the pilot should rollout from the turn to allow the airplane to proceed diagonally to a point tangent on the downwind side of the second pylon. The pilot should complete the rollout with the proper wind correction angle to correct for wind drift,

so that the airplane arrives at a point downwind from the second pylon that is equal in distance from the pylon as the corresponding point was from the first pylon at the beginning of the maneuver.

At this point, the pilot should begin a turn in the opposite direction by lowering the upwind wing to the point where the visual reference line aligns with the pylon. The pilot should then continue the turn the same way the corresponding turn was performed around the first pylon but in the opposite direction.

With prompt correction, and a very fine control pressures, it is possible to hold the visual reference line directly on the pylon even in strong winds. The pilot may make corrections for temporary variations, such as those caused by gusts or inattention by reducing the bank angle slightly to fly relatively straight to bring forward a lagging visual reference line or by increasing the bank angle temporarily to turn back a visual reference line that has moved ahead. With practice, these corrections may become slight enough to be barely noticeable. It is important to understand that variations in pylon position are according to the apparent movement of the visual reference line. Attempting to correct pivotal altitude by the use of the altimeter is ineffective.

Eights-on-pylons are performed at bank angles ranging from shallow to steep. *[Figure 6-14]* The pilot should understand that the bank chosen does not alter the pivotal altitude. As proficiency is gained, the instructor should increase the complexity of the maneuver by directing the student to enter at a distance from the pylon that results in a specific bank angle at the steepest point in the pylon turn.

The most common error in attempting to hold a pylon is incorrect use of the rudder. When the projection of the visual reference line moves forward with respect to the pylon, many pilots tend to apply inside rudder pressure to yaw the wing backward. When the reference line moves behind the pylon, they tend to apply outside rudder pressure to yaw the wing forward. The pilot should use the rudder only for coordination.

Other common errors in the performance of eights-on-pylons are:

- Failure to adequately clear the area above, below, and on either side of the airplane for safety hazards, initially and throughout the maneuver.

- Poor selection of ground references.

- Failure to establish a constant, level altitude prior to entering the maneuver.

- Failure to maintain adequate altitude control during the maneuver.

- Failure to properly assess wind direction.

- Failure to properly execute constant radius turns.

- Failure to manipulate the flight controls in a smooth and continuous manner.

- Failure to establish the appropriate wind correction angles.

- Failure to apply coordinated aileron and rudder pressure, resulting in slips or skids.

- Failure to maintain orientation as the maneuver progresses.

Chapter Summary

At the completion of ground reference maneuvers, the pilot should not only be able to command the airplane to specific pitch, roll, and yaw attitudes but, while correcting for the effects of wind drift, also control the airplane's orientation in relation to ground-based references. It should be reinforced that safety is paramount in all aspects of flying. Ground reference maneuvers require planning and high levels of vigilance to ensure that the practice and performance of these maneuvers are executed where the safety to groups of people, livestock, communities, and the pilot is not compromised. To master ground reference maneuvers, a pilot must develop coordination, timing, and division of attention to accurately maneuver the airplane in reference to flight attitudes and specific ground references. With these enhanced skills, the pilot significantly strengthens their competency in everyday flight maneuvers, such as straight-and-level, turns, climbs, and descents.

Figure 6-14. *Bank angle versus pivotal altitude.*

Chapter 7

Airport Traffic Patterns

Traffic pattern indicators (indicates location of base leg)

Windsock

Segmented circle

Introduction

Airport traffic patterns are developed to ensure that air traffic is flown into and out of an airport safely. Each airport traffic pattern is established based on the local conditions, including the direction and placement of the pattern, the altitude at which it is to be flown, and the procedures for entering and exiting the pattern. It is imperative that pilots are taught correct traffic pattern procedures and exercise constant vigilance in the vicinity of airports when entering and exiting the traffic pattern. Information regarding the procedures for a specific airport can be found in the Chart Supplements. Additional information on airport operations and traffic patterns can be found in the Aeronautical Information Manual (AIM).

Distribution of Mid-air Collisions in the Airport Pattern

30%	34%		34%
20%			
16%		16%	
10%			
0%			
Downwind	Final	Short final	Runway

Crosswind

Entry

WIND

Right-Hand Traffic

Downwind

Base

Final

0%

0%

6%

10%

82%

Airport Traffic Patterns and Operation

Just as roads and streets are essential for operating automobiles, airports or airstrips are essential for operating airplanes. Every flight begins and ends at an airport or other suitable landing field; therefore, it is essential that pilots learn the traffic rules, traffic procedures, and traffic pattern layouts that may be in use at various airports.

When an automobile is driven on congested city streets, it can be brought to a stop to give way to conflicting traffic; however, an airplane can only speed up, climb, descend, and be slowed down. Consequently, traffic patterns and traffic control procedures have been established for use at airports. Traffic patterns provide procedures for takeoffs, departures, arrivals, and landings. The exact nature of each airport traffic pattern is dependent on the runway in use, wind conditions (which determine the runway in use), obstructions, and other factors.

Control towers and radar facilities provide a means of adjusting the flow of arriving and departing aircraft and render assistance to pilots in busy terminal areas. Airport lighting and runway marking systems are used frequently to alert pilots to abnormal conditions and hazards so arrivals and departures can be made safely.

Airports vary in complexity from small grass or sod strips to major terminals with paved runways and taxiways. Regardless of the type of airport, a pilot must know and abide by the rules and general operating procedures applicable to the airport being used. The objective is to keep air traffic moving with maximum safety and efficiency. Information on traffic patterns and operating procedures for an airport is documented in the Chart Supplements, as well as visual markings on the airport itself. The use of any traffic pattern, service, or procedure does not diminish the pilot's responsibility to see and avoid other aircraft during flight.

Standard Airport Traffic Pattern

To assure that air traffic flows into and out of an airport in an orderly manner, an airport traffic pattern is established based on the local conditions, to include the direction and altitude of the pattern and the procedures for entering and leaving the pattern. Unless the airport displays approved visual markings indicating that turns should be made to the right, the pilot should make all turns in the pattern to the left.

When operating at an airport with an operating control tower, the pilot receives a clearance to approach or depart, as well as pertinent information about the traffic pattern by radio. If there is not a control tower, it is the pilot's responsibility to determine the direction of the traffic pattern, to comply with the appropriate traffic rules, and to display common courtesy toward other pilots operating in the area.

A pilot is not expected to have extensive knowledge of all traffic patterns at all airports, but if the pilot is familiar with the basic rectangular pattern, it is easy to make proper approaches and departures from most airports, regardless of whether or not they have control towers. At airports with operating control towers, the tower operator can instruct pilots to enter the traffic pattern at any point or to make a straight-in approach without flying the usual rectangular pattern. Many other deviations are possible if the tower operator and the pilot work together in an effort to keep traffic moving smoothly. Jets or heavy airplanes will frequently fly wider and/or higher patterns than lighter airplanes, and in many cases, will make a straight-in approach for landing.

Compliance with the basic rectangular traffic pattern reduces the possibility of conflicts at airports without an operating control tower. It is imperative that a pilot form the habit of exercising constant vigilance in the vicinity of airports even when the air traffic appears to be light. Midair collisions usually occur on clear days with unlimited visibility. Never assume you have found all of the air traffic and stop scanning.

Figure 7-1 shows a standard rectangular traffic pattern. The traffic pattern altitude is usually 1,000 feet above the elevation of the airport surface. The use of a common altitude at a given airport is the key factor in minimizing the risk of collisions at airports without operating control towers.

When operating in the traffic pattern at an airport without an operating control tower, the pilot should maintain an airspeed of no more than 200 knots (230 miles per hour (mph)) as required by Title 14 of the Code of Federal Regulations (14 CFR) part 91. In any case, the pilot should adjust the airspeed, when necessary, so that it is compatible with the airspeed of the other airplanes in the pattern.

When entering the traffic pattern at an airport without an operating control tower, inbound pilots are expected to observe other aircraft already in the pattern and to conform to the traffic pattern in use. If there are no other aircraft present, the pilot should check traffic indicators on the ground and wind indicators to determine which runway and traffic pattern direction to use. *[Figure 7-2]* Many airports have L-shaped traffic pattern indicators displayed with a segmented circle adjacent to the runway. The short member of the L shows the direction in which the traffic pattern turns are made when using the runway parallel to the long member. The pilot should check the indicators from a distance or altitude well away from any other airplanes that may be flying in the traffic pattern. Upon identifying the proper traffic pattern, the pilot should enter into the traffic pattern at a point well clear of the other airplanes.

Figure 7-1. *Traffic patterns.*

When approaching an airport for landing, the traffic pattern is normally entered at a 45° angle to the downwind leg, headed toward a point abeam the midpoint of the runway to be used for landing. When arriving, the pilot should be aware of the proper traffic pattern altitude before entering the pattern and remain clear of the traffic flow until established on the entry leg. Entries into traffic patterns while descending create specific collision hazards and should always be avoided.

The pilot should ensure that the entry leg is of sufficient length to provide a clear view of the entire traffic pattern

Figure 7-2. *Traffic pattern indicators.*

and to allow adequate time for planning the intended path in the pattern and the landing approach.

The downwind leg is a course flown parallel to the landing runway, but in a direction opposite to the intended landing direction. This leg is flown approximately ½ to 1 mile out from the landing runway and at the specified traffic pattern altitude. When flying on the downwind leg, the pilot should complete all before landing checks and extend the landing gear if the airplane is equipped with retractable landing gear. Pattern altitude is maintained until at least abeam the approach end of the landing runway. At this point, the pilot should reduce power and begin a descent. The pilot should continue the downwind leg past a point abeam the approach end of the runway to a point approximately 45° from the approach end of the runway, and make a medium bank turn onto the base leg. Pilots should consider tailwinds and not descend too much on the downwind, so as to have a very low base leg altitude.

The base leg is the transitional part of the traffic pattern between the downwind leg and the final approach leg. Depending on the wind condition, the pilot should establish the base leg at a sufficient distance from the approach end of the landing runway to permit a gradual descent to the intended touchdown point. The ground track of the airplane while on the base leg is perpendicular to the extended centerline of the landing runway, although the longitudinal axis of the airplane may not be aligned with the ground track when it is necessary to turn into the wind to counteract drift. While on the base leg, the pilot must ensure, before turning onto the final approach, that there is no danger of colliding

with another aircraft that is already established on the final approach. Pilots must not attempt an overly steep turn to final, especially uncoordinated! If in doubt, go around.

The final approach leg is a descending flightpath starting from the completion of the base-to-final turn and extending to the point of touchdown. This is probably the most important leg of the entire pattern, because of the sound judgment and precision required to accurately control the airspeed and descent angle while approaching the intended touchdown point.

14 CFR part 91, states that aircraft, while on final approach to land or while landing, have the right-of-way over other aircraft in flight or operating on the surface. When two or more aircraft are approaching an airport for the purpose of landing, the aircraft at the lower altitude has the right-of-way. Pilots should not take advantage of this rule to cut in front of another aircraft that is on final approach to land or to overtake that aircraft.

The upwind leg is a course flown parallel to the landing runway in the same direction as landing traffic. The upwind leg is flown at controlled airports and after go-arounds.

When necessary, the upwind leg is the part of the traffic pattern in which the pilot will transition from the final approach to the climb altitude to initiate a go-around. When a safe altitude is attained, the pilot should commence a shallow bank turn to the upwind side of the airport. This allows better visibility of the runway for departing aircraft.

The departure leg of the rectangular pattern is a straight course aligned with, and leading from, the takeoff runway. This leg begins at the point the airplane leaves the ground and continues until the pilot begins the 90° turn onto the crosswind leg.

On the departure leg after takeoff, the pilot should continue climbing straight ahead and, if remaining in the traffic pattern, commence a turn to the crosswind leg beyond the departure end of the runway within 300 feet of the traffic pattern altitude. If departing the traffic pattern, the pilot should continue straight out or exit with a 45° turn (to the left when in a left-hand traffic pattern; to the right when in a right-hand traffic pattern) beyond the departure end of the runway after reaching the traffic pattern altitude.

The crosswind leg is the part of the rectangular pattern that is horizontally perpendicular to the extended centerline of the takeoff runway. The pilot should enter the crosswind leg by making approximately a 90° turn from the upwind leg. The pilot should continue on the crosswind leg, to the downwind leg position.

Since in most cases the takeoff is made into the wind, the wind will now be approximately perpendicular to the airplane's flightpath. As a result, the pilot should turn or head the airplane slightly into the wind while on the crosswind leg to maintain a ground track that is perpendicular to the runway centerline extension.

Non-Towered Airports

Non towered airports traffic patterns are always entered at pattern altitude. How you enter the pattern depends upon the direction of arrival. The preferred method for entering from the downwind leg side of the pattern is to approach the pattern on a course 45° to the downwind leg and join the pattern at midfield.

There are several ways to enter the pattern if you are coming from the upwind legs side of the airport. One method of entry from the opposite side of the pattern is to announce your intentions and cross over midfield at least 500 feet above pattern altitude (normally 1,500 feet AGL.) However, if large or turbine aircraft operate at your airport, it is best to remain 2,000 feet AGL so you're not in conflict with their traffic pattern. When well clear of the pattern—approximately 2 miles—scan carefully for traffic, descend to pattern altitude, then turn right to enter at 45° to the downwind leg at midfield. *[Figure 7-4A]* An alternate method is to enter on a midfield crosswind at pattern altitude, carefully scan for traffic, announce your intentions and then turned down downwind. *[Figure 7-4B]* This technique should not be used if the pattern is busy.

Always remember to give way to aircraft on the preferred 45° entry and to aircraft already established on downwind. In either case, it is vital to announce your intentions, and remember to scan outside. Before joining the downwind leg, adjust your course or speed to blend into the traffic. Adjust power on the downwind leg, or sooner, to fit into the flow of traffic. Avoid flying too fast or too slow. Speeds recommended by the airplane manufacturer should be used. They will generally fall between 70 to 80 knots for fixed-gear singles, and 80 to 90 knots for high-performance retractable.

Safety Considerations

According to the National Transportation Safety Board (NTSB), the most probable cause of mid-air collisions is the pilot failing to see and avoid other aircraft. When in the traffic, pilots must continue to scan for other aircraft and check blind spots caused by fixed aircraft structures,

Figure 7-4. *Preferred entry from upwind leg side of airport (A). Alternate midfield entry from upwind leg side of airport (B).*

such as doorposts and wings. High-wing airplanes have restricted visibility above while low-wing airplanes have limited visibility below. The worst-case scenario is a low-wing airplane flying above a high-wing airplane. Banking from time to time can uncover blind spots. The pilot should also occasionally look to the rear of the airplane to check for other aircraft. *Figure 7-5* depicts the greatest threat area for mid-air collisions in the traffic pattern. Listed below are important facts regarding mid-air collisions:

- Mid-air collisions generally occur during daylight hours; 56 percent of the accidents occur in the afternoon, 32 percent occur in the morning, and 2 percent occur at night, dusk, or dawn.

- Most mid-air collisions occur under good visibility.

- A mid-air collision is most likely to occur between two aircraft going in the same direction.

- The majority of pilots involved in mid-air collisions are not on a flight plan.

- Nearly all accidents occur at or near uncontrolled airports and at altitudes below 1,000 feet.

- Pilots of all experience levels are involved in mid-air collisions.

The following are some important procedures that all pilots should be follow when flying in a traffic pattern or in the vicinity of an airport.

- Tune and verify radio frequencies before entering the airport traffic area.

- Report your position 10 miles out and listen for reports from other inbound traffic.

- Report when you are entering downwind, turning downwind to base, and base to final. This is a good practice at a non-towered airport.

- Descend to traffic pattern altitude before entering the pattern.

- Maintain a constant visual scan for other aircraft.

- Tune and monitor the correct Common Traffic Advisory Frequency (CTAF) frequency.

- Be aware that there may be aircraft in the pattern without radios.

- Use exterior lights to improve the chances of being seen.

Chapter Summary

The volume of traffic at an airport can create a hazardous environment. Airport traffic patterns are procedures that improve the flow of traffic at an airport and when properly executed enhance safety. Most reported mid-air collisions occur during the final or short final approach leg of the airport traffic pattern.

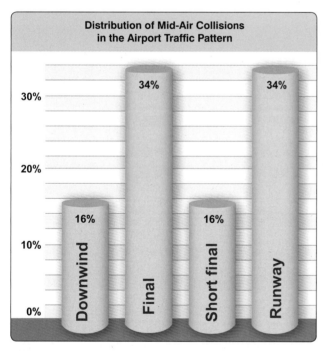

Figure 7-5. *Location distribution of mid-air collisions in the airport traffic pattern.*

Chapter 8

Approaches and Landings

Introduction

There is a saying that while takeoff is optional, landing is mandatory. Unfortunately, a review of accident statistics indicates that over 45 percent of all general aviation accidents occur during the approach and landing phases of a flight. A closer look shows that the cause of over 90 percent of those cases was pilot related and loss of control was also a major contributing factor in 33 percent of the cases. While the requirement to maneuver close to the ground cannot be eliminated, pilots can develop the skills and follow established procedures to reduce the likelihood of an accident or mishap. This chapter focuses on the approach to landing, factors that affect landings, types of landings, and aspects of faulty landings.

Normal Approach and Landing

A normal approach and landing involves the use of procedures for what is considered a normal situation; that is, when engine power is available, the wind is light, or the final approach is made directly into the wind, the final approach path has no obstacles and the landing surface is firm and of ample length to gradually bring the airplane to a stop. The selected landing point is normally beyond the runway's approach threshold but within the first ⅓ portion of the runway.

The factors involved and the procedures described for the normal approach and landing also have applications to the other-than-normal approaches and landings and are discussed later in this chapter. This being the case, the principles of normal operations are explained first and must be understood before proceeding to the more complex operations. To help the pilot better understand the factors that influence judgment and procedures, the last part of the approach pattern and the actual landing is divided into five phases:

1. the base leg

2. the final approach

3. the round out (flare)

4. the touchdown

5. the after-landing roll

It must be remembered that the manufacturer's recommended procedures, including airplane configuration and airspeeds, and other information relevant to approaches and landings in a specific make and model airplane are contained in the Federal Aviation Administration (FAA)-approved Airplane Flight Manual and/or Pilot's Operating Handbook (AFM/POH) for that airplane. If any of the information in this chapter differs from the airplane manufacturer's recommendations as contained in the AFM/POH, the airplane manufacturer's recommendations take precedence.

Base Leg

The placement of the base leg is one of the more important judgments made by the pilot in any landing approach. *[Figure 8-1]* The pilot must accurately judge the altitude and distance from which a gradual, stabilized descent results in landing at the desired spot. The distance depends on the altitude of the base leg, the effect of wind, and the amount of wing flaps used. When there is a strong wind on final approach or the flaps are used to produce a steep angle of descent, the base leg must be positioned closer to the approach end of the runway than would be required with a light wind or no flaps. Normally, the landing gear is extended and the before-landing check completed prior to reaching the base leg.

After turning onto the base leg, start the descent with reduced power and airspeed of approximately 1.4 V_{SO}, which is the

Figure 8-1. *Base leg and final approach.*

stalling speed with power off, landing gear and flaps down. For example, if V_{SO} is 60 knots, the speed should be 1.4 times 60 or 84 knots. Landing flaps may be partially lowered, if desired, at this time. Full flaps are not recommended until the final approach is established. A drift correction is established and maintained to follow a ground track perpendicular to the extension of the centerline of the runway on which the landing is to be made. Since the final approach and landing are normally made into the wind, there is somewhat of a crosswind during the base leg. This requires that the airplane be angled sufficiently into the wind to prevent drifting farther away from the intended landing spot.

The base leg is continued to the point where a medium to shallow-banked turn aligns the airplane's path directly with the centerline of the landing runway. This descending turn is completed at a safe altitude and dependent upon the height of the terrain and any obstructions along the ground track. The turn to the final approach is sufficiently above the airport elevation to permit a final approach long enough to accurately estimate the resultant point of touchdown while maintaining the proper approach airspeed. This requires careful planning as to the starting point and the radius of the turn. Normally, it is recommended that the angle of bank not exceed a medium bank because the steeper the angle of bank, the higher the airspeed at which the airplane stalls. Since the base-to-final turn is made at a relatively low altitude, it is important that a stall not occur at this point. If an extremely steep bank is needed to prevent overshooting the proper final approach path, it is advisable to discontinue the approach, go around, and plan to start the turn earlier on the next approach rather than risk a hazardous situation.

Final Approach

After the base-to-final approach turn is completed, the longitudinal axis of the airplane is aligned with the centerline of the runway or landing surface so that drift (if any) is recognized immediately. On a normal approach, with no wind drift, the longitudinal axis is kept aligned with the runway centerline throughout the approach and landing. (The proper way to correct for a crosswind is explained under the section, Crosswind Approach and Landing. For now, only an approach and landing where the wind is straight down the runway are discussed.)

After aligning the airplane with the runway centerline, the final flap setting is completed and the pitch attitude adjusted as required for the desired rate of descent. Slight adjustments in pitch and power may be necessary to maintain the descent attitude and the desired approach airspeed. In the absence of the manufacturer's recommended airspeed, a speed equal to 1.3 V_{SO} should be used. If V_{SO} is 60 knots, the speed should be 78 knots. When the pitch attitude and airspeed have been stabilized, the airplane is re-trimmed to relieve the pressures being held on the controls.

A stabilized descent angle is controlled throughout the approach so that the airplane lands in the center of the first third of the runway. The descent angle is affected by all four fundamental forces that act on an airplane (lift, drag, thrust, and weight). If all the forces are constant, the descent angle is constant in a no-wind condition. The pilot controls these forces by adjusting the airspeed, attitude, power, and drag (flaps or forward slip). The wind also plays a prominent part in the gliding distance over the ground *[Figure 8-2]*; the pilot does not have control over the wind but corrects for its effect on the airplane's descent by appropriate pitch and power adjustments.

Considering the factors that affect the descent angle on the final approach, for all practical purposes at a given pitch attitude there is only one power setting for one airspeed, one

Figure 8-2. *Effect of headwind on final approach.*

flap setting, and one wind condition. A change in any one of these variables requires an appropriate coordinated change in the other controllable variables. For example, if the pitch attitude is raised too high without an increase of power, the airplane settles very rapidly and touches down short of the desired spot. For this reason, never try to stretch a glide by applying back-elevator pressure alone to reach the desired landing spot. This shortens the gliding distance if power is not added simultaneously. The proper angle of descent and airspeed is maintained by coordinating pitch attitude changes and power changes.

The objective of a good, stabilized final approach is to descend at an angle and airspeed that permits the airplane to reach the desired touchdown point at an airspeed that results in minimum floating just before touchdown; in essence, a semi-stalled condition. To accomplish this, it is essential that both the descent angle and the airspeed be accurately controlled. Since on a normal approach the power setting is not fixed as in a power-off approach, the power and pitch attitude are adjusted simultaneously as necessary to control the airspeed and the descent angle, or to attain the desired altitudes along the approach path. By lowering the nose and reducing power to keep approach airspeed constant, a descent at a higher rate can be made to correct for being too high in the approach. This is one reason for performing approaches with partial power; if the approach is too high, merely lower the nose and reduce the power. When the approach is too low, add power and raise the nose.

Use of Flaps

The lift/drag factors are varied by the pilot to adjust the descent through the use of landing flaps. *[Figures 8-3 and 8-4]* Flap extension during landings provides several advantages by:

- Producing greater lift and permitting lower landing speed,

- Producing greater drag, permitting a steeper descent angle without airspeed increase, and

- Reducing the length of the landing roll.

Flap extension has a definite effect on the airplane's pitch behavior. The increased camber from flap deflection produces lift primarily on the rear portion of the wing. This produces a nose-down pitching moment; however, the change in tail loads from the downwash deflected by the flaps over the horizontal tail has a significant influence on the pitching moment. Consequently, pitch behavior depends on the design features of the particular airplane.

Flap deflection of up to 15° primarily produces lift with minimal drag. The airplane has a tendency to balloon up with initial flap deflection because of the lift increase. The nose-down pitching moment, however, tends to offset the balloon. Flap deflection beyond 15° produces a large increase in drag. Also, deflection beyond 15° produces a significant nose-up pitching moment in high-wing airplanes because the resulting downwash increases the airflow over the horizontal tail.

The time of flap extension and the degree of deflection are related. Large flap deflections at one single point in the landing pattern produce large lift changes that require significant pitch and power changes in order to maintain airspeed and descent angle. Consequently, there is an advantage to extending flaps in increments while in the landing pattern. Incremental deflection of flaps on downwind, base leg, and final approach allow smaller adjustments of pitch and power compared to extension of full flaps all at one time.

When the flaps are lowered, the airspeed decreases unless the power is increased or the pitch attitude lowered. On final approach, the pilot must estimate where the airplane lands

Figure 8-3. *Effect of flaps on the landing point.*

Figure 8-4. *Effect of flaps on the approach angle.*

through judgment of the descent angle. If it appears that the airplane is going to overshoot the desired landing spot, more flaps are used, if not fully extended, or the power reduced further and the pitch attitude lowered. This results in a steeper approach. If the desired landing spot is being undershot and a shallower approach is needed, both power and pitch attitude are increased to readjust the descent angle. Never retract the flaps to correct for undershooting since that suddenly decreases the lift and causes the airplane to sink rapidly.

The airplane must be re-trimmed on the final approach to compensate for the change in aerodynamic forces. With the reduced power and with a slower airspeed, the airflow produces less lift on the wings and less downward force on the horizontal stabilizer resulting in a significant nose-down tendency. The elevator must then be trimmed more nose-up.

The round out, touchdown, and landing roll are much easier to accomplish when they are preceded by a proper final approach consisting of precise control of airspeed, attitude, power, and drag resulting in a stabilized descent angle.

Estimating Height and Movement

During the approach, round out, and touchdown; vision is of prime importance. To provide a wide scope of vision and to foster good judgment of height and movement, the pilot's head should assume a natural, straight-ahead position. Visual focus is not fixed on any one side or any one spot ahead of the airplane. Instead, it is changed slowly from a point just over the airplane's nose to the desired touchdown zone and back again. This is done while maintaining a deliberate awareness of distance from either side of the runway using your peripheral field of vision.

Accurate estimation of distance is, besides being a matter of practice, dependent upon how clearly objects are seen. It requires that the vision be focused properly in order that the important objects stand out as clearly as possible.

Speed blurs objects at close range. For example, most everyone has noted this in an automobile moving at high speed. Nearby objects seem to merge together in a blur, while objects farther away stand out clearly. The driver subconsciously focuses the eyes sufficiently far ahead of the automobile to see objects distinctly.

The distance at which the pilot's vision is focused should be proportionate to the speed at which the airplane is traveling over the ground. Thus, as speed is reduced during the round out, the distance ahead of the airplane at which it is possible to focus is brought closer accordingly.

If the pilot attempts to focus on a reference that is too close or looks directly down, the reference becomes blurred, *[Figure 8-5]* and the reaction is either too abrupt or too late. In this case, the pilot's tendency is to over-control, round out high, and make full-stall, drop-in landings. If the pilot focuses too far ahead, accuracy in judging the closeness of the ground is lost and the consequent reaction is too slow since there does not appear to be a necessity for action. This results in the airplane flying into the ground nose first. The change of visual focus from a long distance to a short distance requires a definite time interval and, even though the time is brief, the airplane's speed during this interval is such that the airplane travels an appreciable distance, both forward and downward toward the ground.

If the focus is changed gradually, being brought progressively closer as speed is reduced, the time interval and the pilot's reaction are reduced and the whole landing process smoothed out.

Figure 8-5. *Focusing too close blurs vision.*

Round Out (Flare)

The round out is a slow, smooth transition from a normal approach attitude to a landing attitude, gradually rounding out the flightpath to one that is parallel with, and within a very few inches above, the runway. When the airplane, in a normal descent, approaches within what appears to be 10 to 20 feet above the ground, the round out or flare is started. This is a continuous process until the airplane touches down on the ground.

As the airplane reaches a height above the ground where a change into the proper landing attitude can be made, back-elevator pressure is gradually applied to slowly increase the pitch attitude and angle of attack (AOA). *[Figure 8-6]* This causes the airplane's nose to gradually rise toward the desired landing attitude. The AOA is increased at a rate that allows the airplane to continue settling slowly as forward speed decreases.

When the AOA is increased, the lift is momentarily increased and this decreases the rate of descent. Since power normally is reduced to idle during the round out, the airspeed also gradually decreases. This causes lift to decrease again and necessitates raising the nose and further increasing the AOA. During the round out, the airspeed is decreased to touchdown speed while the lift is controlled so the airplane settles gently onto the landing surface. The round out is executed at a rate that the proper landing attitude and the proper touchdown airspeed are attained simultaneously just as the wheels contact the landing surface.

The rate at which the round out is executed depends on the airplane's height above the ground, the rate of descent, and the pitch attitude. A round out started excessively high must be executed more slowly than one from a lower height to allow the airplane to descend to the ground while the proper landing attitude is being established. The rate of rounding out must also be proportionate to the rate of closure with the ground. When the airplane appears to be descending very slowly, the increase in pitch attitude must be made at a correspondingly slow rate.

Visual cues are important in flaring at the proper altitude and maintaining the wheels a few inches above the runway until eventual touchdown. Flare cues are primarily dependent on the angle at which the pilot's central vision intersects the ground (or runway) ahead and slightly to the side. Proper depth perception is a factor in a successful flare, but the visual cues used most are those related to changes in runway or terrain perspective and to changes in the size of familiar objects near the landing area, such as fences, bushes, trees, hangars, and even sod or runway texture. Focus direct central vision at a shallow downward angle from 10° to 15° toward the runway as the round out/flare is initiated. *[Figure 8-7]* Maintaining the same viewing angle causes the point of visual interception with

Figure 8-6. *Changing angle of attack during roundout.*

Figure 8-7. *To obtain necessary visual cues, the pilot should look toward the runway at a shallow angle.*

the runway to move progressively rearward as the airplane loses altitude. This is an important visual cue in assessing the rate of altitude loss. Conversely, forward movement of the visual interception point indicates an increase in altitude and means that the pitch angle was increased too rapidly, resulting in an over flare. Location of the visual interception point in conjunction with assessment of flow velocity of nearby off-runway terrain, as well as the similarity of appearance of height above the runway ahead of the airplane (in comparison to the way it looked when the airplane was taxied prior to takeoff), is also used to judge when the wheels are just a few inches above the runway.

The pitch attitude of the airplane in a full-flap approach is considerably lower than in a no-flap approach. To attain the proper landing attitude before touching down, the nose must travel through a greater pitch change when flaps are fully extended. Since the round out is usually started at approximately the same height above the ground regardless of the degree of flaps used, the pitch attitude must be increased at a faster rate when full flaps are used; however, the round out is still be executed at a rate proportionate to the airplane's downward motion.

Once the actual process of rounding out is started, do not push the elevator control forward. If too much back-elevator pressure was exerted, this pressure is either slightly relaxed or held constant, depending on the degree of the error. In some cases, it may be necessary to advance the throttle slightly to prevent an excessive rate of sink or a stall, either of which results in a hard, drop-in type landing.

It is recommended that a pilot form the habit of keeping one hand on the throttle throughout the approach and landing should a sudden and unexpected hazardous situation require an immediate application of power.

Touchdown

The touchdown is the gentle settling of the airplane onto the landing surface. The round out and touchdown are normally made with the engine idling and the airplane at minimum controllable airspeed so that the airplane touches down on the main gear at approximately stalling speed. As the airplane settles, the proper landing attitude is attained by application of whatever back-elevator pressure is necessary.

Some pilots try to force or fly the airplane onto the ground without establishing the proper landing attitude. The airplane should never be flown on the runway with excessive speed. A common technique to making a smooth touchdown is to actually focus on holding the wheels of the aircraft a few inches off the ground as long as possible using the elevators while the power is smoothly reduced to idle. In most cases, when the wheels are within 2 or 3 feet off the ground, the airplane is still settling too fast for a gentle touchdown; therefore, this descent must be retarded by increasing back-elevator pressure. Since the airplane is already close to its stalling speed and is settling, this added back-elevator pressure only slows the settling instead of stopping it. At the same time, it results in the airplane touching the ground in the proper landing attitude and the main wheels touching down first so that little or no weight is on the nose wheel. *[Figure 8-8]*

After the main wheels make initial contact with the ground, back-elevator pressure is held to maintain a positive AOA for aerodynamic braking and to hold the nose wheel off the ground until the airplane decelerates. As the airplane's momentum decreases, back-elevator pressure is gradually relaxed to allow the nose wheel to gently settle onto the runway. This permits steering with the nose wheel. At the same time, it decreases the AOA and reduces lift on the wings

Figure 8-8. *A well-executed roundout results in attaining the proper landing attitude.*

to prevent floating or skipping and allows the full weight of the airplane to rest on the wheels for better braking action.

It is extremely important that the touchdown occur with the airplane's longitudinal axis exactly parallel to the direction in which the airplane is moving along the runway. Failure to accomplish this imposes severe side loads on the landing gear. To avoid these side stresses, do not allow the airplane to touch down while turned into the wind or drifting.

After-Landing Roll

The landing process must never be considered complete until the airplane decelerates to the normal taxi speed during the landing roll or has been brought to a complete stop when clear of the landing area. Numerous accidents occur as a result of pilots abandoning their vigilance and failing to maintain positive control after getting the airplane on the ground.

A pilot must be alert for directional control difficulties immediately upon and after touchdown due to the ground friction on the wheels. Loss of directional control may lead to an aggravated, uncontrolled, tight turn on the ground, or a ground loop. The combination of centrifugal force acting on the center of gravity (CG) and ground friction of the main wheels resisting it during the ground loop may cause the airplane to tip or lean enough for the outside wingtip to contact the ground. This imposes a sideward force that could collapse the landing gear.

The rudder serves the same purpose on the ground as it does in the air—it controls the yawing of the airplane. The effectiveness of the rudder is dependent on the airflow, which depends on the speed of the airplane. As the speed decreases and the nose wheel has been lowered to the ground, the steerable nose provides more positive directional control.

The brakes of an airplane serve the same primary purpose as the brakes of an automobile—to reduce speed on the ground. In airplanes, they are also used as an aid in directional control when more positive control is required than could be obtained with rudder or nose wheel steering alone.

To use brakes, on an airplane equipped with toe brakes, the pilot slides the toes or feet up from the rudder pedals to the brake pedals. If rudder pressure is being held at the time braking action is needed, that pressure is not to be released as the feet or toes are being slid up to the brake pedals because control may be lost before brakes can be applied.

Putting maximum weight on the wheels after touchdown is an important factor in obtaining optimum braking performance. During the early part of rollout, some lift continues to be generated by the wing. After touchdown, the nose wheel is lowered to the runway to maintain directional control. During deceleration, the nose may pitch down by braking and the weight transferred to the nose wheel from the main wheels. This does not aid in braking action, so back pressure is applied to the controls without lifting the nose wheel off the runway. This enables directional control while keeping weight on the main wheels.

Careful application of the brakes is initiated after the nose wheel is on the ground and directional control is established. Maximum brake effectiveness is just short of the point where skidding occurs. If the brakes are applied so hard that skidding takes place, braking becomes ineffective. Skidding is stopped by releasing the brake pressure. Braking effectiveness is not enhanced by alternately applying, releasing, and reapplying brake pressure. The brakes are applied firmly and smoothly as necessary.

During the ground roll, the airplane's direction of movement can be changed by carefully applying pressure on one brake or uneven pressures on each brake in the desired direction. Caution must be exercised when applying brakes to avoid overcontrolling.

The ailerons serve the same purpose on the ground as they do in the air—they change the lift and drag components of the wings. During the after-landing roll, they are used to keep the wings level in much the same way they are used in flight. If a wing starts to rise, aileron control is applied toward that wing to lower it. The amount required depends on speed because as the forward speed of the airplane decreases, the ailerons become less effective. Procedures for using ailerons in crosswind conditions are explained further in this chapter, in the Crosswind Approach and Landing section.

After the airplane is on the ground, back-elevator pressure is gradually relaxed to place weight on the nose wheel to aid in better steering. If available runway permits, the speed of the airplane is allowed to dissipate in a normal manner. Once the airplane has slowed sufficiently and has turned on to the taxiway and stopped, retract the flaps and perform the after-landing checklist. Many accidents have occurred as a result of the pilot unintentionally operating the landing gear control and retracting the gear instead of the flap control when the airplane was still rolling. The habit of positively identifying both of these controls, before actuating them, must be formed from the very beginning of flight training and continued in all future flying activities.

Stabilized Approach Concept

A stabilized approach is one in which the pilot establishes and maintains a constant angle glide path towards a predetermined point on the landing runway. It is based on the pilot's judgment of certain visual clues and depends on the maintenance of a constant final descent airspeed and configuration.

An airplane descending on final approach at a constant rate and airspeed is traveling in a straight line toward a spot on the ground ahead. This spot is not the spot on which the airplane touches down because some float occurs during the round out (flare). *[Figure 8-9]* Neither is it the spot toward which the airplane's nose is pointed because the airplane is flying at a fairly high AOA, and the component of lift exerted parallel to the Earth's surface by the wings tends to carry the airplane forward horizontally.

The point toward which the airplane is progressing is termed the "aiming point." *[Figure 8-9]* It is the point on the ground at which, if the airplane maintains a constant glide path and was not flared for landing, it would strike the ground. To a pilot moving straight ahead toward an object, it appears to be stationary. It does not appear to move under the nose of the aircraft and does not appear to move forward away from the aircraft. This is how the aiming point can be distinguished—it does not move. However, objects in front of and beyond the aiming point do appear to move as the distance is closed, and they appear to move in opposite directions. During instruction in landings, one of the most important skills a pilot must acquire is how to use visual cues to accurately determine the true aiming point from any distance out on final approach. From this, the pilot is not only able to determine if the glide path results in either an under or overshoot but, taking into account float during round out, the pilot is able to predict the touchdown point to within a few feet.

For a constant angle glide path, the distance between the horizon and the aiming point remains constant. If a final approach descent is established and the distance between the perceived aiming point and the horizon appears to increase (aiming point moving down away from the horizon), then the true aiming point, and subsequent touchdown point, is farther down the runway. If the distance between the perceived aiming point and the horizon decreases, meaning that the aiming point is moving up toward the horizon, the true aiming point is closer than perceived.

Figure 8-9. *Stabilized approach.*

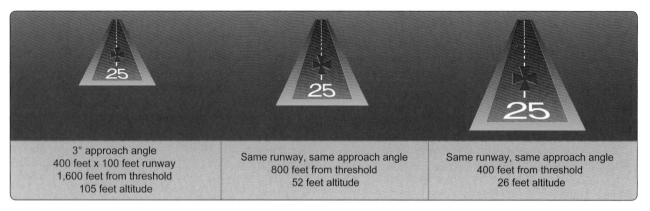

3° approach angle 400 feet x 100 feet runway 1,600 feet from threshold 105 feet altitude	Same runway, same approach angle 800 feet from threshold 52 feet altitude	Same runway, same approach angle 400 feet from threshold 26 feet altitude

Figure 8-10. *Runway shape during stabilized approach.*

When the airplane is established on final approach, the shape of the runway image also presents clues as to what must be done to maintain a stabilized approach to a safe landing.

Obviously, runway is normally shaped in the form of an elongated rectangle. When viewed from the air during the approach, the phenomenon known as perspective causes the runway to assume the shape of a trapezoid with the far end looking narrower than the approach end and the edge lines converging ahead.

As an airplane continues down the glide path at a constant angle (stabilized), the image the pilot sees is still trapezoidal but of proportionately larger dimensions. In other words, during a stabilized approach, the runway shape does not change. *[Figure 8-10]*

If the approach becomes shallow, the runway appears to shorten and become wider. Conversely, if the approach is steepened, the runway appears to become longer and narrower. *[Figure 8-11]*

The objective of a stabilized approach is to select an appropriate touchdown point on the runway, and adjust the glide path so that the true aiming point and the desired touchdown point basically coincide. Immediately after rolling out on final approach, adjust the pitch attitude and power so that the airplane is descending directly toward the aiming point at the appropriate airspeed, in the landing configuration, and trimmed for "hands off" flight. With the approach set up in this manner, the pilot is free to devote full attention toward outside references. Do not stare at any one place, but rather scan from one point to another, such as from the aiming point to the horizon, to the trees and bushes along the runway, to an area well short of the runway, and back to the aiming point. This makes it easier to perceive a deviation from the desired glide path and determine if the airplane is proceeding directly toward the aiming point.

If there is any indication that the aiming point on the runway is not where desired, an adjustment must be made to the glide path. This in turn moves the aiming point. For instance, if the aiming point is short of the desired touchdown point and results in an undershoot, an increase in pitch attitude and engine power is warranted. A constant airspeed must be maintained. The pitch and power change, therefore, must be made smoothly and simultaneously. This results in a shallowing of the glide path with the aiming point moving towards the desired touchdown point. Conversely, if the aiming point is farther down the runway than the desired touchdown point resulting in an overshoot, the glide path is steepened by a simultaneous decrease in pitch attitude and

Figure 8-11. *Change in runway shape if approach becomes narrow or steep.*

power. Once again, the airspeed must be held constant. It is essential that deviations from the desired glide path be detected early so that only slight and infrequent adjustments to glide path are required.

The closer the airplane gets to the runway, the larger and more frequent the required corrections become, resulting in an unstable approach. Common errors in the performance of normal approaches and landings are:

- Inadequate wind drift correction on the base leg.

- Overshooting or undershooting the turn onto final approach resulting in too steep or too shallow a turn onto final approach.

- Flat or skidding turns from base leg to final approach as a result of overshooting/inadequate wind drift correction.

- Poor coordination during turn from base to final approach.

- Failure to complete the landing checklist in a timely manner.

- Unstable approach.

- Failure to adequately compensate for flap extension.

- Poor trim technique on final approach.

- Attempting to maintain altitude or reach the runway using elevator alone.

- Focusing too close to the airplane resulting in a too high round out.

- Focusing too far from the airplane resulting in a too low round out.

- Touching down prior to attaining proper landing attitude.

- Failure to hold sufficient back-elevator pressure after touchdown.

- Excessive braking after touchdown.

- Loss of aircraft control during touchdown and roll out.

Intentional Slips

A slip occurs when the bank angle of an airplane is too steep for the existing rate of turn. Unintentional slips are most often the result of uncoordinated rudder/aileron application. Intentional slips, however, are used to dissipate altitude without increasing airspeed and/or to adjust airplane ground track during a crosswind. Intentional slips are especially useful in forced landings and in situations where obstacles must be cleared during approaches to confined areas. A slip can also be used as an emergency means of rapidly reducing airspeed in situations where wing flaps are inoperative or not installed.

A slip is a combination of forward movement and sideward (with respect to the longitudinal axis of the airplane) movement, the lateral axis being inclined and the sideward movement being toward the low end of this axis (low wing). An airplane in a slip is in fact flying sideways, which results in a change in the direction that the relative wind strikes the airplane. Slips are characterized by a marked increase in drag and corresponding decrease in airplane climb, cruise, and glide performance. It is the increase in drag, however, that makes it possible for an airplane in a slip to descend rapidly without an increase in airspeed.

Most airplanes exhibit the characteristic of positive static directional stability and, therefore, have a natural tendency to compensate for slipping. An intentional slip, therefore, requires deliberate cross-controlling ailerons and rudder throughout the maneuver.

A "sideslip" is entered by lowering a wing and applying just enough opposite rudder to prevent a turn. In a sideslip, the airplane's longitudinal axis remains parallel to the original flightpath, but the airplane no longer flies straight ahead. Instead, the horizontal component of wing lift forces the airplane also to move somewhat sideways toward the low wing. [Figure 8-12] The amount of slip, and therefore the

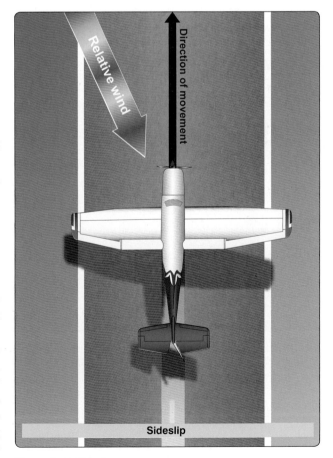

Figure 8-12. *Sideslip.*

rate of sideward movement, is determined by the bank angle. The steeper the bank is, the greater the degree of slip. As bank angle is increased additional opposite rudder is required to prevent turning. Sideslips are frequently used when landing with a crosswind to keep the aircraft aligned with the runway centerline while stopping any drift left or right of the centerline.

A "forward slip" is one in which the airplane's direction of motion continues the same as before the slip was begun. Assuming the airplane is originally in straight flight, the wing on the side toward which the slip is to be made should be lowered by use of the ailerons. Simultaneously, the airplane's nose must be yawed in the opposite direction by applying opposite rudder so that the airplane's longitudinal axis is at an angle to its original flightpath. [Figure 8-13] The degree to which the nose is yawed in the opposite direction from the bank should be such that the original ground track is maintained. In a forward slip, the amount of slip, and therefore the sink rate, is determined by the bank angle. The steeper the bank is, the steeper the descent.

In most light airplanes, the steepness of a slip is limited by the amount of rudder travel available. In both sideslips and forward slips, the point may be reached where full rudder is required to maintain heading even though the ailerons are capable of further steepening the bank angle. This is the practical slip limit because any additional bank would cause the airplane to turn even though full opposite rudder is being applied. If there is a need to descend more rapidly, even though the practical slip limit has been reached, lowering the nose not only increases the sink rate but also increases airspeed. The increase in airspeed increases rudder effectiveness permitting a steeper slip. Conversely, when the nose is raised, rudder effectiveness decreases and the bank angle must be reduced.

Discontinuing a slip is accomplished by leveling the wings and simultaneously releasing the rudder pressure while readjusting the pitch attitude to the normal glide attitude. If the pressure on the rudder is released abruptly, the nose swings too quickly into line and the airplane tends to acquire excess speed. Because of the location of the pitot tube and static vents, airspeed indicators in some airplanes may have considerable error when the airplane is in a slip. The pilot must be aware of this possibility and recognize a properly performed slip by the attitude of the airplane, the sound of the airflow, and the feel of the flight controls. Unlike skids, however, if an airplane in a slip is made to stall, it displays very little of the yawing tendency that causes a skidding stall to develop into a spin. The airplane in a slip may do little more than tend to roll into a wings level attitude. In fact, in some airplanes stall characteristics may even be improved.

Go-Arounds (Rejected Landings)

Whenever landing conditions are not satisfactory, a go-around is warranted. There are many factors that can contribute to unsatisfactory landing conditions. Situations such as air traffic control (ATC) requirements, unexpected appearance of hazards on the runway, overtaking another airplane, wind shear, wake turbulence, mechanical failure, and/or an unstable approach are all examples of reasons to discontinue a landing approach and make another approach under more favorable conditions. The assumption that an aborted landing is invariably the consequence of a poor approach, which in turn is due to insufficient experience or skill, is a fallacy. The go-around is not strictly an emergency procedure. It is a normal maneuver that is also used in an emergency situation. Like any other normal maneuver, the go-around must be practiced and perfected. The flight instructor needs to emphasize early on, and the pilot must be made to understand, that the go-around maneuver is an alternative to any approach and/or landing.

Although the need to discontinue a landing may arise at any point in the landing process, the most critical go-around is one started when very close to the ground. The earlier a condition that warrants a go-around is recognized, the safer

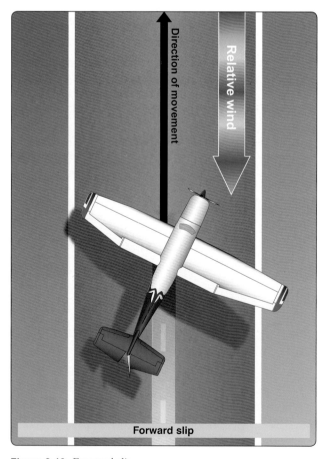

Figure 8-13. *Forward slip.*

the go-around/rejected landing is. The go-around maneuver is not inherently dangerous in itself. It becomes dangerous only when delayed unduly or executed improperly. Delay in initiating the go-around normally stems from two sources:

1. Landing expectancy or set—the anticipatory belief that conditions are not as threatening as they are and that the approach is surely terminated with a safe landing,

2. Pride—the mistaken belief that the act of going around is an admission of failure—failure to execute the approach properly. The improper execution of the go-around maneuver stems from a lack of familiarity with the three cardinal principles of the procedure: power, attitude, and configuration.

Power

Power is the pilot's first concern. The instant a pilot decides to go around, full or maximum allowable takeoff power must be applied smoothly and without hesitation and held until flying speed and controllability are restored. Applying only partial power in a go-around is never appropriate. The pilot must be aware of the degree of inertia that must be overcome before an airplane that is settling towards the ground can regain sufficient airspeed to become fully controllable and capable of climbing or turning safely. The application of power is smooth, as well as positive. Abrupt movements of the throttle in some airplanes causes the engine to falter. Carburetor heat is turned off to obtain maximum power.

Attitude

Attitude is always critical when close to the ground, and when power is added, a deliberate effort on the part of the pilot is required to keep the nose from pitching up prematurely. The airplane executing a go-around must be maintained in an attitude that permits a buildup of airspeed well beyond the stall point before any effort is made to gain altitude or to execute a turn. Raising the nose too early could result in

a stall from which the airplane could not be recovered if the go-around is performed at a low altitude.

A concern for quickly regaining altitude during a go-around produces a natural tendency to pull the nose up. A pilot executing a go-around must accept the fact that an airplane cannot climb until it can fly, and it cannot fly below stall speed. In some circumstances, it is desirable to lower the nose briefly to gain airspeed. As soon as the appropriate climb airspeed and pitch attitude are attained, "rough trim" the airplane to relieve any adverse control pressures. More precise trim adjustments can be made when flight conditions have stabilized.

Configuratio

After establishing the proper climb attitude and power settings, be concerned first with flaps and secondly with the landing gear (if retractable). When the decision is made to perform a go-around, takeoff power is applied immediately and the pitch attitude changed so as to slow or stop the descent. After the descent has been stopped, the landing flaps are partially retracted or placed in the takeoff position as recommended by the manufacturer. Caution must be used in retracting the flaps. Depending on the airplane's altitude and airspeed, it is wise to retract the flaps intermittently in small increments to allow time for the airplane to accelerate progressively as they are being raised. A sudden and complete retraction of the flaps could cause a loss of lift resulting in the airplane settling into the ground. *[Figure 8-14]*

Unless otherwise specified in the AFM/POH, it is generally recommended that the flaps be retracted (at least partially) before retracting the landing gear for two reasons. First, on most airplanes full flaps produce more drag than the landing gear; and second, in case the airplane inadvertently touches down as the go-around is initiated; it is most desirable to

Figure 8-14. *Go-around procedure.*

have the landing gear in the down-and-locked position. After a positive rate of climb is established, the landing gear is retracted.

When takeoff power is applied, it is usually necessary to hold considerable pressure on the controls to maintain straight flight and a safe climb attitude. Since the airplane is trimmed for the approach (a low power and low airspeed condition), application of maximum allowable power requires considerable control pressure to maintain a climb pitch attitude. The addition of power tends to raise the airplane's nose suddenly and veer to the left. Forward elevator pressure must be anticipated and applied to hold the nose in a safe climb attitude. Right rudder pressure must be increased to counteract torque and P-factor and to keep the nose straight. The airplane must be held in the proper flight attitude regardless of the amount of control pressure that is required. Trim is applied to relieve adverse control pressures and assist in maintaining a proper pitch attitude. On airplanes that produce high control pressures when using maximum power on go-arounds, use caution when reaching for the flap handle. Airplane control is critical during this high-workload phase.

The landing gear is retracted only after the initial or rough trim is accomplished and when it is certain the airplane will remain airborne. During the initial part of an extremely low go-around, it is possible for the airplane to settle onto the runway and bounce. This situation is not particularly dangerous provided the airplane is kept straight and a constant, safe pitch attitude is maintained. With the application of power, the airplane attains a safe flying speed rapidly and the advanced power cushions any secondary touchdown.

If the pitch attitude is increased excessively in an effort to keep the airplane from contacting the runway, it may cause the airplane to stall. This is likely to occur if no trim correction is made and the flaps remain fully extended. Do not attempt to retract the landing gear until after a rough trim is accomplished and a positive rate of climb is established.

Ground Effect

Ground effect is a factor in every landing and every takeoff in fixed-wing airplanes. Ground effect can also be an important factor in go-arounds. If the go-around is made close to the ground, the airplane may be in the ground effect area. Pilots are often lulled into a sense of false security by the apparent "cushion of air" under the wings that initially assists in the transition from an approach descent to a climb. This "cushion of air," however, is imaginary. The apparent increase in airplane performance is, in fact, due to a reduction in induced drag in the ground effect area. It is "borrowed" performance that must be repaid when the airplane climbs out of the ground effect area. The pilot must factor in ground effect when initiating a go-around close to the ground. An attempt to climb prematurely may result in the airplane not being able to climb or even maintain altitude at full power.

Common errors in the performance of go-arounds (rejected landings) are:

- Failure to recognize a condition that warrants a rejected landing
- Indecision
- Delay in initiating a go-around
- Failure to apply maximum allowable power in a timely manner
- Abrupt power application
- Improper pitch attitude
- Failure to configure the airplane appropriately
- Attempting to climb out of ground effect prematurely
- Failure to adequately compensate for torque/P factor
- Loss of aircraft control

Crosswind Approach and Landing

Many runways or landing areas are such that landings must be made while the wind is blowing across rather than parallel to the landing direction. All pilots must be prepared to cope with these situations when they arise. The same basic principles and factors involved in a normal approach and landing apply to a crosswind approach and landing; therefore, only the additional procedures required for correcting for wind drift are discussed here.

Crosswind landings are a little more difficult to perform than crosswind takeoffs, mainly due to different problems involved in maintaining accurate control of the airplane while its speed is decreasing rather than increasing as on takeoff.

There are two usual methods of accomplishing a crosswind approach and landing—the crab method and the wing-low (sideslip) method. Although the crab method may be easier for the pilot to maintain during final approach, it requires a high degree of judgment and timing in removing the crab immediately prior to touchdown. The wing-low method is recommended in most cases, although a combination of both methods may be used.

Crosswind Final Approach

The crab method is executed by establishing a heading (crab) toward the wind with the wings level so that the airplane's ground track remains aligned with the centerline of the runway. [Figure 8-15] This crab angle is maintained until just prior to touchdown, when the longitudinal axis of the

Figure 8-15. *Crabbed approach.*

airplane must be aligned with the runway to avoid sideward contact of the wheels with the runway. If a long final approach is being flown, one option is to use the crab method until just before the round out is started and then smoothly change to the wing-low method for the remainder of the landing.

The wing-low (sideslip) method compensates for a crosswind from any angle, but more important, it keeps the airplane's ground track and longitudinal axis aligned with the runway centerline throughout the final approach, round out, touchdown, and after-landing roll. This prevents the airplane from touching down in a sideward motion and imposing damaging side loads on the landing gear.

To use the wing-low method, align the airplane's heading with the centerline of the runway, note the rate and direction of drift, and promptly apply drift correction by lowering the upwind wing. *[Figure 8-16]* The amount the wing must be lowered depends on the rate of drift. When the wing is lowered, the airplane tends to turn in that direction. To compensate for the turn, it is necessary to simultaneously apply sufficient opposite rudder pressure to keep the airplane's

Figure 8-16. *Sideslip approach.*

longitudinal axis aligned with the runway. In other words, the drift is controlled with aileron and the heading with rudder. The airplane is now side slipping into the wind just enough that both the resultant flightpath and the ground track are aligned with the runway. If the crosswind diminishes, this crosswind correction is reduced accordingly, or the airplane begins slipping away from the desired approach path. *[Figure 8-17]*

To correct for strong crosswind, the slip into the wind is increased by lowering the upwind wing a considerable amount. As a consequence, this results in a greater tendency of the airplane to turn. Since turning is not desired, considerable opposite rudder must be applied to keep the airplane's longitudinal axis aligned with the runway. In some airplanes, there may not be sufficient rudder travel available to compensate for the strong turning tendency caused by the steep bank. If the required bank is such that full opposite rudder does not prevent a turn, the wind is too strong to safely land the airplane on that particular runway with those wind conditions. Since the airplane's capability is exceeded, it is imperative that the landing be made on a more favorable runway either at that airport or at an alternate airport.

Flaps are used during most approaches since they tend to have a stabilizing effect on the airplane. The degree to which flaps are extended vary with the airplane's handling characteristics, as well as the wind velocity.

Crosswind Round Out (Flare)
Generally, the round out is made like a normal landing approach, but the application of a crosswind correction is continued as necessary to prevent drifting.

Since the airspeed decreases as the round out progresses, the flight controls gradually become less effective. As a result, the crosswind correction being held becomes inadequate. When using the wing-low method, it is necessary to gradually increase the deflection of the rudder and ailerons to maintain the proper amount of drift correction.

Do not level the wings and keep the upwind wing down throughout the round out. If the wings are leveled, the airplane begins drifting and the touchdown occurs while drifting. Remember, the primary objective is to land the airplane without subjecting it to any side loads that result from touching down while drifting.

Crosswind Touchdown
If the crab method of drift correction is used throughout the final approach and round out, the crab must be removed the instant before touchdown by applying rudder to align the airplane's longitudinal axis with its direction of movement. This requires timely and accurate action. Failure to

Figure 8-17. *Crosswind approach and landing.*

accomplish this results in severe side loads being imposed on the landing gear.

If the wing-low method is used, the crosswind correction (aileron into the wind and opposite rudder) is maintained throughout the round out, and the touchdown made on the upwind main wheel. During gusty or high wind conditions, prompt adjustments must be made in the crosswind correction to assure that the airplane does not drift as the airplane touches down. As the forward momentum decreases after initial contact, the weight of the airplane causes the downwind main wheel to gradually settle onto the runway.

In those airplanes having nose-wheel steering interconnected with the rudder, the nose wheel is not aligned with the runway as the wheels touch down because opposite rudder is being held in the crosswind correction. To prevent swerving in the direction the nose wheel is offset, the corrective rudder pressure must be promptly relaxed just as the nose wheel touches down.

Crosswind After-Landing Roll

Particularly during the after-landing roll, special attention must be given to maintaining directional control by the use

of rudder or nose-wheel steering, while keeping the upwind wing from rising by the use of aileron. When an airplane is airborne, it moves with the air mass in which it is flying regardless of the airplane's heading and speed. When an airplane is on the ground, it is unable to move with the air mass (crosswind) because of the resistance created by ground friction on the wheels.

Characteristically, an airplane has a greater profile or side area behind the main landing gear than forward of the gear. With the main wheels acting as a pivot point and the greater surface area exposed to the crosswind behind that pivot point, the airplane tends to turn or weathervane into the wind.

Wind acting on an airplane during crosswind landings is the result of two factors. One is the natural wind, which acts in the direction the air mass is traveling, while the other is induced by the forward movement of the airplane and acts parallel to the direction of movement. Consequently, a crosswind has a headwind component acting along the airplane's ground track and a crosswind component acting 90° to its track. The resultant or relative wind is somewhere between the two components. As the airplane's forward speed decreases during the after landing roll, the headwind component decreases and the relative wind has more of a crosswind component. The greater the crosswind component, the more difficult it is to prevent weathervaning.

Maintaining control on the ground is a critical part of the after-landing roll because of the weathervaning effect of the wind on the airplane. Additionally, tire side load from runway contact while drifting frequently generates roll-overs in tricycle-geared airplanes. The basic factors involved are cornering angle and side load.

Cornering angle is the angular difference between the heading of a tire and its path. Whenever a load bearing tire's path and heading diverge, a side load is created. It is accompanied by tire distortion. Although side load differs in varying tires and air pressures, it is completely independent of speed, and through a considerable range, is directly proportional to the cornering angle and the weight supported by the tire. As little as 10° of cornering angle creates a side load equal to half the supported weight; after 20°, the side load does not increase with increasing cornering angle. For each high-wing, tricycle-geared airplane, there is a cornering angle at which roll-over is inevitable. The roll-over axis is the line linking the nose and main wheels. At lesser angles, the roll-over may be avoided by use of ailerons, rudder, or steerable nose wheel but not brakes.

While the airplane is decelerating during the after-landing roll, more and more aileron is applied to keep the upwind wing from rising. Since the airplane is slowing down, there is less airflow around the ailerons and they become less effective. At the same time, the relative wind becomes more of a crosswind and exerting a greater lifting force on the upwind wing. When the airplane is coming to a stop, the aileron control must be held fully toward the wind.

Maximum Safe Crosswind Velocities

Takeoffs and landings in certain crosswind conditions are inadvisable or even dangerous. *[Figure 8-18]* If the crosswind is great enough to warrant an extreme drift correction, a hazardous landing condition may result. Therefore, the takeoff and landing capabilities with respect to the reported surface wind conditions and the available landing directions must be considered.

Before an airplane is type certificated by the Federal Aviation Administration (FAA), it must be flight tested and meet certain requirements. Among these is the demonstration of being satisfactorily controllable with no exceptional degree of skill or alertness on the part of the pilot in 90° crosswinds up to a velocity equal to 0.2 V_{SO}. This means a windspeed of two-tenths of the airplane's stalling speed with power off and landing gear/flaps down. Regulations require that the

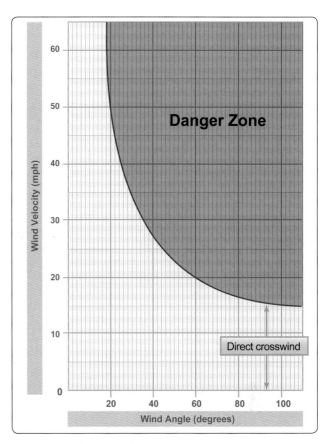

Figure 8-18. *Crosswind chart.*

demonstrated crosswind velocity be included on a placard in airplanes certificated after May 3, 1962.

The headwind component and the crosswind component for a given situation is determined by reference to a crosswind component chart. *[Figure 8-19]* It is imperative that pilots determine the maximum crosswind component of each airplane they fly and avoid operations in wind conditions that exceed the capability of the airplane.

Common errors in the performance of crosswind approaches and landings are:

- Attempting to land in crosswinds that exceed the airplane's maximum demonstrated crosswind component

- Inadequate compensation for wind drift on the turn from base leg to final approach, resulting in undershooting or overshooting

- Inadequate compensation for wind drift on final approach

- Unstable approach

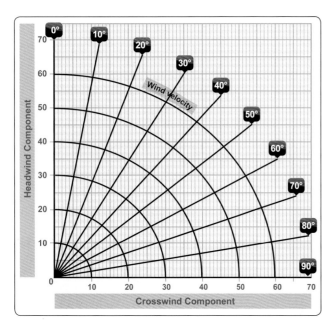

Figure 8-19. *Crosswind component chart.*

- Failure to compensate for increased drag during sideslip resulting in excessive sink rate and/or too low an airspeed

- Touchdown while drifting

- Excessive airspeed on touchdown

- Failure to apply appropriate flight control inputs during rollout

- Failure to maintain direction control on rollout

- Excessive braking

- Loss of aircraft control

Turbulent Air Approach and Landing

For landing in turbulent conditions, use a power-on approach at an airspeed slightly above the normal approach speed. This provides for more positive control of the airplane when strong horizontal wind gusts, or up and down drafts, are experienced. Like other power-on approaches, a coordinated combination of both pitch and power adjustments is usually required. As in most other landing approaches, the proper approach attitude and airspeed require a minimum round out and should result in little or no floating during the landing.

To maintain control during an approach in turbulent air with gusty crosswind, use partial wing flaps. With less than full flaps, the airplane is in a higher pitch attitude. Thus, it requires less of a pitch change to establish the landing attitude and touchdown at a higher airspeed to ensure more positive control. Excessive speed causes the airplane to float past the desired landing area.

One procedure is to use the normal approach speed plus one-half of the wind gust factors. If the normal speed is 70 knots, and the wind gusts are 15 knots, an increase of airspeed to 77 knots is appropriate. In any case, the airspeed and the number of flaps used should conform to airplane manufacturer recommendations in the AFM/POH.

Use an adequate amount of power to maintain the proper airspeed and descent path throughout the approach, and retard the throttle to idling position only after the main wheels contact the landing surface. Care must be exercised in closing the throttle before the pilot is ready for touchdown. In turbulent conditions, the sudden or premature closing of the throttle may cause a sudden increase in the descent rate that results in a hard landing.

When landing from power approaches in turbulence, the touchdown is made with the airplane in approximately level flight attitude. The pitch attitude at touchdown would be only enough to prevent the nose wheel from contacting the surface before the main wheels have touched the surface. After touchdown, avoid the tendency to apply forward pressure on the yoke, as this may result in wheel barrowing and possible loss of control. Allow the airplane to decelerate normally, assisted by careful use of wheel brakes. Avoid heavy braking until the wings are devoid of lift and the airplane's full weight is resting on the landing gear.

Short-Field Approach and Landing

Short-field approaches and landings require the use of procedures for approaches and landings at fields with a relatively short landing area or where an approach is made over obstacles that limit the available landing area. *[Figures 8-20 and 8-21]* As in short-field takeoffs, it is one of the most critical of the maximum performance operations. Short field operations require the pilot fly the airplane at one of its crucial performance capabilities while close to the ground in order to safely land within confined areas. This low-speed type of power-on approach is closely related to the performance of flight at minimum controllable airspeeds.

To land within a short-field or a confined area, the pilot must have precise, positive control of the rate of descent and airspeed to produce an approach that clears any obstacles, result in little or no floating during the round out, and permit the airplane to be stopped in the shortest possible distance.

The procedures for landing in a short-field or for landing approaches over obstacles as recommended in the AFM/POH should be used. A stabilized approach is essential. *[Figures 8-22 and 8-23]* These procedures generally involve the use of full flaps and the final approach started from an

Figure 8-20. *Landing over an obstacle.*

Figure 8-21. *Landing on a short-field.*

Figure 8-22. *Stabilized approach.*

Figure 8-23. *Unstabilized approach.*

altitude of at least 500 feet higher than the touchdown area. A wider than normal pattern is normally used so that the airplane can be properly configured and trimmed. In the absence of the manufacturer's recommended approach speed, a speed of not more than 1.3 V_{SO} is used. For example, in an airplane that stalls at 60 knots with power off, and flaps and landing gear extended, an approach speed no higher than 78 knots is used. In gusty air, no more than one-half the gust factor is added. An excessive amount of airspeed could result in a touchdown too far from the runway threshold or an after-landing roll that exceeds the available landing area.

After the landing gear and full flaps have been extended, simultaneously adjust the power and the pitch attitude to establish and maintain the proper descent angle and airspeed. A coordinated combination of both pitch and power adjustments is required. When this is done properly, very little change in the airplane's pitch attitude and power setting is necessary to make corrections in the angle of descent and airspeed.

The short-field approach and landing is in reality an accuracy approach to a spot landing. The procedures previously outlined in the section on the stabilized approach concept are used. If it appears that the obstacle clearance is excessive and touchdown occurs well beyond the desired spot leaving insufficient room to stop, power is reduced while lowering the pitch attitude to steepen the descent path and increase the rate of descent. If it appears that the descent angle does not ensure safe clearance of obstacles, power is increased while simultaneously raising the pitch attitude to shallow the descent path and decrease the rate of descent. Care must be taken to avoid an excessively low airspeed. If the speed is allowed to become too slow, an increase in pitch and application of full power may only result in a further rate of descent. This occurs when the AOA is so great and creating so much drag that the maximum available power is insufficient to overcome it. This is generally referred

to as operating in the region of reversed command or operating on the back side of the power curve. When there is doubt regarding the outcome of the approach, make a go around and try again or divert to a more suitable landing area.

Because the final approach over obstacles is made at a relatively steep approach angle and close to the airplane's stalling speed, the initiation of the round out or flare must be judged accurately to avoid flying into the ground or stalling prematurely and sinking rapidly. A lack of floating during the flare with sufficient control to touch down properly is verification that the approach speed was correct.

Touchdown should occur at the minimum controllable airspeed with the airplane in approximately the pitch attitude that results in a power-off stall when the throttle is closed. Care must be exercised to avoid closing the throttle too rapidly, as closing the throttle may result in an immediate increase in the rate of descent and a hard landing.

Upon touchdown, the airplane is held in this positive pitch attitude as long as the elevators remain effective. This provides aerodynamic braking to assist in deceleration. Immediately upon touchdown and closing the throttle, appropriate braking is applied to minimize the after-landing roll. The airplane is normally stopped within the shortest possible distance consistent with safety and controllability. If the proper approach speed has been maintained, resulting in minimum float during the round out and the touchdown made at minimum control speed, minimum braking is required.

Common errors in the performance of short-field approaches and landings are:

- Failure to allow enough room on final to set up the approach, necessitating an overly steep approach and high sink rate

- Unstable approach

- Undue delay in initiating glide path corrections

- Too low an airspeed on final resulting in inability to flare properly and landing hard

- Too high an airspeed resulting in floating on round out

- Prematurely reducing power to idle on round out resulting in hard landing

- Touchdown with excessive airspeed

- Excessive and/or unnecessary braking after touchdown

- Failure to maintain directional control

- Failure to recognize and abort a poor approach that cannot be completed safely

Soft-Field Approach and Landing

Landing on fields that are rough or have soft surfaces, such as snow, sand, mud, or tall grass, require unique procedures. When landing on such surfaces, the objective is to touch down as smooth as possible and at the slowest possible landing speed. A pilot must control the airplane in a manner that the wings support the weight of the airplane as long as practical to minimize drag and stresses imposed on the landing gear by the rough or soft surface.

The approach for the soft-field landing is similar to the normal approach used for operating into long, firm landing areas. The major difference between the two is that during the soft-field landing, the airplane is held 1 to 2 feet off the surface in ground effect as long as possible. This permits a more gradual dissipation of forward speed to allow the wheels to touch down gently at minimum speed. This technique minimizes the nose-over forces that suddenly affect the airplane at the moment of touchdown. Power is used throughout the level-off and touchdown to ensure touchdown at the slowest possible airspeed, and the airplane is flown onto the ground with the weight fully supported by the wings. *[Figure 8-24]*

The use of flaps during soft-field landings aids in touching down at minimum speed and is recommended whenever practical. In low-wing airplanes, the flaps may suffer damage from mud, stones, or slush thrown up by the wheels. If flaps are used, it is generally inadvisable to retract them during the after-landing roll because the need for flap retraction is less important than the need for total concentration on maintaining full control of the airplane.

The final-approach airspeed used for short-field landings is equally appropriate to soft-field landings. The use of higher approach speeds may result in excessive float in ground effect, and floating makes a smooth, controlled touchdown even more difficult. There is no reason for a steep angle of descent unless obstacles are present in the approach path.

Touchdown on a soft or rough field is made at the lowest possible airspeed with the airplane in a nose-high pitch attitude. In nose-wheel type airplanes, after the main wheels touch the surface, hold sufficient back-elevator pressure to keep the nose wheel off the surface. Using back-elevator pressure and engine power, the pilot can control the rate at which the weight of the airplane is transferred from the wings to the wheels.

Field conditions may warrant that the pilot maintain a flight condition in which the main wheels are just touching the surface but the weight of the airplane is still being supported by the wings until a suitable taxi surface is reached. At any time during this transition phase, before the weight of the airplane is being supported by the wheels, and before the nose wheel is on the surface, the ability is retained to apply full power and perform a safe takeoff (obstacle clearance and field length permitting) should the pilot elect to abandon the landing. Once committed to a landing, the pilot should gently lower the nose wheel to the surface. A slight addition of power usually aids in easing the nose wheel down.

Figure 8-24. *Soft/rough field approach and landing.*

The use of brakes on a soft field is not needed and should be avoided as this may tend to impose a heavy load on the nose gear due to premature or hard contact with the landing surface, causing the nose wheel to dig in. The soft or rough surface itself provides sufficient reduction in the airplane's forward speed. Often upon landing on a very soft field, an increase in power is required to keep the airplane moving and from becoming stuck in the soft surface.

Common errors in the performance of soft-field approaches and landings are:

- Excessive descent rate on final approach
- Excessive airspeed on final approach
- Unstable approach
- Round out too high above the runway surface
- Poor power management during round out and touchdown
- Hard touchdown
- Inadequate control of the airplane weight transfer from wings to wheels after touchdown
- Allowing the nose wheel to "fall" to the runway after touchdown rather than controlling its descent

Power-Off Accuracy Approaches

Power-off accuracy approaches are approaches and landings made by gliding with the engine idling, through a specific pattern to a touchdown beyond and within 200 feet of a designated line or mark on the runway. The objective is to instill in the pilot the judgment and procedures necessary for accurately flying the airplane, without power, to a safe landing.

The ability to estimate the distance an airplane glides to a landing is the real basis of all power-off accuracy approaches and landings. This largely determines the amount of maneuvering that may be done from a given altitude. In addition to the ability to estimate distance, it requires the ability to maintain the proper glide while maneuvering the airplane.

With experience and practice, altitudes up to approximately 1,000 feet can be estimated with fair accuracy; while above this level the accuracy in judgment of height above the ground decreases, since all features tend to merge. The best aid in perfecting the ability to judge height above this altitude is through the indications of the altimeter and associating them with the general appearance of the Earth.

The judgment of altitude in feet, hundreds of feet, or thousands of feet is not as important as the ability to estimate gliding angle and its resultant distance. A pilot who knows the normal glide angle of the airplane can estimate with reasonable accuracy, the approximate spot along a given ground path at which the airplane lands, regardless of altitude. A pilot who also has the ability to accurately estimate altitude, can judge how much maneuvering is possible during the glide, which is important to the choice of landing areas in an actual emergency.

The objective of a good final approach is to descend at an angle that permits the airplane to reach the desired landing area and at an airspeed that results in minimum floating just before touchdown. To accomplish this, it is essential that both the descent angle and the airspeed be accurately controlled.

Unlike a normal approach when the power setting is variable, on a power-off approach the power is fixed at the idle setting. Pitch attitude is adjusted to control the airspeed. This also changes the glide or descent angle. By lowering the nose to keep the approach airspeed constant, the descent angle steepens. If the airspeed is too high, raise the nose, and when the airspeed is too low, lower the nose. If the pitch attitude is raised too high, the airplane settles rapidly due to a slow airspeed and insufficient lift. For this reason, never try to stretch a glide to reach the desired landing spot.

Uniform approach patterns, such as the 90°, 180°, or 360° power-off approaches are described further in this chapter. Practice in these approaches provides a pilot with a basis on which to develop judgment in gliding distance and in planning an approach.

The basic procedure in these approaches involves closing the throttle at a given altitude and gliding to a key position. This position, like the pattern itself, must not be allowed to become the primary objective; it is merely a convenient point in the air from which the pilot can judge whether the glide safely terminates at the desired spot. The selected key position should be one that is appropriate for the available altitude and the wind condition. From the key position, the pilot must constantly evaluate the situation.

It must be emphasized that, although accurate spot touchdowns are important, safe and properly executed approaches and landings are vital. A pilot must never sacrifice a good approach or landing just to land on the desired spot.

90° Power-Off Approach

The 90° power-off approach is made from a base leg and requires only a 90° turn onto the final approach. The approach path may be varied by positioning the base leg closer to or farther out from the approach end of the runway according to wind conditions. [Figure 8-25] The glide from the key position on the base leg through the 90° turn to the final approach is the final part of all accuracy landing maneuvers. The 90° power-off approach usually begins from a

Figure 8-25. *Plan the base leg for wind conditions.*

Within the figure:

1. Strong Wind
Set up closest base for steeper glideslope on final

2. Medium Wind
Set up closer base for steeper glideslope on final

3. Light Wind
Set up normal base for normal final

rectangular pattern at approximately 1,000 feet above the ground or at normal traffic pattern altitude. The airplane is flown on a downwind leg at the same distance from the landing surface as in a normal traffic pattern. The before landing checklist should be completed on the downwind leg, including extension of the landing gear if the airplane is equipped with retractable gear.

After a medium-banked turn onto the base leg is completed, the throttle is retarded slightly and the airspeed allowed to decrease to the normal base-leg speed. *[Figure 8-26]* On the base leg, the airspeed, wind drift correction, and altitude are maintained while proceeding to the 45° key position. At this position, the intended landing spot appears to be on a 45° angle from the airplane's nose.

The pilot can determine the strength and direction of the wind from the amount of crab necessary to hold the desired ground track on the base leg. This helps in planning the turn onto the final approach and in lowering the correct number of flaps.

At the 45° key position, the throttle is closed completely, the propeller control (if equipped) advanced to the full increase revolution per minute (rpm) position, and altitude maintained until the airspeed decreases to the manufacturer's recommended glide speed. In the absence of a recommended speed, use 1.4 V_{SO}. When this airspeed is attained, the nose is lowered to maintain the gliding speed and the controls trimmed. The base-to-final turn is planned and accomplished so that upon rolling out of the turn, the airplane is aligned with the runway centerline. When on final approach, the wing flaps are lowered and the pitch attitude adjusted, as necessary, to establish the proper descent angle and airspeed (1.3 V_{SO}), then the controls trimmed. Slight adjustments in pitch attitude or flaps setting are used as necessary to control the glide angle and airspeed. However, never try to stretch the glide or retract the flaps to reach the desired landing spot. The final approach may be made with or without the use of slips.

After the final-approach glide has been established, full attention is then given to making a good, safe landing rather than concentrating on the selected landing spot. The base-leg position and the flap setting already determined the probability of landing on the spot. In any event, it is better to execute a good landing 200 feet from the spot than to make a poor landing precisely on the spot.

180° Power-Off Approach

The 180° power-off approach is executed by gliding with the power off from a given point on a downwind leg to a preselected landing spot. *[Figure 8-27]* It is an extension of the principles involved in the 90° power-off approach just described. The objective is to further develop judgment in estimating distances and glide ratios, in that the airplane is flown without power from a higher altitude and through a 90° turn to reach the base-leg position at a proper altitude for executing the 90° approach.

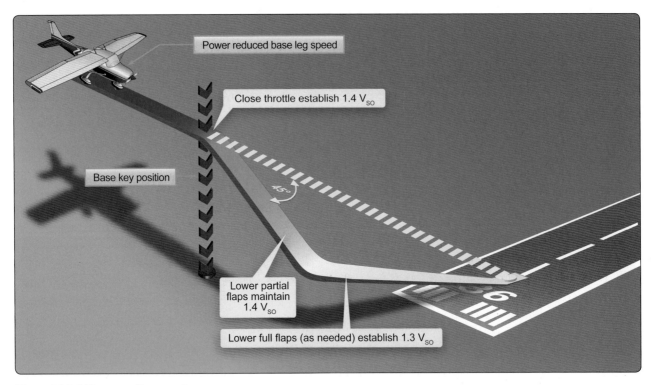

Figure 8-26. *90° power-off approach.*

The 180° power-off approach requires more planning and judgment than the 90° power-off approach. In the execution of 180° power-off approaches, the airplane is flown on a downwind heading parallel to the landing runway. The altitude from which this type of approach is started varies with the type of airplane, but should usually not exceed 1,000 feet above the ground, except with large airplanes. Greater accuracy in judgment and maneuvering is required at higher altitudes.

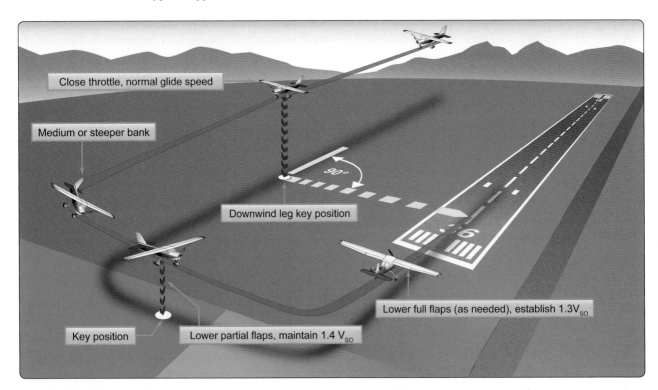

Figure 8-27. *180° power-off approach.*

When abreast of or opposite the desired landing spot, the throttle is closed and altitude maintained while decelerating to the manufacturer's recommended glide speed or 1.4 V_{SO}. The point at which the throttle is closed is the downwind key position.

The turn from the downwind leg to the base leg is a uniform turn with a medium or slightly steeper bank. The degree of bank and amount of this initial turn depend upon the glide angle of the airplane and the velocity of the wind. Again, the base leg is positioned as needed for the altitude or wind condition. Position the base leg to conserve or dissipate altitude so as to reach the desired landing spot.

The turn onto the base leg is made at an altitude high enough and close enough to permit the airplane to glide to what would normally be the base key position in a 90° power-off approach.

Although the key position is important, it must not be overemphasized nor considered as a fixed point on the ground. Many inexperienced pilots may gain a conception of it as a particular landmark, such as a tree, crossroad, or other visual reference, to be reached at a certain altitude. This misconception leaves the pilot at a total loss any time such objects are not present. Both altitude and geographical location should be varied as much as is practical to eliminate any such misconceptions. After reaching the base key position, the approach and landing are the same as in the 90° power-off approach.

360° Power-Off Approach

The 360° power-off approach is one in which the airplane glides through a 360° change of direction to the preselected landing spot. The entire pattern is designed to be circular, but the turn may be shallow, steepened, or discontinued at any point to adjust the accuracy of the flightpath.

The 360° approach is started from a position over the approach end of the landing runway or slightly to the side of it, with the airplane headed in the proposed landing direction and the landing gear and flaps retracted. *[Figure 8-28]* It is usually initiated from approximately 2,000 feet or more above the ground—where the wind may vary significantly from that at lower altitudes. This must be taken into account when maneuvering the airplane to a point from which a 90° or 180° power-off approach can be completed.

After the throttle is closed over the intended point of landing, the proper glide speed is immediately established, and a medium-banked turn made in the desired direction so as to arrive at the downwind key position opposite the intended landing spot. At or just beyond the downwind key position, the landing gear is extended if the airplane is equipped with retractable gear. The altitude at the downwind key position should be approximately 1,000 to 1,200 feet above the ground.

After reaching that point, the turn is continued to arrive at a base-leg key position, at an altitude of about 800 feet above the terrain. Flaps may be used at this position, as necessary,

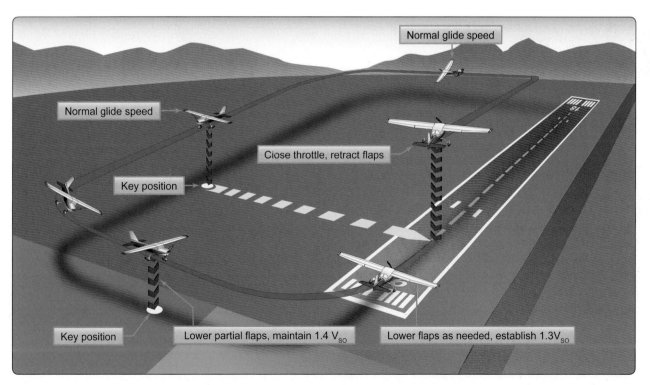

Figure 8-28. *360° power-off approach.*

but full flaps are not used until established on the final approach. The angle of bank is varied as needed throughout the pattern to correct for wind conditions and to align the airplane with the final approach. The turn-to-final should be completed at a minimum altitude of 300 feet above the terrain.

Common errors in the performance of power-off accuracy approaches are:

- Downwind leg is too far from the runway/landing area
- Overextension of downwind leg resulting from a tailwind
- Inadequate compensation for wind drift on base leg
- Skidding turns in an effort to increase gliding distance
- Failure to lower landing gear in retractable gear airplanes
- Attempting to "stretch" the glide during an undershoot
- Premature flap extension/landing gear extension
- Use of throttle to increase the glide instead of merely clearing the engine
- Forcing the airplane onto the runway in order to avoid overshooting the designated landing spot

Emergency Approaches and Landings (Simulated)

During dual training flights, the instructor should give simulated emergency landings by retarding the throttle and calling "simulated emergency landing." The objective of these simulated emergency landings is to develop a pilot's accuracy, judgment, planning, procedures, and confidence when little or no power is available. A simulated emergency landing may be given with the airplane in any configuration. When the instructor calls "simulated emergency landing," immediately establish a glide attitude and ensure that the flaps and landing gear are in the proper configuration for the existing situation. When the proper glide speed is attained, the nose can then be lowered and the airplane trimmed to maintain that speed.

A constant gliding speed is maintained because variations of gliding speed nullify all attempts at accuracy in judgment of gliding distance and the landing spot. The many variables, such as altitude, obstruction, wind direction, landing direction, landing surface and gradient, and landing distance requirements of the airplane, determines the pattern and approach procedures to use.

Use any combination of normal gliding maneuvers, from wings level to spirals to eventually arrive at the normal key position at a normal traffic pattern altitude for the selected landing area. From the key point on, the approach is a normal power-off approach. [Figure 8-29]

With the greater choice of fields afforded by higher altitudes, the inexperienced pilot may be inclined to delay making a decision, and with considerable altitude in which to maneuver, errors in maneuvering and estimation of glide distance may develop.

All pilots must learn to determine the wind direction and estimate its speed from the windsock at the airport, smoke from factories or houses, dust, brush fires, and windmills.

Once a field has been selected, a pilot should always be required to indicate the proposed landing area to the instructor. Normally, the pilot should be required to plan and fly a pattern for landing on the field first elected until the instructor terminates the simulated emergency landing. This provides the instructor an opportunity to explain and correct any errors; it also gives the pilot an opportunity to see the results of the errors. However, if the pilot realizes during the approach that a poor field has been selected—one that would obviously result in disaster if a landing were to be made—and there is a more advantageous field within gliding distance, a change to the better field should be permitted. The hazards involved in these last-minute decisions, such as excessive maneuvering at very low altitudes, must be thoroughly explained by the instructor.

Instructors must stress slipping the airplane, using flaps, varying the position of the base leg, and varying the turn onto final approach as ways of correcting for misjudgment of altitude and glide angle.

Eagerness to get down is one of the most common faults of inexperienced pilots during simulated emergency landings. They forget about speed and arrive at the edge of the field with too much speed to permit a safe landing. Too much speed is just as dangerous as too little; it results in excessive floating and overshooting the desired landing spot. Instructors must stress during their instruction that pilots cannot dive at a field and expect to land on it.

During all simulated emergency landings, keep the engine warm and cleared. During a simulated emergency landing, either the instructor or the pilot should have complete control of the throttle. There must be no doubt as to who has control since many near accidents have occurred from such misunderstandings.

Every simulated emergency landing approach is terminated as soon as it can be determined whether a safe landing could have been made. In no case should it be continued to a point

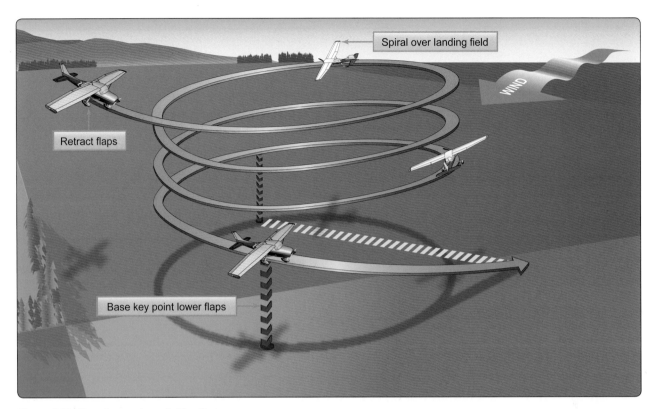

Figure 8-29. *Remain over intended landing area.*

where it creates an undue hazard or an annoyance to persons or property on the ground.

In addition to flying the airplane from the point of simulated engine failure to where a reasonable safe landing could be made, a pilot should also receive instruction on certain emergency cockpit procedures. The habit of performing these cockpit procedures must be developed to such an extent that, when an engine failure actually occurs, a pilot checks the critical items that are necessary to get the engine operating again while selecting a field and planning an approach. Combining the two operations—accomplishing emergency procedures and planning and flying the approach—are difficult during the early training in emergency landings.

There are definite steps and procedures to be followed in a simulated emergency landing. Although they may differ somewhat from the procedures used in an actual emergency, they must be learned thoroughly and each step called out to the instructor. The use of a checklist is strongly recommended. Most airplane manufacturers provide a checklist of the appropriate items. *[Figure 8-30]*

Critical items to be checked include the position of the fuel tank selector, the quantity of fuel in the tank selected, the fuel pressure gauge to see if the electric fuel pump is needed, the position of the mixture control, the position of the magneto switch, and the use of carburetor heat. Many actual emergency landings have been made and later found to be the result of the fuel selector valve being positioned to an empty tank while the other tank had plenty of fuel. It may be wise to change the position of the fuel selector valve even though the fuel gauge indicates fuel in all tanks because fuel gauges can be inaccurate. Many actual emergency landings could have been prevented if the pilots had developed the habit of checking these critical items during flight training to the extent that it carried over into later flying.

Instruction in emergency procedures is not limited to simulated emergency landings caused by power failures. Other emergencies associated with the operation of the airplane should be explained, demonstrated, and practiced if practicable. Among these emergencies are fire in flight, electrical or hydraulic system malfunctions, unexpected severe weather conditions, engine overheating, imminent fuel exhaustion, and the emergency operation of airplane systems and equipment.

Faulty Approaches and Landings

Low Final Approach

When the base leg is too low, insufficient power is used, landing flaps are extended prematurely or the velocity of the wind is misjudged, sufficient altitude is lost, which causes the airplane to be well below the proper final approach path. In such a situation, the pilot would have to apply considerable

65 KIAS (flaps DOWN)

2. Mixture—IDLE CUT-OFF
3. Fuel selector valve—OFF
4. Ignition switch—OFF
5. Wing flaps—AS REQUIRED
6. Master switch—OFF

ENGINE FAILURE DURING FLIGHT (RESTART PROCEDURES)

1. Airspeed—70 KIAS
2. Carburetor heat—ON
3. Fuel selector valve—BOTH
4. Mixture—RICH
5. Ignition switch—BOTH (or START if propeller is stopped)
6. Primer—IN and LOCKED

FORCED LANDINGS

EMERGENCY LANDING WITHOUT ENGINE POWER

1. Airspeed—79 KIAS (flaps UP)
   65 KIAS (flaps DOWN)
2. Mixture—IDLE CUT-OFF
3. Fuel selector valve—OFF
4. Ignition switch—OFF
5. Wing flaps—AS REQUIRED (30° RECOMMENDED)
6. Master switch—OFF
7. Doors—UNLATCH PRIOR TO TOUCHDOWN
8. Touchdown—SLIGHTLY TAIL LOW
9. Brakes—APPLY HEAVILY

PRECAUTIONARY LANDING WITH ENGINE POWER

Figure 8-30. *Sample emergency checklist.*

power to fly the airplane (at an excessively low altitude) up to the runway threshold. When it is realized the runway cannot be reached unless appropriate action is taken, power must be applied immediately to maintain the airspeed while the pitch attitude is raised to increase lift and stop the descent. When the proper approach path has been intercepted, the correct approach attitude is reestablished and the power reduced and a stabilized approach maintained. *[Figure 8-31]* Do not increase the pitch attitude without increasing the power because the airplane decelerates rapidly and may approach the critical AOA and stall. Do not retract the flaps; this suddenly decreases lift and causes the airplane to sink more rapidly. If there is any doubt about the approach being safely completed, it is advisable to execute an immediate go-around.

High Final Approach

When the final approach is too high, lower the flaps as required. Further reduction in power may be necessary,

while lowering the nose simultaneously to maintain approach airspeed and steepen the approach path. *[Figure 8-32]* When the proper approach path is intercepted, adjust the power as required to maintain a stabilized approach. When steepening the approach path, care must be taken that the descent does not result in an excessively high sink rate. If a high sink rate is continued close to the surface, it may be difficult to slow to a proper rate prior to ground contact. Any sink rate in excess of 800–1,000 feet per minute (fpm) is considered excessive. A go-around should be initiated if the sink rate becomes excessive.

Slow Final Approach

On the final approach, when the airplane is flown at a slower than normal airspeed, the pilot's judgment of the rate of sink (descent) and the height of round out is difficult. During an excessively slow approach, the wing is operating near the critical AOA and, depending on the pitch attitude changes

Figure 8-31. *Right and wrong methods of correction for low final approach.*

and control usage, the airplane may stall or sink rapidly, contacting the ground with a hard impact.

Whenever a slow speed approach is noted, apply power to accelerate the airplane and increase the lift to reduce the sink rate and to prevent a stall. This is done while still at a high enough altitude to reestablish the correct approach airspeed and attitude. If too slow and too low, it is best to execute a go-around.

Use of Power

Power can be used effectively during the approach and round out to compensate for errors in judgment. Power is added to accelerate the airplane to increase lift without increasing the AOA and the descent slowed to an acceptable rate. If the proper landing attitude is attained and the airplane is only slightly high, the landing attitude is held constant and sufficient power applied to help ease the airplane onto the ground. After the airplane has touched down, close the throttle so the additional thrust and lift are removed and the airplane remains on the ground.

High Round Out

Sometimes when the airplane appears to temporarily stop moving downward, the round out has been made too rapidly and the airplane is flying level, too high above the runway. Continuing the round out further reduces the airspeed and increases the AOA to the critical angle. This results in the airplane stalling and dropping hard onto the runway. To prevent this, the pitch attitude is held constant until the airplane decelerates enough to again start descending. Then the round out is continued to establish the proper landing attitude. This procedure is only used when there is adequate airspeed. It may be necessary to add a slight amount of power to keep the airspeed from decreasing excessively and to avoid losing lift too rapidly.

Figure 8-32. *Change in glidepath and increase in descent rate for high final approach.*

Although back-elevator pressure may be relaxed slightly, the nose should not be lowered to make the airplane descend when fairly close to the runway unless some power is added momentarily. The momentary decrease in lift that results from lowering the nose and decreasing the AOA might cause the airplane to contact the ground with the nose wheel first and result in the nose wheel collapsing.

When the proper landing attitude is attained, the airplane is approaching a stall because the airspeed is decreasing and the critical AOA is being approached, even though the pitch attitude is no longer being increased. *[Figure 8-33]*

It is recommended that a go-around be executed any time it appears the nose must be lowered significantly or that the landing is in any other way uncertain.

Late or Rapid Round Out

Starting the round out too late or pulling the elevator control back too rapidly to prevent the airplane from touching down prematurely can impose a heavy load factor on the wing and cause an accelerated stall.

Suddenly increasing the AOA and stalling the airplane during a round out is a dangerous situation since it may cause the airplane to land extremely hard on the main landing gear and then bounce back into the air. As the airplane contacts the ground, the tail is forced down very rapidly by the back-elevator pressure and by inertia acting downward on the tail.

Recovery from this situation requires prompt and positive application of power prior to occurrence of the stall. This may be followed by a normal landing if sufficient runway is available—otherwise the pilot should execute a go-around immediately.

If the round out is late, the nose wheel may strike the runway first, causing the nose to bounce upward. Do not attempt to force the airplane back onto the ground; execute a go-around immediately.

Floating During Round Out

If the airspeed on final approach is excessive, it usually results in the airplane floating. *[Figure 8-34]* Before touchdown can be made, the airplane may be well past the desired landing point and the available runway may be insufficient. When diving the airplane on final approach to land at the proper point, there is an appreciable increase in airspeed. The proper touchdown attitude cannot be established without producing an excessive AOA and lift. This causes the airplane to gain altitude or balloon.

Any time the airplane floats, judgment of speed, height, and rate of sink must be especially acute. The pilot must smoothly and gradually adjust the pitch attitude as the airplane decelerates to touchdown speed and starts to settle, so the proper landing attitude is attained at the moment of touchdown. The slightest error in judgment and timing results in either ballooning or bouncing.

The recovery from floating is dependent upon the amount of floating and the effect of any crosswind, as well as the amount of runway remaining. Since prolonged floating utilizes considerable runway length, it must be avoided especially on short runways or in strong crosswinds. If a landing cannot be made on the first third of the runway, or the airplane drifts sideways, execute a go-around.

Ballooning During Round Out

If the pilot misjudges the rate of sink during a landing and thinks the airplane is descending faster than it should, there is a tendency to increase the pitch attitude and AOA too rapidly.

Figure 8-33. *Rounding out too high.*

Figure 8-34. *Floating during roundout.*

This not only stops the descent, but actually starts the airplane climbing. This climbing during the round out is known as ballooning. *[Figure 8-35]* Ballooning is dangerous because the height above the ground is increasing and the airplane is rapidly approaching a stalled condition. The altitude gained in each instance depends on the airspeed or the speed with which the pitch attitude is increased.

Depending on the severity of ballooning, the use of throttle is helpful in cushioning the landing. By adding power, thrust is increased to keep the airspeed from decelerating too rapidly and the wings from suddenly losing lift, but throttle must be closed immediately after touchdown. Remember that torque is created as power is applied, and it is necessary to use rudder pressure to keep the airplane straight as it settles onto the runway.

When ballooning is excessive, it is best to execute a go-around immediately; do not attempt to salvage the landing.

Power must be applied before the airplane enters a stalled condition.

The pilot must be extremely cautious of ballooning when there is a crosswind present because the crosswind correction may be inadvertently released or it may become inadequate. Because of the lower airspeed after ballooning, the crosswind affects the airplane more. Consequently, the wing has to be lowered even further to compensate for the increased drift. It is imperative that the pilot makes certain that the appropriate wing is down and that directional control is maintained with opposite rudder. If there is any doubt, or the airplane starts to drift, execute a go-around.

Bouncing During Touchdown

When the airplane contacts the ground with a sharp impact as the result of an improper attitude or an excessive rate of sink, it tends to bounce back into the air. Though the airplane's tires and shock struts provide some springing action, the airplane

Figure 8-35. *Ballooning during roundout.*

does not bounce like a rubber ball. Instead, it rebounds into the air because the wing's AOA was abruptly increased, producing a sudden addition of lift. *[Figure 8-36]*

The abrupt change in AOA is the result of inertia instantly forcing the airplane's tail downward when the main wheels contact the ground sharply. The severity of the bounce depends on the airspeed at the moment of contact and the degree to which the AOA or pitch attitude was increased.

Since a bounce occurs when the airplane makes contact with the ground before the proper touchdown attitude is attained, it is almost invariably accompanied by the application of excessive back-elevator pressure. This is usually the result of the pilot realizing too late that the airplane is not in the proper attitude and attempting to establish it just as the second touchdown occurs.

The corrective action for a bounce is the same as for ballooning and similarly depends on its severity. When it is very slight and there is no extreme change in the airplane's pitch attitude, a follow-up landing may be executed by applying sufficient power to cushion the subsequent touchdown and smoothly adjusting the pitch to the proper touchdown attitude.

In the event a very slight bounce is encountered while landing with a crosswind, crosswind correction must be maintained while the next touchdown is made. Remember that since the subsequent touchdown is made at a slower airspeed, the upwind wing has to be lowered even further to compensate for drift.

Extreme caution and alertness must be exercised any time a bounce occurs, but particularly when there is a crosswind. Inexperienced pilots almost invariably release the crosswind correction. When one main wheel of the airplane strikes the runway, the other wheel touches down immediately afterwards, and the wings becomes level. Then, with no crosswind correction as the airplane bounces, the wind causes the airplane to roll with the wind, thus exposing even more surface to the crosswind and drifting the airplane more rapidly.

When a bounce is severe, the safest procedure is to execute a go-around immediately. Do not attempt to salvage the landing. Apply full power while simultaneously maintaining directional control and lowering the nose to a safe climb attitude. The go-around procedure should be continued even though the airplane may descend and another bounce may be encountered. It is extremely foolish to attempt a landing from a bad bounce since airspeed diminishes very rapidly in the nose-high attitude, and a stall may occur before a subsequent touchdown could be made.

Porpoising

In a bounced landing that is improperly recovered, the airplane comes in nose first initiating a series of motions that imitate the jumps and dives of a porpoise. *[Figure 8-37]* The problem is improper airplane attitude at touchdown, sometimes caused by inattention, not knowing where the ground is, miss-trimming or forcing the airplane onto the runway.

Ground effect decreases elevator control effectiveness and increases the effort required to raise the nose. Not enough elevator or stabilator trim can result in a nose low contact with the runway and a porpoise develops.

Porpoising can also be caused by improper airspeed control. Usually, if an approach is too fast, the airplane floats and the pilot tries to force it on the runway when the airplane still wants to fly. A gust of wind, a bump in the runway, or even a slight tug on the control wheel sends the airplane aloft again.

Figure 8-36. *Bouncing during touchdown.*

Figure 8-37. *Porpoising.*

The corrective action for a porpoise is the same as for a bounce and similarly depends on its severity. When it is very slight and there is no extreme change in the airplane's pitch attitude, a follow-up landing may be executed by applying sufficient power to cushion the subsequent touchdown and smoothly adjusting the pitch to the proper touchdown attitude.

When a porpoise is severe, the safest procedure is to execute a go-around immediately. In a severe porpoise, the airplane's pitch oscillations can become progressively worse until the airplane strikes the runway nose first with sufficient force to collapse the nose gear. Attempts to correct a severe porpoise with flight control and power inputs is most likely untimely and out of sequence with the oscillations and only make the situation worse. Do not attempt to salvage the landing. Apply full power while simultaneously maintaining directional control and lowering the nose to a safe climb attitude.

Wheel Barrowing

When a pilot permits the airplane weight to become concentrated about the nose wheel during the takeoff or landing roll, a condition known as wheel barrowing occurs. Wheel barrowing may cause loss of directional control during the landing roll because braking action is ineffective, and the airplane tends to swerve or pivot on the nose wheel, particularly in crosswind conditions. One of the most common causes of wheel barrowing during the landing roll is a simultaneous touchdown of the main and nose wheel with excessive speed, followed by application of forward pressure on the elevator control. Usually, the situation can be corrected by smoothly applying back-elevator pressure.

If wheel barrowing is encountered and runway and other conditions permit, it is advisable to promptly initiate a go-around. Wheel barrowing does not occur if the pilot achieves and maintains the correct landing attitude, touches down at the proper speed, and gently lowers the nose wheel while losing speed on rollout. If the pilot decides to stay on the ground rather than attempt a go-around or if directional control is lost, close the throttle and adjust the pitch attitude smoothly but firmly to the proper landing attitude.

Hard Landing

When the airplane contacts the ground during landings, its vertical speed is instantly reduced to zero. Unless provisions are made to slow this vertical speed and cushion the impact of touchdown, the force of contact with the ground may be so great it could cause structural damage to the airplane.

The purpose of pneumatic tires, shock absorbing landing gear, and other devices is to cushion the impact and to increase the time in which the airplane's vertical descent is stopped. The importance of this cushion may be understood from the computation that a 6-inch free fall on landing is roughly equal to a 340 fpm descent. Within a fraction of a second, the airplane must be slowed from this rate of vertical descent to zero without damage.

During this time, the landing gear, together with some aid from the lift of the wings, must supply whatever force is needed to counteract the force of the airplane's inertia and weight. The lift decreases rapidly as the airplane's forward speed is decreased, and the force on the landing gear increases by the impact of touchdown. When the descent stops, the lift is practically zero, leaving the landing gear alone to carry both the airplane's weight and inertia force. The load imposed at the instant of touchdown may easily be three or four times the actual weight of the airplane depending on the severity of contact.

Touchdown in a Drift or Crab

At times, it is necessary to correct for wind drift by crabbing on the final approach. If the round out and touchdown are made while the airplane is drifting or in a crab, it contacts the ground while moving sideways. This imposes extreme side loads on the landing gear and, if severe enough, may cause structural failure.

The most effective method to prevent drift is the wing-low method. This technique keeps the longitudinal axis of the airplane aligned with both the runway and the direction of motion throughout the approach and touchdown.

There are three factors that cause the longitudinal axis and the direction of motion to be misaligned during touchdown: drifting, crabbing, or a combination of both.

If the pilot does not take adequate corrective action to avoid drift during a crosswind landing, the main wheels' tire tread offers resistance to the airplane's sideward movement in respect to the ground. Consequently, any sidewise velocity of the airplane is abruptly decelerated, resulting in the aircraft being shifted to the right due to the inertia force which is shown in *Figure 8-38*. This creates a moment around the main wheel when it contacts the ground, tending to overturn or tip the airplane. If the windward wingtip is raised by the action of this moment, all the weight and shock of landing is borne by one main wheel. This could cause structural damage. Not only are the same factors present that are attempting to raise a wing, but the crosswind is also acting on the fuselage surface behind the main wheels, tending to yaw (weathervane) the airplane into the wind. This often results in a ground loop.

Ground Loop

A ground loop is an uncontrolled turn during ground operation that may occur while taxiing or taking off, but especially during the after-landing roll. Drift or weathervaning does not always cause a ground loop, although these things may cause the initial swerve. Careless use of the rudder, an uneven ground surface, or a soft spot that retards one main wheel of the airplane may also cause a swerve. In any case, the initial swerve tends to make the airplane ground loop, whether it is a tailwheel-type or nose-wheel type. *[Figure 8-39]*

Nose-wheel type airplanes are somewhat less prone to ground loop than tailwheel-type airplanes. Since the center of gravity (CG) is located forward of the main landing gear

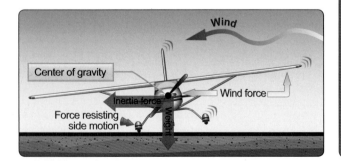

Figure 8-38. *Drifting during touchdown.*

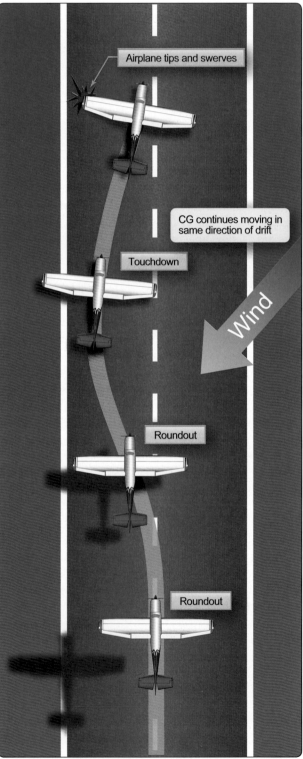

Figure 8-39. *Start of a ground loop.*

on these airplanes, any time a swerve develops, centrifugal force acting on the CG tends to stop the swerving action.

If the airplane touches down while drifting or in a crab, apply aileron toward the high wing and stop the swerve with the rudder. Brakes are used to correct for turns or swerves only when the rudder is inadequate. Exercise caution when applying corrective brake action because it is very easy to over control and aggravate the situation.

If brakes are used, sufficient brake is applied on the low-wing wheel (outside of the turn) to stop the swerve. When the wings are approximately level, the new direction must be maintained until the airplane has slowed to taxi speed or has stopped.

In nose-wheel airplanes, a ground loop is almost always a result of wheel barrowing. A pilot must be aware that even though the nose-wheel type airplane is less prone than the tailwheel-type airplane, virtually every type of airplane, including large multi-engine airplanes, can be made to ground loop when sufficiently mishandled.

Wing Rising After Touchdown
When landing in a crosswind, there may be instances when a wing rises during the after-landing roll. This may occur whether or not there is a loss of directional control, depending on the amount of crosswind and the degree of corrective action.

Any time an airplane is rolling on the ground in a crosswind condition, the upwind wing is receiving a greater force from the wind than the downwind wing. This causes a lift differential. Also, as the upwind wing rises, there is an increase in the AOA, which increases lift on the upwind wing, rolling the airplane downwind.

When the effects of these two factors are great enough, the upwind wing may rise even though directional control is maintained. If no correction is applied, it is possible that the upwind wing rises sufficiently to cause the downwind wing to strike the ground.

In the event a wing starts to rise during the landing roll, immediately apply more aileron pressure toward the high wing and continue to maintain direction. The sooner the aileron control is applied, the more effective it is. The further a wing is allowed to rise before taking corrective action, the more airplane surface is exposed to the force of the crosswind. This diminishes the effectiveness of the aileron.

Hydroplaning

Hydroplaning is a condition that can exist when an airplane has landed on a runway surface contaminated with standing water, slush, and/or wet snow. Hydroplaning can have serious adverse effects on ground controllability and braking efficiency. The three basic types of hydroplaning are dynamic hydroplaning, reverted rubber hydroplaning, and viscous hydroplaning. Any one of the three can render an airplane partially or totally uncontrollable anytime during the landing roll.

Dynamic Hydroplaning
Dynamic hydroplaning is a relatively high-speed phenomenon that occurs when there is a film of water on the runway that is at least one-tenth of an inch deep. As the speed of the airplane and the depth of the water increase, the water layer builds up an increasing resistance to displacement, resulting in the formation of a wedge of water beneath the tire. At some speed, termed the hydroplaning speed (Vp), the water pressure equals the weight of the airplane, and the tire is lifted off the runway surface. In this condition, the tires no longer contribute to directional control and braking action is nil.

Dynamic hydroplaning is related to tire inflation pressure. Data obtained during hydroplaning tests have shown the minimum dynamic hydroplaning speed (Vp) of a tire to be 8.6 times the square root of the tire pressure in pounds per square inch (PSI). For an airplane with a main tire pressure of 24 pounds, the calculated hydroplaning speed would be approximately 42 knots. It is important to note that the calculated speed referred to above is for the start of dynamic hydroplaning. Once hydroplaning has started, it may persist to a significantly slower speed depending on the type being experienced.

Reverted Rubber Hydroplaning
Reverted rubber (steam) hydroplaning occurs during heavy braking that results in a prolonged locked-wheel skid. Only a thin film of water on the runway is required to facilitate this type of hydroplaning. The tire skidding generates enough heat to cause the rubber in contact with the runway to revert to its original uncured state. The reverted rubber acts as a seal between the tire and the runway and delays water exit from the tire footprint area. The water heats and is converted to steam, which supports the tire off the runway.

Reverted rubber hydroplaning frequently follows an encounter with dynamic hydroplaning, during which time the pilot may have the brakes locked in an attempt to slow the airplane. Eventually the airplane slows enough to where the tires make contact with the runway surface and the airplane begins to skid. The remedy for this type of hydroplane is to release the brakes and allow the wheels to spin up and apply moderate braking. Reverted rubber hydroplaning is insidious in that the pilot may not know when it begins, and it can persist to very slow ground speeds (20 knots or less).

Viscous Hydroplaning

Viscous hydroplaning is due to the viscous properties of water. A thin film of fluid no more than one thousandth of an inch in depth is all that is needed. The tire cannot penetrate the fluid and the tire rolls on top of the film. This can occur at a much lower speed than dynamic hydroplane, but requires a smooth or smooth acting surface, such as asphalt or a touchdown area coated with the accumulated rubber of past landings. Such a surface can have the same friction coefficient as wet ice.

When confronted with the possibility of hydroplaning, it is best to land on a grooved runway (if available). Touchdown speed should be as slow as possible consistent with safety. After the nose wheel is lowered to the runway, moderate braking is applied. If deceleration is not detected and hydroplaning is suspected, raise the nose and use aerodynamic drag to decelerate to a point where the brakes do become effective.

Proper braking technique is essential. The brakes are applied firmly until reaching a point just short of a skid. At the first sign of a skid, release brake pressure and allow the wheels to spin up. Directional control is maintained as far as possible with the rudder. Remember that in a crosswind, if hydroplaning occurs, the crosswind causes the airplane to simultaneously weathervane into the wind, as well as slide downwind.

Chapter Summary

Accident statistics show that a pilot is at most risk for an accident during the approach and landing than any other phase of a flight. There are many factors that contribute to accidents in this phase, but an overwhelming percentage of accidents are caused from pilot's lack of proficiency. This chapter presents procedures that, when learned and practiced, are a key to attaining proficiency. Additional information on aerodynamics, airplane performance, and other aspects affecting approaches and landings can be found in the Pilot's Handbook of Aeronautical Knowledge (FAA-H-8083-25, as revised). For information concerning risk assessment as a means of preventing accidents, refer to the Risk Management Handbook (FAA-H-8083-2). Both of these publications are available at www.faa.gov/library/manuals/aviation.

Chapter 9

Performance Maneuvers

Introduction

Flight maneuvers that are initially taught to pilots are designed to be basic and relatively simple: straight-and-level, turns, climbs and descents. However, as a pilot continues through their flight training, additional maneuvers are needed to develop beyond the fundamentals. Performance maneuvers are intended to enhance a pilot's proficiency in flight control application, maneuver planning, situational awareness, and division of attention. To further that intent, performance maneuvers are generally designed so that the application of flight control pressures, attitudes, airspeeds, and orientations are constantly changing throughout the maneuver.

Performance maneuvers also allow for an effective assessment of a pilot's ability to apply the fundamentals; weakness in executing performance maneuvers is likely due to a pilot's lack of understanding or a deficiency of fundamental skills. It is advisable that performance maneuver training should not take place until sufficient competency in the fundamentals is consistently demonstrated by the pilot. Further, initial training for performance maneuvers should always begin with a detailed ground lesson for each maneuver, so that the technicalities are understood prior to flight. In addition, performance maneuver training should be segmented into comprehensible building blocks of instruction so as to allow the pilot an appropriate level of repetition to develop the required skills.

Performance maneuvers, once grasped by the pilot, are very satisfying and rewarding. As the pilot develops skills in executing performance maneuvers, they may likely see an increased smoothness in their flight control application and a higher ability to sense the airplane's attitude and orientation without significant conscious effort.

Steep Turns

Steep turns consist of single to multiple 360° to 720° turns, in either or both directions, using a bank angle between 45° to 60°. The objective of the steep turn is to develop a pilot's skill in flight control smoothness and coordination, an awareness of the airplane's orientation to outside references, division of

attention between flight control application, and the constant need to scan for hazards. *[Figure 9-1]*

When steep turns are first demonstrated, the pilot will be in an unfamiliar environment when compared to what was previously experienced in shallow bank angled turns; however, the fundamental concepts of turns remain the same in the execution of steep turns. When performing steep turns, pilots will be exposed to higher load factors, the airplane's inherent overbanking tendency, the loss of vertical component of lift when the wings are steeply banked, the need for substantial pitch control pressures, and the need for additional power to maintain altitude and airspeed during the turn.

As discussed in previous chapters, when an airplane is banked, the total lift is comprised of a vertical component of lift and a horizontal component of lift. In order to not lose altitude, the pilot must increase the wing's angle of attack (AOA) to ensure that the vertical component of lift is sufficient to maintain altitude. In a steep turn, the pilot will need to increase pitch with elevator back pressures that are greater than what has been previously utilized. Total lift must increase substantially to balance the load factor or G-force (G). The load factor is the vector resultant of gravity and centrifugal force. For example, in a level altitude, 45° banked turn, the resulting load factor is 1.4; in a level altitude, 60° banked turn, the resulting load factor is 2.0. To put this in perspective, with a load factor of 2.0, the effective weight of the aircraft will double. Pilots

Figure 9-1. *Steep turns.*

should realize load factors increase dramatically beyond 60°. Most general aviation airplanes are designed for a load limit of 3.8Gs. Regardless of the airspeed or what airplane is involved, for a given bank angle in a level altitude turn, the same load factor will always be produced. A light, general aviation airplane in a level altitude, 45° angle of bank turn will experience a load factor of 1.4 just as a large commercial airliner will in the same level altitude, 45° angle of bank turn.

Because of the higher load factors, steep turns should be performed at an airspeed that does not exceed the airplane's design maneuvering speed (V_A) or the manufacturer's recommended speed. Maximum turning performance is accomplished when an airplane has both a fast rate of turn and minimum radius of turn, which is effected by both airspeed and angle of bank. Each airplane's turning performance is limited by structural and aerodynamic design, as well as available power. The airplane's limiting load factor determines the maximum bank angle that can be maintained in level flight without exceeding the airplane's structural limitations or stalling. As the load factor increases, so does the stalling speed. For example, if an airplane stalls in level flight at 50 knots, it will stall at 60 knots in a level altitude, 45° banked turn and at 70 knots in a level altitude, 60° banked turn. Stalling speed increases at the square root of the load factor. As the bank angle increases in level flight, the margin between stalling speed and maneuvering speed decreases—an important concept for a pilot to remain cognizant.

In addition to the increased load factors, the airplane will exhibit what is called "overbanking tendency." Recall from a previous chapter on the discussion of overbanking tendency. In most flight maneuvers, bank angles are shallow enough that the airplane exhibits positive or neutral stability about the longitudinal axis; however, as bank angles steepen, the airplane will exhibit the behavior to continue rolling in the direction of the bank unless deliberate and opposite aileron pressure is held against the bank. Also, pilots should be mindful of the various left turning tendencies, such as P-factor, which requires effective rudder aileron coordination.

Before starting any practice maneuver, the pilot must ensure that the area is clear of air traffic and other hazards. Further, distant references such as a mountain peak or road should be chosen to allow the pilot to assess when to begin rollout from the turn. After establishing the manufacturer's recommended entry speed or the design maneuvering speed, the airplane should be smoothly rolled into the desired bank angle somewhere between 45° to 60°. As the bank angle is being established, generally prior to 30° of bank, elevator back pressure should be smoothly applied to increase the AOA. After the selected bank angle has been reached, the pilot will find that considerable force is required on the elevator control

to hold the airplane in level flight—to maintain altitude. Pilots should keep in mind that as the AOA increases, so does drag. Consequently, power must be added to maintain altitude and airspeed.

Steep turns can be conducted more easily by the use of elevator trim and power as the maneuver is entered. In many light general aviation airplanes, as the bank angle transitions from medium to steep, increasing elevator up trim and adding a small increase in engine power minimizes control pressure requirements. Pilots must not forget to remove both the trim and power inputs as the maneuver is completed.

To maintain bank angle, altitude, as well as orientation, requires an awareness of the relative position of the horizon to the nose and the wings. The pilot who references the aircraft's attitude by observing only the nose will have difficulty maintaining altitude. A pilot who observes both the nose and the wings relative to the horizon is likely able to maintain altitude within performance standards. Altitude deviations are primary errors exhibited in the execution of steep turns. If the altitude does increase or decrease, changing elevator back pressure could be used to alter the altitude; however, a more effective method is a slight increase or decrease in bank angle to control small altitude deviations. If altitude is decreasing, reducing the bank angle a few degrees helps recover or stop the altitude loss trend; also, if altitude is increasing, increasing the bank angle a few degrees helps recover or stop the altitude increase trend—all bank angle changes should be accomplished with coordinated use of aileron and rudder.

The rollout from the steep turn should be timed so that the wings reach level flight when the airplane is on heading from which the maneuver was started. A good rule of thumb is to begin the rollout at ½ the number of degrees of bank prior to reaching the terminating heading. For example, if a right steep turn was begun on a heading of 270° and if the bank angle is 60°, the pilot should begin the rollout 30° prior or at a heading of 240°. While the rollout is being made, elevator back pressure, trim, and power should be gradually reduced, as necessary, to maintain the altitude and airspeed.

Common errors when performing steep turns are:

- Not clearing the area
- Inadequate pitch control on entry or rollout
- Gaining altitude or losing altitude
- Failure to maintain constant bank angle
- Poor flight control coordination
- Ineffective use of trim

- Ineffective use of power

- Inadequate airspeed control

- Becoming disoriented

- Performing by reference to the flight instrument rather than visual references

- Failure to scan for other traffic during the maneuver

- Attempts to start recovery prematurely

- Failure to stop the turn on designated heading

Steep Spiral

The objective of the steep spiral is to provide a flight maneuver for rapidly dissipating substantial amounts of altitude while remaining over a selected spot. This maneuver is especially effective for emergency descents or landings. A steep spiral is a gliding turn where the pilot maintains a constant radius around a surface-based reference point while rapidly descending—similar to the turns around a point maneuver. Sufficient altitude must be gained prior to practicing the maneuver so that at least three 360° turns are completed. *[Figure 9-2]* The maneuver should not be allowed to continue below 1,500 feet above ground level (AGL) unless an actual emergency exists.

The steep spiral is initiated by properly clearing the airspace for air traffic and hazards. In general, the throttle is closed to idle, carburetor heat is applied if equipped, and gliding speed

is established. Once the proper airspeed is attained, the pitch should be lowered and the airplane rolled to the desired bank angle as the reference point is reached. The steepest bank should not exceed 60°. The gliding spiral should be a turn of constant radius while maintaining the airplane's position to the reference. This can only be accomplished by proper correction for wind drift by steepening the bank on downwind headings and shallowing the bank on upwind headings, just as in the maneuver, turns around a point. During the steep spiral, the pilot must continually correct for any changes in wind direction and velocity to maintain a constant radius.

Operating the engine at idle speed for any prolonged period during the glide may result in excessive engine cooling, spark plug fouling, or carburetor ice. To assist in avoiding these issues, the throttle should be periodically advanced to normal cruise power and sustained for a few seconds. If equipped, monitoring cylinder head temperatures provides a pilot with additional information on engine cooling. When advancing the throttle, the pitch attitude must be adjusted to maintain a constant airspeed and, preferably, this should be done when headed into the wind.

Maintaining a constant airspeed throughout the maneuver is an important skill for a pilot to develop. This is necessary because the airspeed tends to fluctuate as the bank angle is changed throughout the maneuver. The pilot should anticipate pitch corrections as the bank angle is varied throughout the

Figure 9-2. *Steep spiral.*

maneuver. During practice of the maneuver, the pilot should execute three turns and roll out toward a definite object or on a specific heading. During rollout, the smooth and accurate application of the flight controls allow the airplane to recover to a wing's level glide with no change in airspeed. Recovering to normal cruise flight would proceed after the establishment of a wing's level glide.

Common errors when performing steep spirals are:

- Not clearing the area

- Inadequate pitch control on entry or rollout

- Gaining altitude

- Not correcting the bank angle to compensate for wind

- Poor flight control coordination

- Ineffective use of trim

- Inadequate airspeed control

- Becoming disoriented

- Performing by reference to the flight instrument rather than visual references

- Not scanning for other traffic during the maneuver

- Not completing the turn on designated heading or reference

Chandelle

A chandelle is a maximum performance, 180° climbing turn that begins from approximately straight-and-level flight and concludes with the airplane in a wings-level, nose-high attitude just above stall speed. *[Figure 9-3]* The goal is to gain the most altitude possible for a given bank angle and power setting; however, the standard used to judge the maneuver is not the amount of altitude gained, but by the pilot's proficiency as it pertains to maximizing climb performance for the power and bank selected, as well as the skill demonstrated.

A chandelle is best described in two specific phases: the first 90° of turn and the second 90° of turn. The first 90° of turn is described as constant bank and changing pitch; and the second 90° as constant pitch and changing bank. During the first 90°, the pilot will set the bank angle, increase power and pitch at a rate so that maximum pitch-up is set at the completion of the first 90°. If the pitch is not correct, the airplane's airspeed is either above stall speed or the airplane may aerodynamically stall prior to the completion of the maneuver. Starting at the 90° point, the pilot begins a slow and coordinated constant rate rollout so as to have the wings level when the airplane is at the 180° point while maintaining the constant pitch attitude set in the first 90°. If the rate of rollout is too rapid or sluggish, the airplane either does not

1. Complete rollout to wings level at 180° point
2. Airspeed V_{s_1}
3. Maintain coordinated flight
4. Hold airspeed without stalling

1. Continue smooth rollout
2. Hold pitch
3. Maintain coordinated flight

1. Maintain 30° bank until this point and then begin gradual rollout.
2. Maximum pitch should be reached. No further changes in pitch

1. Clear area
2. V_A or manufacturer's recommended speed
3. No lower than 1,500 feet AGL
4. Trim

1. Roll into 30° bank
2. Neuralize ailerons
3. Begin increasing pitch toward climbing attitude
4. Apply full power without exceeding limits
5. Increase rudder pressure

Figure 9-3. *Chandelle.*

complete or exceeds the 180° turn as the wings come level to the horizon.

Prior to starting the chandelle, the flaps and landing gear (if retractable) should be in the UP position. The chandelle is initiated by properly clearing the airspace for air traffic and hazards. The maneuver should be entered from straight-and-level flight or a shallow dive at an airspeed recommended by the manufacturer—in most cases this is the airplane's design maneuvering speed (V_A). [Figure 9-3A] After the appropriate entry airspeed has been established, the chandelle is started by smoothly entering a coordinated turn to the desired angle of bank; once the bank angle is established, which is generally 30°, a climbing turn should be started by smoothly applying elevator back pressure at a constant rate while simultaneously increasing engine power to the recommended setting. In airplanes with a fixed-pitch propeller, the throttle should be set so as to not exceed rotations per minute (rpm) limitations; in airplanes with constant-speed propellers, power may be set at the normal cruise or climb setting as appropriate. [Figure 9-3B]

Since the airspeed is constantly decreasing throughout the chandelle, the effects of left turning tendencies, such as P-factor, becomes more apparent. As airspeed decreases, right-rudder pressure is progressively increased to ensure that the airplane remains in coordinated flight. The pilot should maintain coordinated flight by sensing slipping or skidding pressures applied to the controls and by quick glances to the ball in the turn-and-slip or turn coordinator.

At the 90° point, the pilot should begin to smoothly roll out of the bank at a constant rate while maintaining the pitch attitude set in the first 90°. While the angle of bank is fixed during the first 90°, recall that as airspeed decreases, the overbanking tendency increases. [Figure 9-3C] As a result, proper use of the ailerons allows the bank to remain at a fixed angle until rollout is begun at the start of the final 90°. As the rollout continues, the vertical component of lift increases; therefore, a slight release of elevator back pressure is required to keep the pitch attitude from increasing.

When the airspeed is slowest, near the completion of the chandelle, right rudder pressure is significant, especially when rolling out from a left chandelle due to left adverse yaw and left turning tendencies, such as P-factor. [Figure 9-3D] When rolling out from a right chandelle, the yawing moment is to the right, which partially cancels some of the left turning tendency's effect. Depending on the airplane, either very little left rudder or a reduction in right rudder pressure is required during the rollout from a right chandelle. At the completion of 180° of turn, the wings should be leveled to the horizon, the airspeed should be just above stall speed, and the airplane's pitch high attitude should be held momentarily.

[Figure 9-3E] Once demonstrated that the airplane is in controlled flight, the pitch attitude may be reduced and the airplane returned to straight-and-level cruise flight.

Common errors when performing chandelles are:

- Not clearing the area
- Initial bank is too shallow resulting in a stall
- Initial bank is too steep resulting in failure to gain maximum performance
- Allowing the bank angle to increase after initial establishment
- Not starting the recovery at the 90° point in the turn
- Allowing the pitch attitude to increase as the bank is rolled out during the second 90° of turn
- Leveling the wings prior to the 180° point being reached
- Pitch attitude is low on recovery resulting in airspeed well above stall speed
- Application of flight control pressures is not smooth
- Poor flight control coordination
- Stalling at any point during the maneuver
- Execution of a steep turn instead of a climbing maneuver
- Not scanning for other traffic during the maneuver
- Performing by reference to the flight instrument rather than visual references

Lazy Eight

The lazy eight is a maneuver that is designed to develop the proper coordination of the flight controls across a wide range of airspeeds and attitudes. It is the only standard flight training maneuver that, at no time, flight control pressures are constant. In an attempt to simplify the discussion about this maneuver, the lazy eight can be loosely described by the ground reference maneuver, S-turns across the road. Recall that S-turns across the road are made of opposing 180° turns. For example, first a 180° turn to the right, followed immediately by a 180° turn to the left. The lazy eight adds both a climb and descent to each 180° segment. The first 90° is a climb; the second 90° is a descent. [Figure 9-4]

To aid in the performance of the lazy eight's symmetrical climbing/descending turns, prominent reference points must be selected on the natural horizon. The reference points selected should be at 45°, 90°, and 135° from the direction in which the maneuver is started for each 180° turn. With the general concept of climbing and descending turns grasped, specifics of the lazy eight can then be discussed.

Figure 9-4. *Lazy eight.*

Shown in *Figure 9-4A*, from level flight a gradual climbing turn is begun in the direction of the 45° reference point; the climbing turn should be planned and controlled so that the maximum pitch-up attitude is reached at the 45° point with an approximate bank angle of 15°. *[Figure 9-4B]* As the pitch attitude is raised, the airspeed decreases, which causes the rate of turn to increase. As such, the lazy eight must begin with a slow rate of roll as the combination of increasing pitch and increasing bank may cause the rate of turn to be so rapid that the 45° reference point will be reached before the highest pitch attitude is attained. At the 45° reference point, the pitch attitude should be at the maximum pitch-up selected for the maneuver while the bank angle is slowly increasing. Beyond the 45° reference point, the pitch-up attitude should begin to decrease slowly toward the horizon until the 90° reference point is reached where the pitch attitude should be momentarily level.

The lazy eight requires substantial skill in coordinating the aileron and rudder; therefore, some discussion about coordination is warranted. As pilots understand, the purpose of the rudder is to maintain coordination; slipping or skidding is to be avoided. Pilots should remember that since the airspeed is still decreasing as the airplane is climbing; additional right rudder pressure must be applied to counteract left turning tendencies, such as P-factor. As the airspeed decreases, right rudder pressure must be gradually applied to counteract yaw at the apex of the lazy eight in both the

right and left turns; however, additional right rudder pressure is required when turning or rolling out to the right than left because left adverse yaw augments with the left yawing P-factor in an attempt to yaw the nose to the left. Correction is needed to prevent these additive left yawing moments from decreasing a right turn's rate. In contrast, in left climbing turns or rolling to the left, the left yawing P-factor tends to cancel the effects of adverse yaw to the right; consequently, less right rudder pressure is required. These concepts can be difficult to remember; however, to simplify, rolling right at low airspeeds and high-power settings requires substantial right rudder pressures.

At the lazy eight's 90° reference point, the bank angle should also have reached its maximum angle of approximately 30°. *[Figure 9-4C]* The airspeed should be at its minimum, just about 5 to 10 knots above stall speed, with the airplane's pitch attitude passing through level flight. Coordinated flight at this point requires that, in some flight conditions, a slight amount of opposite aileron pressure may be required to prevent the wings from overbanking while maintaining rudder pressure to cancel the effects of left turning tendencies.

The pilot should not hesitate at the 90° point but should continue to maneuver the airplane into a descending turn. The rollout from the bank should proceed slowly while the airplane's pitch attitude is allowed to decrease. When the airplane has turned 135°, the airplane should be in its lowest pitch attitude. *[Figure 9-4D]* Pilots should remember that the airplane's airspeed is increasing as the airplane's pitch attitude decreases; therefore, to maintain proper coordination will require a decrease in right rudder pressure. As the airplane approaches the 180° point, it is necessary to progressively relax rudder and aileron pressure while simultaneously raising pitch and roll to level flight. As the rollout is being accomplished, the pilot should note the amount of turn remaining and adjust the rate of rollout and pitch change so that the wings and nose are level at the original airspeed just as the 180° point is reached.

Upon arriving at 180° point, a climbing turn should be started immediately in the opposite direction toward the preselected reference points to complete the second half of the lazy eight in the same manner as the first half. *[Figure 9-4E]*

Power should be set so as not to enter the maneuver at an airspeed that would exceed manufacturer's recommendations, which is generally no greater than V_A. Power and bank angle have significant effect on the altitude gained or lost; if excess power is used for a given bank angle, altitude is gained at the completion of the maneuver; however, if insufficient power is used for a given bank angle, altitude is lost.

Common errors when performing lazy eights are:

- Not clearing the area

- Maneuver is not symmetrical across each 180°

- Inadequate or improper selection or use of 45°, 90°, 135° references

- Ineffective planning

- Gain or loss of altitude at each 180° point

- Poor control at the top of each climb segment resulting in the pitch rapidly falling through the horizon

- Airspeed or bank angle standards not met

- Control roughness

- Poor flight control coordination

- Stalling at any point during the maneuver

- Execution of a steep turn instead of a climbing maneuver

- Not scanning for other traffic during the maneuver

- Performing by reference to the flight instrument rather than visual references

Chapter Summary

Performance maneuvers are used to develop a pilot's skills in coordinating the flight control's use and effect while enhancing the pilot's ability to divide attention across the various demands of flight. Performance maneuvers are also designed to further develop a pilot's application and correlation of the fundamentals of flight and integrate developing skills into advanced maneuvers. Developing highly-honed skills in performance maneuvers allows the pilot to effectively progress toward the mastery of flight. Mastery is developed as the mechanics of flight become a subconscious, rather than a conscious, application of the flight controls to maneuver the airplane in attitude, orientation, and position.

Chapter 10
Night Operations

Introduction

The mechanical operation of an airplane at night is no different than operating the same airplane during the day. The airplane does not know if it is being operated in the dark or bright sunlight. It performs and responds to control inputs by the pilot. The pilot, however, is affected by various aspects of night operations and must take them into consideration during night flight operations. Some are actual physical limitations affecting all pilots while others, such as equipment requirements, procedures, and emergency situations, must also be considered.

According to Title 14 of the Code of Federal Regulations (14 CFR) part 1, Definitions and Abbreviations, night is defined as the time between the end of evening civil twilight and the beginning of morning civil twilight. To explain further, morning civil twilight begins when the geometric center of the sun is 6° below the horizon and ends at sunrise. Evening civil twilight begins at sunset and ends when the geometric center of the sun reaches 6° below the horizon.

For 14 CFR part 61 operations, the term night refers to 1 hour after sunset and ending 1 hour before sunrise as 14 CFR part 61 explains that between those hours no person may act as pilot in command (PIC) of an aircraft carrying passengers unless within the preceding 90 days that person has made at least three takeoffs and three landings to a full stop during that night period.

Night flying operations should not be encouraged or attempted except by certificated pilots with knowledge of and experience in the topics discussed in this chapter.

Night Vision

Generally, most pilots are poorly informed about night vision. Human eyes never function as effectively at night as the eyes of animals with nocturnal habits, but if humans learn how to use their eyes correctly and know their limitations, night vision can be improved significantly.

The brain and eyes act as a team for a person to see well; both must be used effectively. Due to the physiology of the eye, limitations on sight are experienced in low light conditions, such as at night. To see at night, the eyes are used differently than during the day. Therefore, it is important to understand the eye's construction and how the eye is affected by darkness. Innumerable light-sensitive nerves called "cones" and "rods" are located at the back of the eye or retina, a layer upon which all images are focused. These nerves connect to the cells of the optic nerve, which transmits messages directly to the brain. The cones are located in the center of the retina, and the rods are concentrated in a ring around the cones. *[Figure 10-1]*

The function of the cones is to detect color, details, and faraway objects. The rods function when something is seen out of the corner of the eye or peripheral vision. They detect objects, particularly those that are moving, but do not give detail or color—only shades of gray. Both the cones and the rods are used for vision during daylight.

Although there is not a clear-cut division of function, the rods make night vision possible. The rods and cones function in daylight and in moonlight, but in the absence of normal light, the process of night vision is placed almost entirely on the rods. The rods are distributed in a band around the cones and do not lie directly behind the pupils, which makes off-center viewing (looking to one side of an object) important during night flight. During daylight, an object can be seen best by looking directly at it, but at night there is a blind spot in the center of the field of vision, the night blind spot. If an object is in this area, it may not be seen. The size of this blind spot increases as the distance between the eye and the object increases as illustrated in *Figure 10-1*. Therefore, the night blind spot can hide

Figure 10-1. *Rods and cones.*

larger objects as the distance between the pilot and an object increases. Use of a scanning procedure to permit off-center viewing of the object is more effective. Consciously practice this scanning procedure to improve night vision.

The eye's adaptation to darkness is another important aspect of night vision. When a dark room is entered, it is difficult to see anything until the eyes become adjusted to the darkness. Almost everyone experiences this when entering a darkened movie theater. In this process, the pupils of the eyes first enlarge to receive as much of the available light as possible. After approximately 5 to 10 minutes, the cones become adjusted to the dim light and the eyes become approximately 100 times more sensitive to the light than they were before the dark room was entered. Much more time, about 30 minutes, is needed for the rods to become adjusted to darkness, but when they do adjust, they are about 100,000 times more sensitive to light than they were in the lighted area. After the adaptation process is complete, much more can be seen, especially if scanning techniques are used correctly.

After the eyes have adapted to the dark, the entire process is reversed when entering a lighted room. The eyes are first dazzled by the brightness, but become completely adjusted in a very few seconds, thereby losing their adaptation to the dark. Now, if the dark room is re-entered, the eyes again go through the long process of adapting to the darkness.

Before and during night flight, the adaptation process of the eyes must be considered. First, adapt to the low level of light and then stay adapted. After the eyes are adapted to the darkness, avoid exposing them for more than one second to any bright white light as that causes temporary blindness. If exposed to a bright light source, such as search lights and landing lights, remember that each eye adapts to the dark independently. By closing or covering one eye when exposed to light, some night vision acuity is retained in the closed eye.

Temporary blindness, caused by an unusually bright light, may result in illusions or after images until the eyes recover from the brightness. The brain creates these illusions reported by the eyes. This results in misjudging or incorrectly identifying objects, such as mistaking slanted clouds for the horizon or populated areas for a landing field. Vertigo is experienced as a feeling of dizziness and imbalance that can create or increase illusions. The illusions seem very real and pilots at every level of experience and skill can be affected. Recognizing that the brain and eyes can play tricks in this manner is the best protection for flying at night.

Good eyesight depends upon physical condition. Fatigue, colds, vitamin deficiency, alcohol, stimulants, smoking, or medication can seriously impair vision. Keep these facts in mind and take adequate precautions to safeguard night vision. In addition to the principles previously discussed, the following items aid in increasing night vision effectiveness.

- Adapt the eyes to darkness prior to flight and keep them adapted. About 30 minutes is needed to adjust the eyes to maximum efficiency after exposure to a bright light.

- If oxygen is available, use it during night flying. Keep in mind that a significant deterioration in night vision can occur at cabin altitudes as low as 5,000 feet.

- Close one eye when exposed to bright light to help avoid the blinding effect.

- Do not wear sunglasses after sunset as this impairs night vision.

- Move the eyes more slowly than in daylight.

- Blink the eyes if they become blurred.

- Concentrate on seeing objects.

- Force the eyes to view off center using scanning techniques.

- Maintain good physical condition.

- Avoid smoking, drinking, and using drugs that may be harmful.

Night Illusions

In addition to night vision limitations, night illusions can cause confusion and distractions during night flying. The following discussion covers some of the common situations that cause illusions associated with night flying.

On a clear night, distant stationary lights can be mistaken for stars or other aircraft. Cloud layers or even the northern lights can confuse a pilot and indicate a false visual horizon. Certain geometrical patterns of ground lights, such as a freeway, runway, approach, or even lights on a moving train, can cause confusion. Dark nights tend to eliminate reference to a visual horizon. As a result, pilots need to rely less on outside references at night and more on flight and navigation instruments.

Visual autokinesis can occur when staring at a single light source for several seconds on a dark night. The result is that the light appears to be moving. The autokinesis effect will not occur if the visual field is expanded through scanning techniques. A good scanning procedure reduces the probability of vision becoming fixed on one source of light.

Distractions and problems can result from a flickering light in the flightdeck, anti-collision light, or other aircraft lights and can cause flicker vertigo. If continuous, the possible physical reactions can be nausea, dizziness, grogginess, unconsciousness, headaches, or confusion. Try to eliminate any light source causing blinking or flickering problems in the flightdeck.

A black-hole approach occurs when the landing is made from over water or non-lighted terrain where the runway lights are the only source of light. Without peripheral visual cues to help, orientation is difficult. The runway can seem out of position (down-sloping or up-sloping) and in the worst case, results in landing short of the runway. If an electronic glide slope or visual approach slope indicator (VASI) is available, it should be used. If navigation aids (NAVAIDs) are unavailable, use the flight instruments to assist in maintaining orientation and a normal approach. Anytime position in relation to the runway or altitude is in doubt, execute a go-around.

Bright runway and approach lighting systems, especially where few lights illuminate the surrounding terrain, may create the illusion of being lower or having less distance to the runway. In this situation, the tendency is to fly a higher approach. Also, flying over terrain with only a few lights makes the runway recede or appear farther away. With this situation, the tendency is to fly a lower-than-normal approach. If the runway has a city in the distance on higher terrain, the tendency is to fly a lower-than-normal approach. A good review of the airfield layout and boundaries before initiating any approach helps maintain a safe approach angle.

Illusions created by runway lights result in a variety of problems. Bright lights or bold colors advance the runway, making it appear closer. Night landings are further complicated by the difficulty of judging distance and the possibility of confusing approach and runway lights. For example, when a double row of approach lights joins the boundary lights of the runway, there can be confusion where the approach lights terminate and runway lights begin. Under certain conditions, approach lights can make the aircraft seem higher in a turn to final, than when its wings are level.

Pilot Equipment

Before beginning a night flight, carefully consider personal equipment that should be readily available during the flight to include a flashlight, aeronautical charts and pertinent data for the flight, and a flightdeck checklist containing procedures for the following tasks, which can be found in 14 CFR part 91:

- Before starting engines
- Before takeoff
- Cruise
- Before landing
- After landing
- Stopping engines
- Emergencies

At least one reliable flashlight is recommended as standard equipment on all night flights. A reliable incandescent or light-emitting diode (LED) flashlight able to produce white/red light and blue for chart reading is preferable. The flash light should be large enough to be easily located in the event it is needed. The white light is used while performing the preflight visual inspection of the airplane, and the red light is used when performing cockpit operations. It is also recommended to have a spare set of batteries for the flashlight readily available.

Since the red light is non-glaring, it will not impair night vision. Some pilots prefer two flashlights, one with a white light for preflight and the other a penlight type with a red light. The latter can be suspended by a string from around the neck to ensure the light is always readily available. One word of caution: if a red light is used for reading an aeronautical chart, the red features of the chart will not show up.

Aeronautical charts are essential for night cross-country flight and, if the intended course is near the edge of the chart, the adjacent chart should also be available. The lights of cities and towns can be seen at surprising distances at night, and if this adjacent chart is not available to identify those landmarks, confusion could result. These checklist items are not just for night flying, they are required for day light flying also. Regardless of the equipment used, organization of the flightdeck eases the burden and enhances safety. Organize equipment and charts and place them within easy reach prior to taxiing.

Airplane Equipment and Lighting

Title 14 of the Code of Federal Regulations (14 CFR) part 91 specifies the basic minimum airplane equipment that is required for night flight. This equipment includes only basic instruments, lights, electrical energy source, and spare fuses.

The standard instruments required by 14 CFR part 91 for instrument flight are a valuable asset for aircraft control at night. Title 14 CFR part 91 specifies that during the period from sunset to sunrise operating aircraft are required to have a functioning anti-collision light system, including a flashing or rotating beacon and position lights. The anti-collision lights however need not be lighted when the pilot in command (PIC) determines that, because of operating conditions, it would be in the interest of safety to turn the lights off. Airplane position lights are arranged similar to those of boats and ships. A red light is positioned on the left wingtip, a green light on the right wingtip, and a white light on the tail. *[Figure 10-2]*

This arrangement provides a means to determine the general direction of movement of other airplanes in flight. If both a red and green light of another aircraft are observed, and the red light is on the left and the green to the right, the airplane is flying the same direction. Care must be taken not to overtake the other aircraft and maintain clearance. If red were on the right and green to the left, the airplane could be on a collision course.

Figure 10-2. *Position lights.*

Landing lights are not only useful for taxi, takeoffs, and landings, but also provide a means by which airplanes can be seen at night by other pilots. Pilots are encouraged to turn on their landing lights when operating within 10 miles of an airport and below 10,000 feet. Operation lights on applies to both day and night or in conditions of reduced visibility. This should also be done in areas where flocks of birds may be expected.

Although turning on aircraft lights supports the "see and be seen" concept, do not become complacent about keeping a sharp lookout for other aircraft. Most aircraft lights blend in with the stars or the lights of the cities at night and go unnoticed unless a conscious effort is made to distinguish them from other lights.

Airport and Navigation Lighting Aids

The lighting systems used for airports, runways, obstructions, and other visual aids at night are other important aspects of night flying. Lighted airports located away from congested areas are identified readily at night by the lights outlining the runways. Airports located near or within large cities are often difficult to identify as the airport lights tend to blend with the city lights. It is important not to only know the exact location of an airport relative to the city, but also to be able to identify these airports by the characteristics of their lighting pattern.

Aeronautical lights are designed and installed in a variety of colors and configurations, each having its own purpose. Although some lights are used only during low ceiling and visibility conditions, this discussion includes only the lights that are fundamental to visual flight rules (VFR) night operation.

It is recommended that prior to a night flight, and particularly a cross-country night flight, that a check of the availability and status of lighting systems at the destination airport is made. This information can be found on aeronautical charts and in the Chart Supplements. The status of each facility can be determined by reviewing pertinent Notices to Airmen (NOTAMs).

Most airports have rotating beacons. The beacon rotates at a constant speed, thus producing a series of light flashes at regular intervals. These flashes may consist of a white flash and one or two different colors that are used to identify various types of landing areas. For example:

- Lighted civilian land airports—alternating white and green lights

- Lighted civilian water airports—alternating white and yellow lights

- Lighted military airports—alternating white and green lights, but are differentiated from civil airports by dual peaked (two quick) white flashes, then green

Beacons producing red flashes indicate obstructions or areas considered hazardous to aerial navigation. Steady-burning red lights are used to mark obstructions on or near airports and sometimes to supplement flashing lights on en route obstructions. High-intensity, flashing white lights are used to mark some supporting structures of overhead transmission lines that stretch across rivers, chasms, and gorges. These high-intensity lights are also used to identify tall structures, such as chimneys and towers.

As a result of technological advancements, runway lighting systems have become quite sophisticated to accommodate takeoffs and landings in various weather conditions. However, if flying is limited to VFR only, it is important to be familiar with the basic lighting of runways and taxiways.

The basic runway lighting system consists of two straight parallel lines of runway edge lights defining the lateral limits of the runway. These lights are aviation white, although aviation yellow may be substituted for a distance of 2,000 feet from the far end of the runway to indicate a caution zone. At some airports, the intensity of the runway edge lights can be activated and adjusted by radio control. The control system consists of a 3-step control responsive to 7, 5, and/or 3 microphone clicks. This 3-step control turns on lighting facilities capable of either 3-step, 2-step, or 1-step operation. The 3-step and 2-step lighting facilities can be altered in intensity, while the 1-step cannot. All lighting is illuminated for a period of 15 minutes from the most recent time of activation and may not be extinguished prior to end of the 15-minute period. Suggested use is to always initially key the mike 7 times; this assures that all controlled lights are turned on to the maximum available intensity. If desired, adjustment can then be made, where the capability is provided, to a lower intensity by keying 5 and/or 3 times. Due to the close proximity of airports using the same frequency, radio-controlled lighting receivers may be set at a low sensitivity requiring the aircraft to be relatively close to activate the system. Consequently, even when lights are on, always key the mike as directed when overflying an airport of intended landing or just prior to entering the final segment of an approach. This assures the aircraft is close enough to activate the system and a full 15-minute lighting duration is available.

The length limits of the runway are defined by straight lines of lights across the runway ends. At some airports, the runway threshold lights are aviation green, and the runway end lights are aviation red. At many airports, the taxiways are also lighted. A taxiway edge lighting system consists of blue lights that outline the usable limits of taxi paths.

Training for Night Flight

Learning to safely fly at night takes time and your proficiency will improve with experience. Pilot's should practice the following maneuvers at night and acquire competency in straight-and-level flight, climbs and descents, level turns, climbing and descending turns, and steep turns. Practicing recovery from unusual attitudes should only be done with a flight instructor. Practice these maneuvers with all the flightdeck lights turned OFF, as well as ON. This blackout training simulates an electrical or instrument light failure. Include using the navigation equipment and local NAVAIDs during the training. In spite of fewer references or checkpoints, night cross-country flights do not present particular problems if pre-planning is adequate. Continuously monitor position, time estimates, and fuel consumed. Use NAVAIDs, if available, to assist in monitoring en route progress.

Preparation and Prefligh

Night flying requires that pilots are aware of, and operate within, their abilities and limitations. Although careful planning of any flight is essential, night flying demands more attention to the details of preflight preparation and planning.

Preparation for a night flight includes a thorough review of the available weather reports and forecasts with particular attention given to temperature/dew point spread. A narrow temperature/dew point spread may indicate the possibility of fog. Emphasis should also be placed on wind direction and speed, since its effect on the airplane cannot be as easily detected at night as during the day.

On night cross-country flights, select and use appropriate aeronautical charts to include the appropriate adjacent charts. Course lines should be drawn in black to be more distinguishable in low-light conditions. Note prominently lighted checkpoints along the prepared course. Rotating beacons at airports, lighted obstructions, lights of cities or towns, and lights from major highway traffic all provide excellent visual checkpoints. If a global positioning system (GPS) is being used for navigation, ensure that it is working properly before the flight. All necessary waypoints should be loaded properly before the flight and the database should be checked for accuracy prior to taking off and then checked again once in flight. The use of radio navigation aids and communication facilities add significantly to the safety and efficiency of night flying.

Check all personal equipment prior to flight to ensure proper functioning and operation. All airplane lights should be checked for operation by turning them on momentarily during the preflight inspection. Position lights can be checked for loose connections by tapping the light fixture. If the lights blink while being tapped, determine the cause prior to flight. Parking ramps should be checked with a flashlight prior to entering the airplane. During the day, it is quite easy to see stepladders, chuckholes, wheel chocks, and other obstructions, but at night, it is more difficult and a check of the area can prevent taxiing mishaps.

Starting, Taxiing, and Runup

Once seated in the airplane and prior to starting the engine, arrange all items and materials to be used during the flight so they will be readily available and convenient to use. Take extra caution at night to assure the propeller area is clear. Turning the rotating beacon ON, or flashing the airplane position lights serves to alert persons nearby to remain clear of the propeller. To avoid excessive drain of electrical current from the battery, it is recommended that unnecessary electrical equipment be turned OFF until after the engine has been started.

After starting the engine and when ready to taxi, turn the taxi or landing light ON. Be aware that continuous use of the landing light with revolutions per minute (rpm) power settings normally used for taxiing may place an excessive drain on the airplane's electrical system. Also, overheating of the landing light is possible because of inadequate airflow to carry the heat away. Use landing lights only as necessary while taxiing. When using lights, consideration should be given to not blinding other pilots. Taxi slowly, particularly in congested areas. If taxi lines are painted on the ramp or taxiway, follow the lines to ensure a proper path along the route.

Use the checklist for the before takeoff and run-up checks and procedures. During the day, forward movement of the airplane can be detected easily. At night, the airplane could creep forward without being noticed unless the pilot is alert for this possibility. Hold or lock the brakes during the run-up and be alert for any forward movement. An instrument check should be done while taxiing to check for proper and correct operation prior to takeoff.

Takeoff and Climb

Night flying is very different from day flying and demands more attention of the pilot. The most noticeable difference is the limited availability of outside visual references. Therefore, flight instruments should be used to a greater degree in controlling the airplane. This is particularly true on night takeoffs and climbs. Adjust the flightdeck lights to a minimum brightness that allow reading the instruments and switches but not hinder outside vision. This also eliminates light reflections on the windshield and windows.

After ensuring that the final approach and runway are clear of other air traffic, or when cleared for takeoff by the air traffic controller, turn the landing and taxi lights ON and line the airplane up with the centerline of the runway. If the runway does not have centerline lighting, use the painted centerline and the runway edge lights. After the airplane is aligned, note the heading indicator and set to correspond to the known runway direction. To begin the takeoff, release the brakes and advance the throttle smoothly to maximum allowable power. As the airplane accelerates, it should be kept moving straight ahead between and parallel to the runway edge lights.

The procedure for night takeoffs is the same as for normal daytime takeoffs except that many of the runway visual cues are not available. Check the flight instruments frequently during the takeoff to ensure the proper pitch attitude, heading, and airspeed are being attained. As the airspeed reaches the normal lift-off speed, adjust the pitch attitude to establish a normal climb. Accomplish this by referring to both outside visual references, such as lights, and to the flight instruments. *[Figure 10-3]*

Figure 10-3. *Establish a positive climb.*

After becoming airborne, the darkness of night often makes it difficult to note whether the airplane is getting closer to or farther from the surface. To ensure the airplane continues in a positive climb, be sure a climb is indicated on the attitude indicator, vertical speed indicator (VSI), and altimeter. It is also important to ensure the airspeed is at best climb speed.

Make necessary pitch and bank adjustments by referencing the attitude and heading indicators. It is recommended that turns not be made until reaching a safe maneuvering altitude. Although the use of the landing lights is helpful during the takeoff, they become ineffective after the airplane has climbed to an altitude where the light beam no longer extends to the surface. The light can cause distortion when it is reflected by haze, smoke, or clouds that might exist in the climb. Therefore, when the landing light is used for the takeoff, turn it off after the climb is well established provided it is not being used for collision avoidance.

Orientation and Navigation

Generally, at night, it is difficult to see clouds and restrictions to visibility, particularly on dark nights or under overcast. When flying under VFR, pilots must exercise caution to avoid flying into clouds. Usually, the first indication of flying into restricted visibility conditions is the gradual disappearance of lights on the ground. If the lights begin to take on an appearance of being surrounded by a halo or glow, use caution in attempting further flight in that same direction. Such a halo or glow around lights on the ground is indicative of ground fog. Remember that if a descent must be made through clouds, smoke, or haze in order to land, the horizontal visibility is considerably less when looking through the restriction than it is when looking straight down through it from above. Under no circumstances should a VFR night flight be made during poor or marginal weather conditions unless both the pilot and aircraft are certificated and equipped for flight under instrument flight rules (IFR).

Crossing large bodies of water at night in single-engine airplanes could be potentially hazardous, in the event of an engine failure, the pilot may not have any option than to land (ditch) the airplane in the water. Another hazard faced by pilots of all aircraft, due to limited or no lighting, is that the horizon blends with the water. During poor visibility conditions over water, the horizon becomes obscure and may result in a loss of orientation. Even on clear nights, the stars may be reflected on the water surface, which could appear as a continuous array of lights, thus making the horizon difficult to identify.

Lighted runways, buildings, or other objects may cause illusions to the pilot when seen from different altitudes. At an altitude of 2,000 feet, a group of lights on an object may be seen individually, while at 5,000 feet or higher, the same lights could appear to be one solid light mass. These illusions may become quite acute with altitude changes and, if not overcome, could present problems in respect to approaches to lighted runways.

Approaches and Landings

When approaching the airport to enter the traffic pattern and land, it is important that the runway lights and other airport lighting be identified as early as possible. If the airport layout is unfamiliar, sighting of the runway may be difficult until very close-in due to the maze of lights observed in the area. [Figure 10-4] Fly toward the rotating beacon until the lights outlining the runway are distinguishable. To fly a traffic pattern of proper size and direction, the runway threshold and runway-edge lights must be positively identified. Once the airport lights are seen, these lights should be kept in sight throughout the approach.

Distance may be deceptive at night due to limited lighting conditions. A lack of intervening references on the ground and the inability to compare the size and location of different ground objects cause this. This also applies to the estimation of altitude and speed. Consequently, more dependence must be placed on flight instruments, particularly the altimeter and the airspeed indicator. When entering the traffic pattern, always give yourself plenty of time to complete the before landing checklist. If the heading indicator contains a heading bug, setting it to the runway heading is an excellent reference for the pattern legs.

Maintain the recommended airspeeds and execute the approach and landing in the same manner as during the day. A low, shallow approach is definitely inappropriate during a night operation. The altimeter and VSI should be constantly cross-checked against the airplane's position along the base leg and final approach. A visual approach slope indicator (VASI) is an indispensable aid in establishing and maintaining a proper glide path. [Figure 10-5]

After turning onto the final approach and aligning the airplane midway between the two rows of runway-edge lights, note and correct for any wind drift. Throughout the final approach, use pitch and power to maintain a stabilized approach. Flaps are used the same as in a normal approach. Usually, halfway through the final approach, the landing light is turned on. Earlier use of the landing light may be necessary because of "Operation Lights ON" or for local traffic considerations. The landing light is sometimes ineffective since the light beam will usually not reach the ground from higher altitudes. The light may even be reflected back into the pilot's eyes by any existing haze, smoke, or fog. This disadvantage is overshadowed by the safety considerations provided by using the "Operation Lights ON" procedure around other traffic.

The round out and touchdown is made in the same manner as in day landings. At night, the judgment of height, speed, and sink rate is impaired by the scarcity of observable objects in the landing area. An inexperienced pilot may have a tendency

Figure 10-4. *Use light patterns for orientation.*

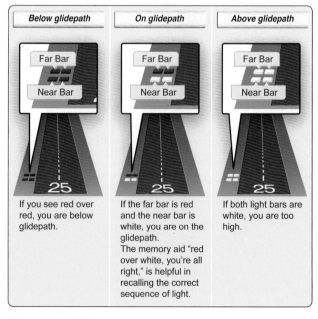

Figure 10-5. *VASI.*

to round out too high until attaining familiarity with the proper height for the correct round out. To aid in determining the proper round out point, continue a constant approach descent until the landing lights reflect on the runway and tire marks on the runway can be seen clearly. At this point, the round out is started smoothly and the throttle gradually reduced to idle as the airplane is touching down. *[Figure 10-6]* During landings without the use of landing lights, the round out may be started when the runway lights at the far end of the runway first appear to be rising higher than the nose of the airplane. This demands a smooth and very timely round out and requires that the pilot feel for the runway surface using power and pitch changes, as necessary, for the airplane to settle slowly to the runway. Blackout landings should always be included in night pilot training as an emergency procedure.

Figure 10-6. *Roundout when tire marks are visible.*

Night Emergencies

Perhaps the greatest concern about flying a single-engine airplane at night is the possibility of a complete engine failure and the subsequent emergency landing. This is a legitimate concern, even though continuing flight into adverse weather and poor pilot judgment account for most serious accidents.

If the engine fails at night, there are several important procedures and considerations to keep in mind. They are as follows:

- Maintain positive control of the airplane and establish the best glide configuration and airspeed. Turn the airplane towards an airport or away from congested areas.

- Check to determine the cause of the engine malfunction, such as the position of fuel selectors, magneto switch, or primer. If possible, the cause of the malfunction should be corrected immediately and the engine restarted.

- Announce the emergency situation to air traffic control (ATC) or Universal Communications (UNICOM). If already in radio contact with a facility, do not change frequencies unless instructed to change.

- If the condition of the nearby terrain is known and is suitable for a forced landing, turn towards an unlighted portion of the area and plan an emergency forced landing to an unlighted portion.

- Consider an emergency landing area close to public access if possible. This may facilitate rescue or help, if needed.

- Maintain orientation with the wind to avoid a downwind landing.

- Complete the before landing checklist, and check the landing lights for operation at altitude and turn ON in sufficient time to illuminate the terrain or obstacles along the flightpath. The landing should be completed in the normal landing attitude at the slowest possible airspeed. If the landing lights are unusable and outside visual references are not available, the airplane should be held in level-landing attitude until the ground is contacted.

- After landing, turn off all switches and evacuate the airplane as quickly as possible.

Chapter Summary

Night operations present additional risks that must be identified and assessed. Night flying operations should not be encouraged or attempted, except by pilots that are certificated, current, and proficient in night flying. Prior to attempting night operations, pilots should receive training and be familiar with the risks associated with night flight and how they differ from daylight operations. Even for experienced pilots, night VFR operations should only be conducted in unrestricted visibility, favorable winds, both on the surface and aloft, and no turbulence. Additional information on pilot vision and illusions can be found in FAA brochure AM-400-98/2 and also Chapters 2 and 17 of the Pilot's Handbook of Aeronautical Knowledge (FAA-H-8083-25A) at www.faa.gov. Additional information on lighting aids can be found in Chapter 2 of the Aeronautical Information Manual (AIM), which can be accessed at www.faa.gov.

Chapter 11
Transition to Complex Airplanes

Introduction

A high-performance airplane is defined as an airplane with an engine capable of developing more than 200 horsepower. A complex airplane is an airplane that has a retractable landing gear, flaps, and a controllable pitch propeller. In lieu of a controllable pitch propeller, the aircraft could also have an engine control system consisting of a digital computer and associated accessories for controlling the engine and the propeller. A seaplane would still be considered complex if it meets the description above except for having floats instead of a retractable landing gear system.

Figure 11-1. *Airfoil types.*

Transition to a complex airplane, or a high-performance airplane, can be demanding for most pilots without previous experience. Increased performance and complexity both require additional planning, judgment, and piloting skills. Transition to these types of airplanes, therefore, should be accomplished in a systematic manner through a structured course of training administered by a qualified flight instructor.

Airplanes can be designed to fly through a wide range of airspeeds. High speed flight requires smaller wing areas and moderately cambered airfoils whereas low speed flight is obtained with airfoils with a greater camber and larger wing area. *[Figure 11-1]* Many compromises are often made by designers to provide for higher speed cruise flight and low speeds for landing. Flaps are a common design effort to increase an airfoil's camber and the wing's surface area for lower speed flight. *[Figure 11-2]*

Since an airfoil cannot have two different cambers at the same time, one of two things must be done. Either the airfoil can be a compromise, or a cruise airfoil can be combined with a device for increasing the camber of the airfoil for low-speed flight. Camber is the asymmetry between the top and the bottom surfaces of an airfoil. One method for varying an airfoil's camber is the addition of trailing-edge flaps. Engineers call these devices a high-lift system.

Function of Flaps

Flaps work primarily by changing the camber of the airfoil which increases the wing's lift coefficient and with some flap designs the surface area of the wing is also increased. Flap deflection does not increase the critical (stall) angle of attack (AOA) and, in some cases, flap deflection actually decreases the critical AOA. Deflection of a wing's control surfaces, such as ailerons and flaps, alters both lift and drag. With aileron deflection, there is asymmetrical lift which imparts a rolling moment about the airplane's longitudinal axis. Wing flaps acts symmetrically about the longitudinal axis producing no rolling moment; however, both lift and drag increase as well as a pitching moment about the lateral axis. Lift is a function of several variables including air density, velocity, surface area, and lift coefficient. Since flaps increase an airfoil's lift coefficient, lift is increased. *[Figure 11-3]*

As flaps are deflected, the aircraft may pitch nose up, nose down or have minimal changes in pitch attitude. Pitching moment is caused by the rearward movement of the wing's center of pressure; however, that pitching behavior depends on several variables including flap type, wing position, downwash behavior, and horizontal tail location.

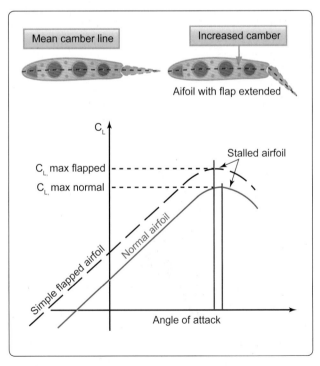

Figure 11-2. *Coefficient of lift comparison for flap extended and retracted positions.*

$$L = \frac{1}{2}pV^2SC_L$$

L = Lift produced

P = Air density

V = Velocity relative to the air

S = Surface area of the wing

C_L = lift coefficient which is determined by the camber of the airfoil used, the chord of the wing and AOA

Figure 11-3. *Lift equation.*

Consequently, pitch behavior depends on the design features of the particular airplane.

Flap deflection of up to 15° primarily produces lift with minimal increases in drag. Deflection beyond 15° produces a large increase in drag. Drag from flap deflection is parasite drag and, as such, is proportional to the square of the speed. Also, deflection beyond 15° produces a significant nose-up pitching moment in most high-wing airplanes because the resulting downwash increases the airflow over the horizontal tail.

Flap Effectiveness

Flap effectiveness depends on a number of factors, but the most noticeable are size and type. For the purpose of this chapter, trailing edge flaps are classified as four basic types: plain (hinge), split, slotted, and Fowler. *[Figure 11-4]*

The plain or hinge flap is a hinged section of the wing. The structure and function are comparable to the other control surfaces—ailerons, rudder, and elevator. The split flap is more complex. It is the lower or underside portion of the wing; deflection of the flap leaves the upper trailing edge of the wing undisturbed. It is, however, more effective than the hinge flap because of greater lift and less pitching moment, but there is more drag. Split flaps are more useful for landing, but the partially deflected hinge flaps have the advantage in takeoff. The split flap has significant drag at small deflections, whereas the hinge flap does not because airflow remains "attached" to the flap.

The slotted flap has a gap between the wing and the leading edge of the flap. The slot allows high-pressure airflow on the wing undersurface to energize the lower pressure over the top, thereby delaying flow separation. The slotted flap has greater lift than the hinge flap but less than the split flap; but, because of a higher lift-drag ratio, it gives better takeoff and climb performance. Small deflections of the slotted flap give a higher drag than the hinge flap but less than the split. This allows the slotted flap to be used for takeoff.

The Fowler flap deflects down and aft to increase the wing area. This flap can be multi-slotted making it the most complex of the trailing-edge systems. This system does, however, give the maximum lift coefficient. Drag characteristics at small

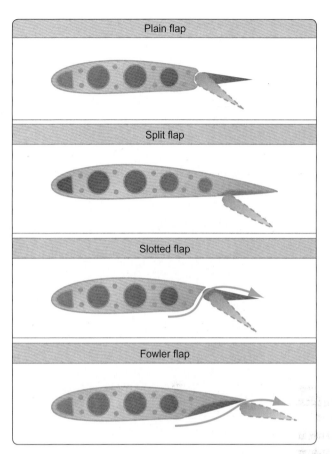

Plain flap

Split flap

Slotted flap

Fowler flap

Figure 11-4. *Four basic types of flaps.*

deflections are much like the slotted flap. Fowler flaps are most commonly used on larger airplanes because of their structural complexity and difficulty in sealing the slots.

Operational Procedures

It would be impossible to discuss all the many airplane design and flap combinations. This emphasizes the importance of the Federal Aviation Administration (FAA) approved Airplane Flight Manual and/or Pilot's Operating Handbook (AFM/POH) for a given airplane. While some AFM/POHs are specific as to operational use of flaps, others leave the use of flaps to pilot discretion. Hence, flap operation makes pilot judgment of critical importance. Since flap operation is used for landings and takeoffs, during which the airplane is in close proximity to the ground where the margin for error is small.

Since the recommendations given in the AFM/POH are based on the airplane and the flap design, the pilot must relate the manufacturer's recommendation to aerodynamic effects of flaps. This requires basic background knowledge of flap aerodynamics and geometry. With this information, a decision as to the degree of flap deflection and time of deflection based on runway and approach conditions relative to the wind conditions can be made.

The time of flap extension and degree of deflection are related. Large flap deflections at one single point in the landing pattern produce large lift changes that require significant pitch and power changes in order to maintain airspeed and glide slope. Incremental deflection of flaps on downwind, base, and final approach allow smaller adjustment of pitch and power compared to extension of full flaps all at one time. This procedure facilitates a more stabilized approach.

While all landings should be accomplished at the slowest speed possible for a given situation, a soft or short-field landing requires minimal speed at touchdown while a short field obstacle approach requires minimum speed and a steep approach angle. Flap extension, particularly beyond 30°, results in significant levels of drag. As such, large angles of flap deployment require higher power settings than used with partial flaps. When steep approach angles and short fields combine with power to offset the drag produced by the flaps, the landing flare becomes critical. The drag produces a high sink rate that must be controlled with power, yet failure to reduce power at a rate so that the power is idle at touchdown allows the airplane to float down the runway. A reduction in power too early can result in a hard landing and damage or loss of control.

Crosswind component is another factor to be considered in the degree of flap extension. The deflected flap presents a surface area for the wind to act on. With flaps extended in a crosswind, the wing on the upwind side is more affected than the downwind wing. The effect is reduced to a slight extent in the crabbed approach since the airplane is more nearly aligned with the wind. When using a wing-low approach, the lowered wing partially blocks the upwind flap. The dihedral of the wing combined with the flap and wind make lateral control more difficult. Lateral control becomes more difficult as flap extension reaches maximum and the crosswind becomes perpendicular to the runway.

With flaps extended, the crosswind effects on the wing become more pronounced as the airplane comes closer to the ground. The wing, flap, and ground form a "container" that is filled with air by the crosswind. Since the flap is located behind the main landing gear when the wind strikes the deflected flap and fuselage side, the upwind wing tends to rise and the airplane tends to turn into the wind. Proper control position is essential for maintaining runway alignment. Depending on the amount of crosswind, it may be necessary to retract the flaps soon after touchdown in order to maintain control of the airplane.

The go-around is another factor to consider when making a decision about degree of flap deflection and about where in the landing pattern to extend flaps. Because of the nose down pitching moment produced with flap extension, trim is used to offset this pitching moment. Application of full power in the go-around increases the airflow over the wing. This produces additional lift causing significant changes in pitch. The pitch-up tendency does not diminish completely with flap retraction because of the trim setting. Expedient retraction of flaps is desirable to eliminate drag; however, the pilot must be prepared for rapid changes in pitch forces as the result of trim and the increase in airflow over the control surfaces. [Figure 11-5]

The degree of flap deflection combined with design configuration of the horizontal tail relative to the wing require carefully monitoring of pitch and airspeed, carefully control flap retraction to minimize altitude loss, and properly use the rudder for coordination. Considering these factors, it is good practice to extend the same degree of flap deflection at the same point in the landing pattern for each landing. This requires that a consistent traffic pattern be used. This allows for a preplanned go-around sequence based on the airplane's position in the landing pattern.

There is no single formula to determine the degree of flap deflection to be used on landing because a landing involves variables that are dependent on each other. The AFM/POH for the particular airplane contains the manufacturer's recommendations for some landing situations. On the other hand, AFM/POH information on flap usage for takeoff is more precise. The manufacturer's requirements are based on the climb performance produced by a given flap design. Under no circumstances should a flap setting given in the AFM/POH be exceeded for takeoff.

Controllable-Pitch Propeller

Fixed-pitch propellers are designed for best efficiency at one speed of rotation and forward speed. This type of propeller provides suitable performance in a narrow range of airspeeds; however, efficiency would suffer considerably outside this range. To provide high-propeller efficiency through a wide range of operation, the propeller blade angle must be controllable. The most effective way of controlling the propeller blade angle is by means of a constant-speed governing system.

Constant-Speed Propeller

The constant-speed propeller keeps the blade angle adjusted for maximum efficiency for most conditions of flight. The pilot controls the engine revolutions per minute (rpm) indirectly by means of a propeller control in the flightdeck, which is connected to a propeller governor. For maximum takeoff power, the propeller control is moved all the way forward to the low pitch/high rpm position, and the throttle is moved forward to the maximum allowable manifold pressure

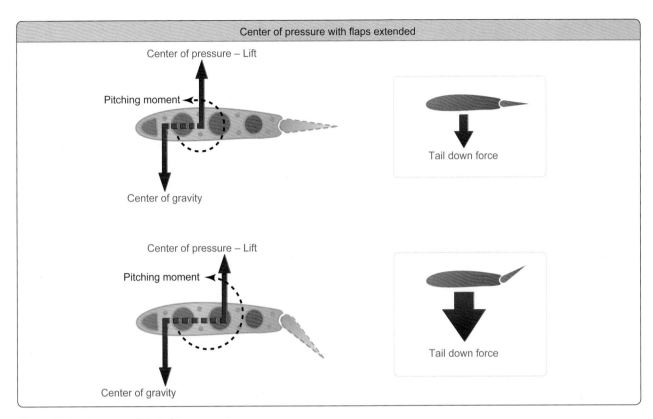

Figure 11-5. *Flaps extended pitching moment.*

position. *[Figure 11-6]* To reduce power for climb or cruise, manifold pressure is reduced to the desired value with the throttle, and the engine rpm is reduced by moving the propeller control back toward the high pitch/low rpm position until the desired rpm is observed on the tachometer. Pulling back on the propeller control causes the propeller blades to move to a higher angle. Increasing the propeller blade angle (of attack) results in an increase in the resistance of the air. This puts a load on the engine so it slows down. In other words, the resistance of the air at the higher blade angle is greater than

Figure 11-6. *Controllable pitch propeller pitch angles.*

the torque, or power, delivered to the propeller by the engine, so it slows down to a point where forces are in balance.

When an aircraft engine is running at constant speed, the torque (power) exerted by the engine at the propeller shaft must equal the opposing load provided by the resistance of the air. The rpm is controlled by regulating the torque absorbed by the propeller—in other words by increasing or decreasing the resistance offered by the air to the propeller. This is accomplished with a constant-speed propeller by means of a governor. The governor, in most cases, is geared to the engine crankshaft and thus is sensitive to changes in engine rpm.

When an airplane is nosed up into a climb from level flight, the engine tends to slow down. Since the governor is sensitive to small changes in engine rpm, it decreases the blade angle just enough to keep the engine speed from falling off. If the airplane is nosed down into a dive, the governor increases the blade angle enough to prevent the engine from over-speeding. This allows the engine to maintain a constant rpm thereby maintaining the power output. Changes in airspeed and power can be obtained by changing rpm at a constant manifold pressure; by changing the manifold pressure at a constant rpm; or by changing both rpm and manifold pressure. The constant-speed propeller makes it possible to obtain an infinite number of power settings.

Takeoff, Climb, and Cruise

During takeoff, when the forward motion of the airplane is at low speeds and when maximum power and thrust are required, the constant-speed propeller sets up a low propeller blade angle (pitch). The low blade angle keeps the AOA, with respect to the relative wind, small and efficient at the low speed. [*Figure 11-7*]

At the same time, it allows the propeller to handle a smaller mass of air per revolution. This light load allows the engine to turn at maximum rpm and develop maximum power. Although the mass of air per revolution is small, the number of rpm is high. Thrust is maximum at the beginning of the takeoff and then decreases as the airplane gains speed and the airplane drag increases. Due to the high slipstream velocity during takeoff, the effective lift of the wing behind the propeller(s) is increased.

As the airspeed increases after lift-off, the load on the engine is lightened because of the small blade angle. The governor senses this and increases the blade angle slightly. Again, the higher blade angle, with the higher speeds, keeps the AOA with respect to the relative wind small and efficient.

For climb after takeoff, the power output of the engine is reduced to climb power by decreasing the manifold pressure and lowering rpm by increasing the blade angle. At the higher (climb) airspeed and the higher blade angle, the propeller is handling a greater mass of air per second at a lower slipstream velocity. This reduction in power is offset by the increase in propeller efficiency. The AOA is again kept small by the increase in the blade angle with an increase in airspeed.

At cruising altitude, when the airplane is in level flight, less power is required to produce a higher airspeed than is used

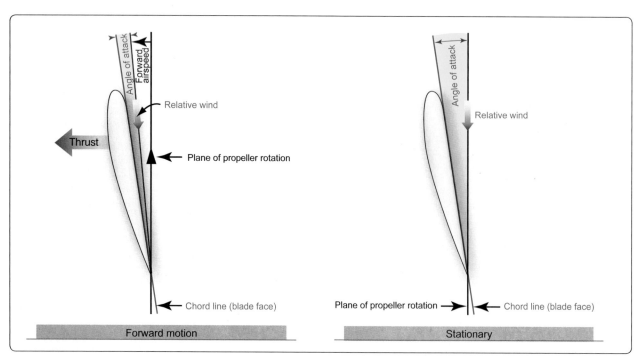

Figure 11-7. *Propeller blade angle.*

in climb. Consequently, engine power is again reduced by lowering the manifold pressure and increasing the blade angle (to decrease rpm). The higher airspeed and higher blade angle enable the propeller to handle a still greater mass of air per second at still smaller slipstream velocity. At normal cruising speeds, propeller efficiency is at or near maximum efficiency.

Blade Angle Control

Once the rpm settings for the propeller are selected, the propeller governor automatically adjusts the blade angle to maintain the selected rpm. It does this by using oil pressure. Generally, the oil pressure used for pitch change comes directly from the engine lubricating system. When a governor is employed, engine oil is used and the oil pressure is usually boosted by a pump that is integrated with the governor. The higher pressure provides a quicker blade angle change. The rpm at which the propeller is to operate is adjusted in the governor head. The pilot changes this setting by changing the position of the governor rack through the flightdeck propeller control.

On some constant-speed propellers, changes in pitch are obtained by the use of an inherent centrifugal twisting moment of the blades that tends to flatten the blades toward low pitch and oil pressure applied to a hydraulic piston connected to the propeller blades which moves them toward high pitch. Another type of constant-speed propeller uses counterweights attached to the blade shanks in the hub. Governor oil pressure and the blade twisting moment move the blades toward the low pitch position, and centrifugal force acting on the counterweights moves them (and the blades) toward the high pitch position. In the first case above, governor oil pressure moves the blades towards high pitch and in the second case, governor oil pressure and the blade twisting moment move the blades toward low pitch. A loss of governor oil pressure, therefore, affects each differently.

Governing Range

The blade angle range for constant-speed propellers varies from about 11½° to 40°. The higher the speed of the airplane, the greater the blade angle range. *[Figure 11-8]*

The range of possible blade angles is termed the propeller's governing range. The governing range is defined by the limits of the propeller blades travel between high and low blade angle pitch stops. As long as the propeller blade angle is within the governing range and not against either pitch stop, a constant engine rpm is maintained. However, once the propeller blade reaches its pitch-stop limit, the engine rpm increases or decreases with changes in airspeed and propeller load similar to a fixed-pitch propeller. For example, once a specific rpm is selected, if the airspeed decreases enough, the propeller blades reduce pitch in an attempt to maintain the selected rpm until they contact their low pitch stops. From that point, any further reduction in airspeed causes the engine rpm to decrease. Conversely, if the airspeed increases, the propeller blade angle increases until the high pitch stop is reached. The engine rpm then begins to increase.

Constant-Speed Propeller Operation

The engine is started with the propeller control in the low pitch/high rpm position. This position reduces the load or drag of the propeller and the result is easier starting and warm-up of the engine. During warm-up, the propeller blade changing mechanism is operated slowly and smoothly through a full cycle. This is done by moving the propeller control (with the manifold pressure set to produce about 1,600 rpm) to the high pitch/low rpm position, allowing the rpm to stabilize, and then moving the propeller control back to the low pitch takeoff position. This is done for two reasons: to determine whether the system is operating correctly and to circulate fresh warm oil through the propeller governor system. Remember the oil has been trapped in the propeller cylinder since the last time the engine was shut down. There is a certain amount of leakage from the propeller cylinder, and the oil tends to congeal, especially if the outside air temperature is low. Consequently, if the propeller is not exercised before takeoff, there is a possibility that the engine may overspeed on takeoff.

An airplane equipped with a constant-speed propeller has better takeoff performance than a similarly powered airplane equipped with a fixed-pitch propeller. This is because with a constant-speed propeller, an airplane can develop its maximum rated horsepower (red line on the tachometer) while motionless. An airplane with a fixed-pitch propeller, on the other hand, must accelerate down the runway to increase airspeed and aerodynamically unload the propeller so that

Aircraft Type	Design Speed (mph)	Blade Angle Range	Pitch	
			Low	High
Fixed gear	160	11½°	10½°	22°
Retractable	180	15°	11°	26°
Turbo retractable	225/240	20°	14°	34°
Turbine retractable	250/300	30°	10°	40°
Transport retractable	325	40°	10/15°	50/55°

Figure 11-8. *Blade angle range (values are approximate).*

rpm and horsepower can steadily build up to their maximum. With a constant-speed propeller, the tachometer reading should come up to within 40 rpm of the red line as soon as full power is applied and remain there for the entire takeoff. Excessive manifold pressure raises the cylinder combustion pressures, resulting in high stresses within the engine. Excessive pressure also produces high-engine temperatures. A combination of high manifold pressure and low rpm can induce damaging detonation. In order to avoid these situations, the following sequence should be followed when making power changes.

- When increasing power, increase the rpm first and then the manifold pressure

- When decreasing power, decrease the manifold pressure first and then decrease the rpm

The cruise power charts in the AFM/POH should be consulted when selecting cruise power settings. Whatever the combinations of rpm and manifold pressure listed in these charts—they have been flight tested and approved by engineers for the respective airframe and engine manufacturer. Therefore, if there are power settings, such as 2,100 rpm and 24 inches manifold pressure in the power chart, they are approved for use. With a constant-speed propeller, a power descent can be made without over-speeding the engine. The system compensates for the increased airspeed of the descent by increasing the propeller blade angles. If the descent is too rapid or is being made from a high altitude, the maximum blade angle limit of the blades is not sufficient to hold the rpm constant. When this occurs, the rpm is responsive to any change in throttle setting.

Although the governor responds quickly to any change in throttle setting, a sudden and large increase in the throttle setting causes a momentary over-speeding of the engine until the blades become adjusted to absorb the increased power. If an emergency demanding full power should arise during approach, the sudden advancing of the throttle causes momentary over-speeding of the engine beyond the rpm for which the governor is adjusted.

Some important points to remember concerning constant-speed propeller operation are:

- The red line on the tachometer not only indicates maximum allowable rpm; it also indicates the rpm required to obtain the engine's rated horsepower.

- A momentary propeller overspeed may occur when the throttle is advanced rapidly for takeoff. This is usually not serious if the rated rpm is not exceeded by 10 percent for more than 3 seconds.

- The green arc on the tachometer indicates the normal operating range. When developing power in this range, the engine drives the propeller. Below the green arc, however, it is usually the windmilling propeller that powers the engine. Prolonged operation below the green arc can be detrimental to the engine.

- On takeoffs from low elevation airports, the manifold pressure in inches of mercury may exceed the rpm. This is normal in most cases, but the pilot should always consult the AFM/POH for limitations.

- All power changes should be made smoothly and slowly to avoid over-boosting and/or over-speeding.

Turbocharging

The turbocharged engine allows the pilot to maintain sufficient cruise power at high altitudes where there is less drag, which means faster true airspeeds and increased range with fuel economy. At the same time, the powerplant has flexibility and can be flown at a low altitude without the increased fuel consumption of a turbine engine. When attached to the standard powerplant, the turbocharger does not take any horsepower from the engine to operate; it is relatively simple mechanically, and some models can pressurize the cabin as well.

The turbocharger is an exhaust-driven device that raises the pressure and density of the induction air delivered to the engine. It consists of two separate components: a compressor and a turbine connected by a common shaft. The compressor supplies pressurized air to the engine for high-altitude operation. The compressor and its housing are between the ambient air intake and the induction air manifold. The turbine and its housing are part of the exhaust system and utilize the flow of exhaust gases to drive the compressor. *[Figure 11-9]*

The turbine has the capability of producing manifold pressure in excess of the maximum allowable for the particular engine. In order not to exceed the maximum allowable manifold pressure, a bypass or waste gate is used so that some of the exhaust is diverted overboard before it passes through the turbine.

The position of the waste gate regulates the output of the turbine and therefore, the compressed air available to the engine. When the waste gate is closed, all of the exhaust gases pass through and drive the turbine. As the waste gate opens, some of the exhaust gases are routed around the turbine through the exhaust bypass and overboard through the exhaust pipe.

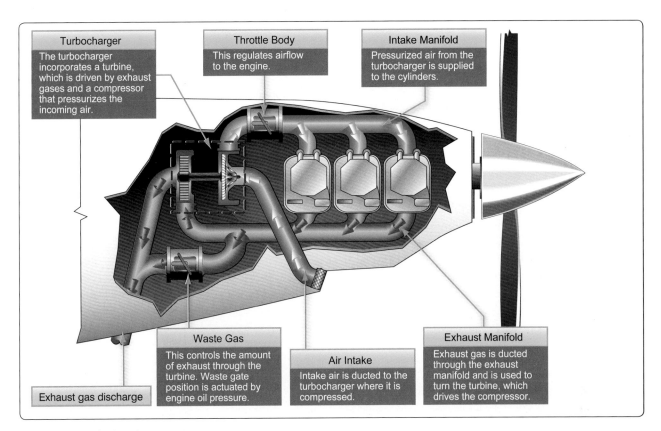

Figure 11-9. *Turbocharging system.*

The waste gate actuator is a spring-loaded piston operated by engine oil pressure. The actuator, which adjusts the waste gate position, is connected to the waste gate by a mechanical linkage.

The control center of the turbocharger system is the pressure controller. This device simplifies turbocharging to one control: the throttle. Once the desired manifold pressure is set, virtually no throttle adjustment is required with changes in altitude. The controller senses compressor discharge requirements for various altitudes and controls the oil pressure to the waste gate actuator, which adjusts the waste gate accordingly. Thus the turbocharger maintains only the manifold pressure called for by the throttle setting.

Ground Boosting Versus Altitude Turbocharging

Altitude turbocharging (sometimes called "normalizing") is accomplished by using a turbocharger that maintains maximum allowable sea level manifold pressure (normally 29–30 "Hg) up to a certain altitude. This altitude is specified by the airplane manufacturer and is referred to as the airplane's critical altitude. Above the critical altitude, the manifold pressure decreases as additional altitude is gained. Ground boosting, on the other hand, is an application of turbocharging where more than the standard 29 inches of manifold pressure is used in flight. In various airplanes using ground boosting, takeoff manifold pressures may go as high as 45 "Hg.

Although a sea level power setting and maximum rpm can be maintained up to the critical altitude, this does not mean that the engine is developing sea level power. Engine power is not determined just by manifold pressure and rpm. Induction air temperature is also a factor. Turbocharged induction air is heated by compression. This temperature rise decreases induction air density, which causes a power loss. Maintaining the equivalent horsepower output requires a somewhat higher manifold pressure at a given altitude than if the induction air were not compressed by turbocharging. If, on the other hand, the system incorporates an automatic density controller which, instead of maintaining a constant manifold pressure, automatically positions the waste gate so as to maintain constant air density to the engine, a near constant horsepower output results.

Operating Characteristics

First and foremost, all movements of the power controls on turbocharged engines should be slow and smooth. Aggressive and/or abrupt throttle movements increase the possibility of over-boosting. Carefully monitor engine indications when making power changes.

When the waste gate is open, the turbocharged engine reacts the same as a normally aspirated engine when the rpm is varied. That is, when the rpm is increased, the manifold pressure decreases slightly. When the engine rpm is decreased, the manifold pressure increases slightly. However, when the waste gate is closed, manifold pressure variation with engine rpm is just the opposite of the normally aspirated engine. An increase in engine rpm results in an increase in manifold pressure, and a decrease in engine rpm results in a decrease in manifold pressure.

Above the critical altitude, where the waste gate is closed, any change in airspeed results in a corresponding change in manifold pressure. This is true because the increase in ram air pressure with an increase in airspeed is magnified by the compressor resulting in an increase in manifold pressure. The increase in manifold pressure creates a higher mass flow through the engine, causing higher turbine speeds and thus further increasing manifold pressure.

When running at high altitudes, aviation gasoline may tend to vaporize prior to reaching the cylinder. If this occurs in the portion of the fuel system between the fuel tank and the engine-driven fuel pump, an auxiliary positive pressure pump may be needed in the tank. Since engine-driven pumps pull fuel, they are easily vapor locked. A boost pump provides positive pressure—pushes the fuel—reducing the tendency to vaporize.

Heat Management

Turbocharged engines must be thoughtfully and carefully operated with continuous monitoring of pressures and temperatures. There are two temperatures that are especially important—turbine inlet temperature (TIT) or, in some installations, exhaust gas temperature (EGT) and cylinder head temperature. TIT or EGT limits are set to protect the elements in the hot section of the turbocharger, while cylinder head temperature limits protect the engine's internal parts.

Due to the heat of compression of the induction air, a turbocharged engine runs at higher operating temperatures than a non-turbocharged engine. Because turbocharged engines operate at high altitudes; their environment is less efficient for cooling. At altitude, the air is less dense and, therefore, cools less efficiently. Also, the less dense air causes the compressor to work harder. Compressor turbine speeds can reach 80,000–100,000 rpm, adding to the overall engine operating temperatures. Turbocharged engines are also operated at higher power settings a greater portion of the time.

High heat is detrimental to piston engine operation. Its cumulative effects can lead to piston, ring, and cylinder head failure and place thermal stress on other operating components. Excessive cylinder head temperature can lead to

detonation, which in turn can cause catastrophic engine failure. Turbocharged engines are especially heat sensitive. The key to turbocharger operation is effective heat management.

Monitor the condition of a turbocharged engine with manifold pressure gauge, tachometer, exhaust gas temperature/turbine inlet temperature gauge, and cylinder head temperature. Manage the "heat system" with the throttle, propeller rpm, mixture, and cowl flaps. At any given cruise power, the mixture is the most influential control over the exhaust gas/TIT. The throttle regulates total fuel flow, but the mixture governs the fuel to air ratio. The mixture, therefore, controls temperature.

Exceeding temperature limits in an after takeoff climb is usually not a problem since a full rich mixture cools with excess fuel. At cruise, power is normally reduced and mixture adjusted accordingly. Under cruise conditions, monitor temperature limits closely because that is when the temperatures are most likely to reach the maximum, even though the engine is producing less power. Overheating in an en route climb, however, may require fully open cowl flaps and a higher airspeed.

Since turbocharged engines operate hotter at altitude than do normally aspirated engines, they are more prone to damage from cooling stress. Gradual reductions in power and careful monitoring of temperatures are essential in the descent phase. Extending the landing gear during the descent may help control the airspeed while maintaining a higher engine power setting. This allows the pilot to reduce power in small increments which allows the engine to cool slowly. It may also be necessary to lean the mixture slightly to eliminate roughness at the lower power settings.

Turbocharger Failure

Because of the high temperatures and pressures produced in the turbine exhaust systems, any malfunction of the turbocharger must be treated with extreme caution. In all cases of turbocharger operation, the manufacturer's recommended procedures should be followed. This is especially so in the case of turbocharger malfunction. However, in those instances where the manufacturer's procedures do not adequately describe the actions to be taken in the event of a turbocharger failure, the following procedures should be used.

Over-Boost Condition

If an excessive rise in manifold pressure occurs during normal advancement of the throttle (possibly owing to faulty operation of the waste gate):

- Immediately retard the throttle smoothly to limit the manifold pressure below the maximum for the rpm and mixture setting

- Operate the engine in such a manner as to avoid a further over-boost condition

Low Manifold Pressure

Although this condition may be caused by a minor fault, it is quite possible that a serious exhaust leak has occurred creating a potentially hazardous situation:

- Shut down the engine in accordance with the recommended engine failure procedures, unless a greater emergency exists that warrants continued engine operation.

- If continuing to operate the engine, use the lowest power setting demanded by the situation and land as soon as practicable.

It is very important to ensure that corrective maintenance is undertaken following any turbocharger malfunction.

Retractable Landing Gear

The primary benefits of being able to retract the landing gear are increased climb performance and higher cruise airspeeds due to the resulting decrease in drag. Retractable landing gear systems may be operated either hydraulically or electrically or may employ a combination of the two systems. Warning indicators are provided in the flightdeck to show the pilot when the wheels are down and locked and when they are up and locked or if they are in intermediate positions. Systems for emergency operation are also provided. The complexity of the retractable landing gear system requires that specific operating procedures be adhered to and that certain operating limitations not be exceeded.

Landing Gear Systems

An electrical landing gear retraction system utilizes an electrically-driven motor for gear operation. The system is basically an electrically-driven jack for raising and lowering the gear. When a switch in the flightdeck is moved to the UP position, the electric motor operates. Through a system of shafts, gears, adapters, an actuator screw, and a torque tube, a force is transmitted to the drag strut linkages. Thus, the gear retracts and locks. Struts are also activated that open and close the gear doors. If the switch is moved to the DOWN position, the motor reverses and the gear moves down and locks. Once activated, the gear motor continues to operate until an up or down limit switch on the motor's gearbox is tripped.

A hydraulic landing gear retraction system utilizes pressurized hydraulic fluid to actuate linkages to raise and lower the gear. When a switch in the flightdeck is moved to the UP position, hydraulic fluid is directed into the gear up line. The fluid flows through sequenced valves and down locks to the gear actuating cylinders. A similar process occurs during gear extension. The pump that pressurizes the fluid in the system can be either engine driven or electrically powered. If an electrically-powered pump is used to pressurize the fluid, the system is referred to as an electrohydraulic system. The system also incorporates a hydraulic reservoir to contain excess fluid and to provide a means of determining system fluid level.

Regardless of its power source, the hydraulic pump is designed to operate within a specific range. When a sensor detects excessive pressure, a relief valve within the pump opens, and hydraulic pressure is routed back to the reservoir. Another type of relief valve prevents excessive pressure that may result from thermal expansion. Hydraulic pressure is also regulated by limit switches. Each gear has two limits switches—one dedicated to extension and one dedicated to retraction. These switches de-energize the hydraulic pump after the landing gear has completed its gear cycle. In the event of limit switch failure, a backup pressure relief valve activates to relieve excess system pressure.

Controls and Position Indicators

Landing gear position is controlled by a switch on the flightdeck panel. In most airplanes, the gear switch is shaped like a wheel in order to facilitate positive identification and to differentiate it from other flightdeck controls.

Landing gear position indicators vary with different make and model airplanes. However, the most common types of landing gear position indicators utilize a group of lights. One type consists of a group of three green lights, which illuminate when the landing gear is down and locked. *[Figure 11-10]* Another type consists of one green light to indicate when the landing gear is down and an amber light to indicate when the gear is up. *[Figure 11-11]* Still other systems incorporate a red or amber light to indicate when the gear is in transit or unsafe for landing. *[Figure 11-12]* The lights are usually of the "press to test" type, and the bulbs are interchangeable. *[Figure 11-10]*

Other types of landing gear position indicators consist of tab-type indicators with markings "UP" to indicate the gear is up and locked, a display of red and white diagonal stripes to show when the gear is unlocked, or a silhouette of each gear to indicate when it locks in the DOWN position.

Landing Gear Safety Devices

Most airplanes with a retractable landing gear have a gear warning horn that sounds when the airplane is configured for landing and the landing gear is not down and locked. Normally, the horn is linked to the throttle or flap position and/or the airspeed indicator so that when the airplane is below a certain airspeed, configuration, or power setting with the gear retracted, the warning horn sounds.

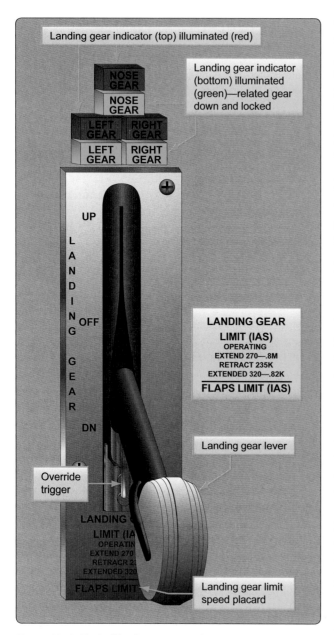

Figure 11-10. *Typical landing gear switch with three light indicator.*

Labels in figure:
- Landing gear indicator (top) illuminated (red)
- Landing gear indicator (bottom) illuminated (green)—related gear down and locked
- NOSE GEAR
- LEFT GEAR / RIGHT GEAR
- UP
- LANDING GEAR
- OFF
- DN
- Override trigger
- LANDING GEAR LEVER
- Landing gear lever
- Landing gear limit speed placard
- LANDING GEAR LIMIT (IAS) OPERATING EXTEND 270—.8M RETRACT 235K EXTENDED 320—.82K FLAPS LIMIT (IAS)

this way, if the landing gear switch in the flightdeck is placed in the RETRACT position when weight is on the gear, the gear remains extended, and the warning horn may sound as an alert to the unsafe condition. Once the weight is off the gear, however, such as on takeoff, the safety switch releases and the gear retracts.

Many airplanes are equipped with additional safety devices to prevent collapse of the gear when the airplane is on the ground. These devices are called ground locks. One common type is a pin installed in aligned holes drilled in two or more units of the landing gear support structure. Another type is a spring-loaded clip designed to fit around and hold two or more units of the support structure together. All types of ground locks usually have red streamers permanently attached to them to readily indicate whether or not they are installed.

Emergency Gear Extension Systems
The emergency extension system lowers the landing gear if the main power system fails. Some airplanes have an emergency release handle in the flightdeck, which is connected through a mechanical linkage to the gear up locks. When the handle is operated, it releases the up locks and allows the gear to free fall or extend under their own weight. *[Figure 11-13]*

On other airplanes, release of the up lock is accomplished using compressed gas, which is directed to up lock release cylinders. In some airplanes, design configurations make emergency extension of the landing gear by gravity and air loads alone impossible or impractical. In these airplanes, provisions are included for forceful gear extension in an emergency. Some installations are designed so that either hydraulic fluid or compressed gas provides the necessary pressure, while others use a manual system, such as a hand crank for emergency gear extension. *[Figure 11-14]* Hydraulic pressure for emergency operation of the landing gear may be provided by an auxiliary hand pump, an accumulator, or an electrically-powered hydraulic pump depending on the design of the airplane.

Operational Procedures
Preflight
Because of their complexity, retractable landing gear demands a close inspection prior to every flight. The inspection should begin inside the flightdeck. First, make certain that the landing gear selector switch is in the GEAR DOWN position. Then, turn on the battery master switch and ensure that the landing gear position indicators show that the gear is DOWN and locked.

External inspection of the landing gear consists of checking individual system components. *[Figure 11-14]* The landing gear, wheel well, and adjacent areas should be clean and free

Accidental retraction of a landing gear may be prevented by such devices as mechanical down locks, safety switches, and ground locks. Mechanical down locks are built-in components of a gear retraction system and are operated automatically by the gear retraction system. To prevent accidental operation of the down locks and inadvertent landing gear retraction while the airplane is on the ground, electrically-operated safety switches are installed.

A landing gear safety switch, sometimes referred to as a squat switch, is usually mounted in a bracket on one the main gear shock struts. *[Figure 11-12]* When the strut is compressed by the weight of the airplane, the switch opens the electrical circuit to the motor or mechanism that powers retraction. In

Figure 11-11. *Landing gear handles and single and multiple light indictor.*

of mud and debris. Dirty switches and valves may cause false safe light indications or interrupt the extension cycle before the landing gear is completely down and locked. The wheel wells should be clear of any obstructions, as foreign objects may damage the gear or interfere with its operation. Bent gear doors may be an indication of possible problems with normal gear operation.

Ensure shock struts are properly inflated and that the pistons are clean. Check main gear and nose gear up lock and down lock mechanisms for general condition. Power sources and retracting mechanisms are checked for general condition, obvious defects, and security of attachment. Check hydraulic lines for signs of chafing and leakage at attach points. Warning system micro switches (squat switches) are checked for cleanliness and security of attachment. Actuating cylinders, sprockets, universal joints, drive gears, linkages, and any other accessible components are checked for condition and obvious defects. The airplane structure to which the landing gear is attached is checked for distortion, cracks, and general condition. All bolts and rivets should be intact and secure.

Takeoff and Climb

Normally, the landing gear is retracted after lift-off when the airplane has reached an altitude where, in the event of an engine failure or other emergency requiring an aborted takeoff,

Figure 11-12. *Landing gear safety switch.*

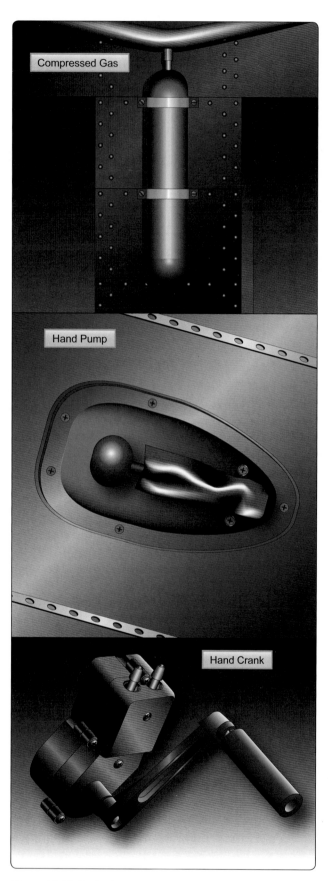

Figure 11-13. *Typical emergency gear extension systems.*

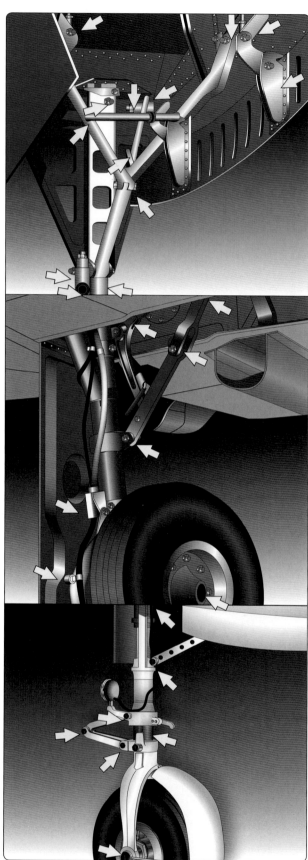

Figure 11-14. *Retractable landing gear inspection checkpoints.*

the airplane could no longer be landed on the runway. This procedure, however, may not apply to all situations. Preplan landing gear retraction taking into account the following:

- Length of the runway
- Climb gradient
- Obstacle clearance requirements
- The characteristics of the terrain beyond the departure end of the runway
- The climb characteristics of the particular airplane.

For example, in some situations it may be preferable, in the event of an engine failure, to make an off airport forced landing with the gear extended in order to take advantage of the energy absorbing qualities of the terrain (see Chapter 19, "Emergency Procedures"). In which case, a delay in retracting the landing gear after takeoff from a short runway may be warranted. In other situations, obstacles in the climb path may warrant a timely gear retraction after takeoff. Also, in some airplanes the initial climb pitch attitude is such that any view of the runway remaining is blocked, making an assessment of the feasibility of touching down on the remaining runway difficult.

Avoid premature landing gear retraction and do not retract the landing gear until a positive rate of climb is indicated on the flight instruments. If the airplane has not attained a positive rate of climb, there is always the chance it may settle back onto the runway with the gear retracted. This is especially so in cases of premature lift-off. Remember that leaning forward to reach the landing gear selector may result in inadvertent forward pressure on the yoke, which causes the airplane to descend.

As the landing gear retracts, airspeed increases and the airplane's pitch attitude may change. The gear may take several seconds to retract. Gear retraction and locking (and gear extension and locking) is accompanied by sound and feel that are unique to the specific make and model airplane. Become familiar with the sound and feel of normal gear retraction so that any abnormal gear operation can be readily recognized. Abnormal landing gear retraction is most often a clear sign that the gear extension cycle will also be abnormal.

Approach and Landing

The operating loads placed on the landing gear at higher airspeeds may cause structural damage due to the forces of the airstream. Limiting speeds, therefore, are established for gear operation to protect the gear components from becoming overstressed during flight. These speeds may not be found on the airspeed indicator.

They are published in the AFM/POH for the particular airplane and are usually listed on placards in the flightdeck. [Figure 11-15] The maximum landing extended speed (V_{LE}) is the maximum speed at which the airplane can be flown with the landing gear extended. The maximum landing gear operating speed (V_{LO}) is the maximum speed at which the landing gear may be operated through its cycle.

The landing gear is extended by placing the gear selector switch in the GEAR DOWN position. As the landing gear extends, the airspeed decreases and the pitch attitude may change. During the several seconds it takes for the gear to extend, be attentive to any abnormal sounds or feel. Confirm that the landing gear has extended and locked by the normal sound and feel of the system operation, as well as by the gear position indicators in the flightdeck. Unless the landing gear has been previously extended to aid in a descent to traffic pattern altitude, the landing gear should be extended by the time the airplane reaches a point on the downwind leg that is opposite the point of intended landing. Establish a standard procedure consisting of a specific position on the downwind leg at which to lower the landing gear. Strict adherence to this procedure aids in avoiding unintentional gear up landings.

Operation of an airplane equipped with a retractable landing gear requires the deliberate, careful, and continued use of an appropriate checklist. When on the downwind leg, make it a habit to complete the before landing checklist for that airplane. This accomplishes two purposes. It ensures that action has been taken to lower the gear and establishes awareness so that the gear down indicators can be rechecked prior to landing.

Unless good operating practices dictate otherwise, the landing roll should be completed and the airplane should be clear of the runway before any levers or switches are operated.

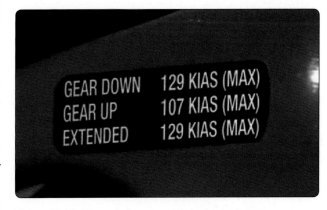

Figure 11-15. *Placarded gear speeds in the cockpit.*

This technique greatly reduces the chance of inadvertently retracting the landing gear while on the ground. Wait until after rollout and clearing the runway to focus attention on the after landing checklist. This practice allows for positive identification of the proper controls.

When transitioning to retractable gear airplanes, it is important to consider some frequent pilot errors. These include pilots that have:

- Neglected to extend landing gear

- Inadvertently retracted landing gear

- Activated gear but failed to check gear position

- Misused emergency gear system

- Retracted gear prematurely on takeoff

- Extended gear too late

These mistakes are not only committed by pilots who have just transitioned to complex aircraft, but also by pilots who have developed a sense of complacency over time. In order to minimize the chances of a landing gear-related mishap:

- Use an appropriate checklist. (A condensed checklist mounted in view is a reminder for its use and easy reference can be especially helpful.)

- Be familiar with, and periodically review, the landing gear emergency extension procedures for the particular airplane.

- Be familiar with the landing gear warning horn and warning light systems for the particular airplane. Use the horn system to cross-check the warning light system when an unsafe condition is noted.

- Review the procedure for replacing light bulbs in the landing gear warning light displays for the particular airplane, so that you can properly replace a bulb to

determine if the bulb(s) in the display is good. Check to see if spare bulbs are available in the airplane spare bulb supply as part of the preflight inspection.

- Be familiar with and aware of the sounds and feel of a properly operating landing gear system.

Transition Training

Transition to a complex airplane or a high-performance airplane should be accomplished through a structured course of training administered by a competent and qualified flight instructor. The training should be accomplished in accordance with a ground and flight training syllabus. *[Figure 11-16]*

This sample syllabus for transition training is an example. The arrangement of the subject matter may be changed and the emphasis shifted to fit the qualifications of the transitioning pilot, the airplane involved, and the circumstances of the training situation. The goal is to ensure proficiency standards are achieved. These standards are contained in the Practical Test Standards (PTS) or Airmen Certification Standard (ACS) as appropriate for the certificate that the transitioning pilot holds or is working towards.

The training times indicated in the syllabus are for illustration purposes. Actual times must be based on the capabilities of the pilot. The time periods may be minimal for pilots with higher qualifications or increased for pilots who do not meet certification requirements or have had little recent flight experience.

Chapter Summary

Flying a complex or high-performance airplane requires a pilot to further divide his or her attention during the most critical phases of flight: take-off and landing. The knowledge, judgment, and piloting skills required to fly these airplanes must be developed. It is essential that adequate training is

Ground Instruction	Flight Instruction
One hour	**One hour**
1. Operations sections of flight manual	1. Flight training maneuvers
2. Line inspection	2. Takeoffs, landings and go-arounds
3. Cockpit familiarization	
One hour	**One hour**
1. Aircraft loading, limitations and servicing	1. Emergency operations
2. instruments, radio and special equipment	2. Control by reference to instruments
3. Aircraft systems	3. Use of radio and autopilot
One hour	**One hour**
1. Performance section of flight manual	1. Short and soft-field takeoffs and landings
2. Cruise control	2. Maximum performance operations
3. Review	

Figure 11-16. *Sample transition training syllabus.*

received to ensure a complete understanding of the systems, their operation (both normal and emergency), and operating limitations.

Chapter 12
Transition to Multiengine Airplanes

Introduction

This chapter is devoted to the factors associated with the operation of small multiengine airplanes. For the purpose of this handbook, a "small" multiengine airplane is a reciprocating or turbopropeller-powered airplane with a maximum certificated takeoff weight of 12,500 pounds or less. This discussion assumes a conventional design with two engines—one mounted on each wing. Reciprocating engines are assumed unless otherwise noted. The term "light-twin," although not formally defined in the regulations, is used herein as a small multiengine airplane with a maximum certificated takeoff weight of 6,000 pounds or less.

There are several unique characteristics of multiengine airplanes that make them worthy of a separate class rating. Knowledge of these factors and proficient flight skills are a key to safe flight in these airplanes. This chapter deals extensively with the numerous aspects of one engine inoperative (OEI) flight. However, pilots are strongly cautioned not to place undue emphasis on mastery of OEI flight as the sole key to flying multiengine airplanes safely. The inoperative engine information that follows is extensive only because this chapter emphasizes the differences between flying multiengine airplanes as contrasted to single-engine airplanes.

The modern, well-equipped multiengine airplane can be remarkably capable under many circumstances. But, as with single-engine airplanes, it must be flown prudently by a current and competent pilot to achieve the highest possible level of safety.

This chapter contains information and guidance on the performance of certain maneuvers and procedures in small multiengine airplanes for the purposes of flight training and pilot certification testing. The airplane manufacturer is the final authority on the operation of a particular make and model airplane. Flight instructors and students should use the Federal Aviation Administration's Approved Flight Manual (AFM) and/or the Pilot's Operating Handbook (POH) but realize that the airplane manufacturer's guidance and procedures take precedence.

General

The basic difference between operating a multiengine airplane and a single-engine airplane is the potential problem involving an engine failure. The penalties for loss of an engine are twofold: performance and control. The most obvious problem is the loss of 50 percent of power, which reduces climb performance 80 to 90 percent, sometimes even more. The other is the control problem caused by the remaining thrust, which is now asymmetrical. Attention to both these factors is crucial to safe OEI flight. The performance and systems redundancy of a multiengine airplane is a safety advantage only to a trained and proficient pilot.

Terms and Definition

Pilots of single-engine airplanes are already familiar with many performance "V" speeds and their definitions. Twin-engine airplanes have several additional V-speeds unique to OEI operation. These speeds are differentiated by the notation "SE" for single engine. A review of some key V-speeds and several new V-speeds unique to twin-engine airplanes are listed below.

- V_R—rotation speed—speed at which back pressure is applied to rotate the airplane to a takeoff attitude.

- V_{LOF}—lift-off speed—speed at which the airplane leaves the surface. (NOTE: Some manufacturers reference takeoff performance data to V_R, others to V_{LOF}.)

- V_X—best angle of climb speed—speed at which the airplane gains the greatest altitude for a given distance of forward travel.

- V_{XSE}—best angle-of-climb speed with OEI.

- V_Y—best rate of climb speed—speed at which the airplane gains the most altitude for a given unit of time.

- V_{YSE}—best rate of climb speed with OEI. Marked with a blue radial line on most airspeed indicators.

Above the single-engine absolute ceiling, V_{YSE} yields the minimum rate of sink.

- V_{SSE}—safe, intentional OEI speed—originally known as safe single-engine speed, now formally defined in Title 14 of the Code of Federal Regulations (14 CFR) part 23, Airworthiness Standards, and required to be established and published in the AFM/POH. It is the minimum speed to intentionally render the critical engine inoperative.

- V_{REF}—reference landing speed—an airspeed used for final approach, which adjust the normal approach speed for winds and gusty conditions. V_{REF} is 1.3 times the stall speed in the landing configuration.

- V_{MC}—minimum control speed with the critical engine inoperative—marked with a red radial line on most airspeed indicators. The minimum speed at which directional control can be maintained under a very specific set of circumstances outlined in 14 CFR part 23, Airworthiness Standards. Under the small airplane certification regulations currently in effect, the flight test pilot must be able to (1) stop the turn that results when the critical engine is suddenly made inoperative within 20° of the original heading, using maximum rudder deflection and a maximum of 5° bank, and (2) thereafter, maintain straight flight with not more than a 5° bank. There is no requirement in this determination that the airplane be capable of climbing at this airspeed. V_{MC} only addresses directional control. Further discussion of V_{MC} as determined during airplane certification and demonstrated in pilot training follows in minimum control airspeed (V_{MC}) demonstration. *[Figure 12-1]*

Unless otherwise noted, when V-speeds are given in the AFM/POH, they apply to sea level, standard day conditions at maximum takeoff weight. Performance speeds vary with aircraft weight, configuration, and atmospheric conditions. The speeds may be stated in statute miles per hour (mph) or knots (kt), and they may be given as calibrated airspeeds (CAS) or indicated airspeeds (IAS). As a general rule, the newer AFM/POHs show V-speeds in knots indicated airspeed (KIAS). Some V-speeds are also stated in knots calibrated airspeed (KCAS) to meet certain regulatory requirements. Whenever available, pilots should operate the airplane from published indicated airspeeds.

With regard to climb performance, the multiengine airplane, particularly in the takeoff or landing configuration, may be considered to be a single-engine airplane with its powerplant divided into two units. There is nothing in 14 CFR part 23 that requires a multiengine airplane to maintain altitude while in the takeoff or landing configuration with OEI. In fact,

Figure 12-1. *Airspeed indicator markings for a multiengine airplane.*

many twins are not required to do this in any configuration, even at sea level.

The current 14 CFR part 23 single-engine climb performance requirements for reciprocating engine-powered multiengine airplanes are as follows.

- More than 6,000 pounds maximum weight and/or V_{SO} more than 61 knots: the single-engine rate of climb in feet per minute (fpm) at 5,000 feet mean sea level (MSL) must be equal to at least .027 V_{SO} 2. For airplanes type certificated February 4, 1991, or thereafter, the climb requirement is expressed in terms of a climb gradient, 1.5 percent. The climb gradient is not a direct equivalent of the .027 V_{SO} 2 formula. Do not confuse the date of type certification with the airplane's model year. The type certification basis of many multiengine airplanes dates back to the Civil Aviation Regulations (CAR) 3.

- 6,000 pounds or less maximum weight and V_{SO} 61 knots or less: the single-engine rate of climb at 5,000 feet MSL must simply be determined. The rate of climb could be a negative number. There is no requirement for a single-engine positive rate of climb at 5,000 feet or any other altitude. For light-twins type certificated February 4, 1991, or thereafter, the single-engine climb gradient (positive or negative) is simply determined.

Rate of climb is the altitude gain per unit of time, while climb gradient is the actual measure of altitude gained per 100 feet of horizontal travel, expressed as a percentage. An altitude gain of 1.5 feet per 100 feet of travel (or 15 feet per 1,000, or 150 feet per 10,000) is a climb gradient of 1.5 percent.

There is a dramatic performance loss associated with the loss of an engine, particularly just after takeoff. Any airplane's climb performance is a function of thrust horsepower, which is in excess of that required for level flight. In a hypothetical twin with each engine producing 200 thrust horsepower, assume that the total level flight thrust horsepower required is 175. In this situation, the airplane would ordinarily have a reserve of 225 thrust horsepower available for climb. Loss of one engine would leave only 25 (200 minus 175) thrust horsepower available for climb, a drastic reduction. Sea level rate of climb performance losses of at least 80 to 90 percent, even under ideal circumstances, are typical for multiengine airplanes in OEI flight.

Operation of Systems

This section deals with systems that are generally found on multiengine airplanes. Multiengine airplanes share many features with complex single-engine airplanes. There are certain systems and features covered that are generally unique to airplanes with two or more engines.

Propellers

The propellers of the multiengine airplane may outwardly appear to be identical in operation to the constant-speed propellers of many single-engine airplanes, but this is not the case. The propellers of multiengine airplanes are featherable, to minimize drag in the event of an engine failure. Depending upon single-engine performance, this feature often permits continued flight to a suitable airport following an engine failure. To feather a propeller is to stop engine rotation with the propeller blades streamlined with the airplane's relative wind, thus to minimize drag. *[Figure 12-2]*

Feathering is necessary because of the change in parasite drag with propeller blade angle. *[Figure 12-3]* When the propeller blade angle is in the feathered position, the change in parasite drag is at a minimum and, in the case of a typical multiengine airplane, the added parasite drag from a single feathered propeller is a relatively small contribution to the airplane total drag.

At the smaller blade angles near the flat pitch position, the drag added by the propeller is very large. At these small blade angles, the propeller windmilling at high rates per minute (rpm) can create such a tremendous amount of drag that the airplane may be uncontrollable. The propeller windmilling at high speed in the low range of blade angles can produce an increase in parasite drag, which may be as great as the parasite drag of the basic airplane.

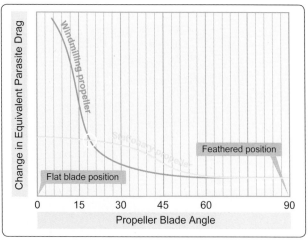

Figure 12-3. *Propeller drag contribution.*

Figure 12-2. *Feathered propeller.*

As a review, the constant-speed propellers on almost all single-engine airplanes are of the non-feathering, oil-pressure-to-increase-pitch design. In this design, increased oil pressure from the propeller governor drives the blade angle towards high pitch, low rpm.

In contrast, the constant-speed propellers installed on most multiengine airplanes are full feathering, counterweighted, oil-pressure-to-decrease-pitch designs. In this design, increased oil pressure from the propeller governor drives the blade angle towards low pitch, high rpm—away from the feather blade angle. In effect, the only thing that keeps these propellers from feathering is a constant supply of high-pressure engine oil. This is a necessity to enable propeller feathering in the event of a loss of oil pressure or a propeller governor failure.

The aerodynamic forces alone acting upon a windmilling propeller tend to drive the blades to low pitch, high rpm. Counterweights attached to the shank of each blade tend to drive the blades to high pitch, low rpm. Inertia, or apparent force (called centrifugal force) acting through

the counterweights, is generally slightly greater than the aerodynamic forces. Oil pressure from the propeller governor is used to counteract the counterweights and drives the blade angles to low pitch, high rpm. A reduction in oil pressure causes the rpm to be reduced from the influence of the counterweights. *[Figure 12-4]*

To feather the propeller, the propeller control is brought fully aft. All oil pressure is dumped from the governor, and the counterweights drive the propeller blades towards feather. As centrifugal force acting on the counterweights decays from decreasing rpm, additional forces are needed to completely feather the blades. This additional force comes from either a spring or high-pressure air stored in the propeller dome, which forces the blades into the feathered position. The entire process may take up to 10 seconds.

Feathering a propeller only alters blade angle and stops engine rotation. To completely secure the engine, the pilot must still turn off the fuel (mixture, electric boost pump, and fuel selector), ignition, alternator/generator, and close the cowl flaps. If the airplane is pressurized, there may also be an air bleed to close for the failed engine. Some airplanes are equipped with firewall shutoff valves that secure several of these systems with a single switch.

Completely securing a failed engine may not be necessary or even desirable depending upon the failure mode, altitude, and time available. The position of the fuel controls, ignition, and alternator/generator switches of the failed engine has no effect on aircraft performance. There is always the distinct possibility of manipulating the incorrect switch under conditions of haste or pressure.

To unfeather a propeller, the engine must be rotated so that oil pressure can be generated to move the propeller blades

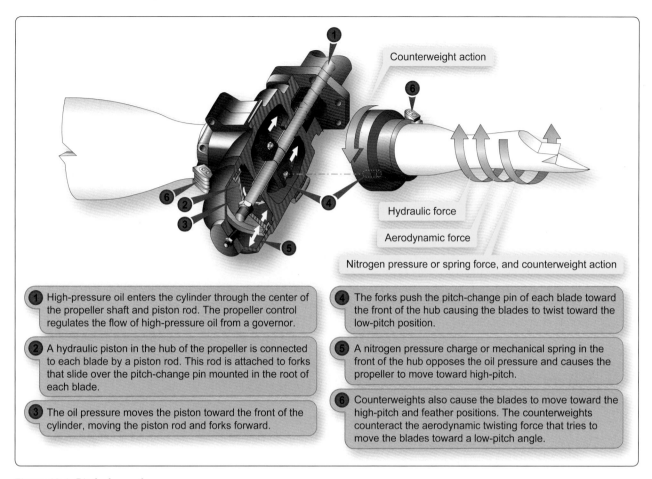

Figure 12-4. *Pitch change forces.*

Counterweight action

Hydraulic force

Aerodynamic force

Nitrogen pressure or spring force, and counterweight action

1 High-pressure oil enters the cylinder through the center of the propeller shaft and piston rod. The propeller control regulates the flow of high-pressure oil from a governor.

2 A hydraulic piston in the hub of the propeller is connected to each blade by a piston rod. This rod is attached to forks that slide over the pitch-change pin mounted in the root of each blade.

3 The oil pressure moves the piston toward the front of the cylinder, moving the piston rod and forks forward.

4 The forks push the pitch-change pin of each blade toward the front of the hub causing the blades to twist toward the low-pitch position.

5 A nitrogen pressure charge or mechanical spring in the front of the hub opposes the oil pressure and causes the propeller to move toward high-pitch.

6 Counterweights also cause the blades to move toward the high-pitch and feather positions. The counterweights counteract the aerodynamic twisting force that tries to move the blades toward a low-pitch angle.

from the feathered position. The ignition is turned on prior to engine rotation with the throttle at low idle and the mixture rich. With the propeller control in a high rpm position, the starter is engaged. The engine begins to windmill, start, and run as oil pressure moves the blades out of feather. As the engine starts, the propeller rpm should be immediately reduced until the engine has had several minutes to warm up; the pilot should monitor cylinder head and oil temperatures.

In any event, the AFM/POH procedures should be followed for the exact unfeathering procedure. Both feathering and starting a feathered reciprocating engine on the ground are strongly discouraged by manufacturers due to the excessive stress and vibrations generated.

As just described, a loss of oil pressure from the propeller governor allows the counterweights, spring, and/or dome charge to drive the blades to feather. Logically then, the propeller blades should feather every time an engine is shut down as oil pressure falls to zero. Yet, this does not occur. Preventing this is a small pin in the pitch changing mechanism of the propeller hub that does not allow the propeller blades to feather once rpm drops below approximately 800. The pin senses a lack of centrifugal force from propeller rotation and falls into place, preventing the blades from feathering. Therefore, if a propeller is to be feathered, it must be done before engine rpm decays below approximately 800. On one popular model of turboprop engine, the propeller blades do, in fact, feather with each shutdown. This propeller is not equipped with such centrifugally-operated pins due to a unique engine design.

An unfeathering accumulator is a device that permits starting a feathered engine in flight without the use of the electric starter. An accumulator is any device that stores a reserve of high pressure. On multiengine airplanes, the unfeathering accumulator stores a small reserve of engine oil under pressure from compressed air or nitrogen. To start a feathered engine in flight, the pilot moves the propeller control out of the feather position to release the accumulator pressure. The oil flows under pressure to the propeller hub and drives the blades toward the high rpm, low pitch position, whereupon the propeller usually begins to windmill. (On some airplanes, an assist from the electric starter may be necessary to initiate rotation and completely unfeather the propeller.) If fuel and ignition are present, the engine starts and runs. For airplanes used in training, this saves much electric starter and battery wear. High oil pressure from the propeller governor recharges the accumulator just moments after engine rotation begins.

Propeller Synchronization

Many multiengine airplanes have a propeller synchronizer (prop sync) installed to eliminate the annoying "drumming" or "beat" of propellers whose rpm are close, but not precisely the same. To use prop sync, the propeller rpm is coarsely matched by the pilot and the system is engaged. The prop sync adjusts the rpm of the "slave" engine to precisely match the rpm of the "master" engine and then maintains that relationship.

The prop sync should be disengaged when the pilot selects a new propeller rpm and then re-engaged after the new rpm is set. The prop sync should always be off for takeoff, landing, and single-engine operation. The AFM/POH should be consulted for system description and limitations.

A variation on the propeller synchronizer is the propeller synchrophaser. Prop synchrophase acts much like a synchronizer to precisely match rpm, but the synchrophaser goes one step further. It not only matches rpm but actually compares and adjusts the positions of the individual blades of the propellers in their arcs. There can be significant propeller noise and vibration reductions with a propeller synchrophaser. From the pilot's perspective, operation of a propeller synchronizer and a propeller synchrophaser are very similar. A synchrophaser is also commonly referred to as prop sync, although that is not entirely correct nomenclature from a technical standpoint.

As a pilot aid to manually synchronizing the propellers, some twins have a small gauge mounted in or by the tachometer(s) with a propeller symbol on a disk that spins. The pilot manually fine tunes the engine rpm so as to stop disk rotation, thereby synchronizing the propellers. This is a useful backup to synchronizing engine rpm using the audible propeller beat. This gauge is also found installed with most propeller synchronizer and synchrophase systems. Some synchrophase systems use a knob for the pilot to control the phase angle.

Fuel Crossfeed

Fuel crossfeed systems are also unique to multiengine airplanes. Using crossfeed, an engine can draw fuel from a fuel tank located in the opposite wing.

On most multiengine airplanes, operation in the crossfeed mode is an emergency procedure used to extend airplane range and endurance in OEI flight. There are a few models that permit crossfeed as a normal, fuel balancing technique in normal operation, but these are not common. The AFM/ POH describes crossfeed limitations and procedures that vary significantly among multiengine airplanes.

Checking crossfeed operation on the ground with a quick repositioning of the fuel selectors does nothing more than ensure freedom of motion of the handle. To actually check crossfeed operation, a complete, functional crossfeed system check should be accomplished. To do this, each engine should be operated from its crossfeed position during the run-up. The engines should be checked individually and allowed to run at moderate power (1,500 rpm minimum) for at least 1 minute to ensure that fuel flow can be established from the crossfeed source. Upon completion of the check, each engine should be operated for at least 1 minute at moderate power from the main (takeoff) fuel tanks to reconfirm fuel flow prior to takeoff.

This suggested check is not required prior to every flight. Crossfeed lines are ideal places for water and debris to accumulate unless they are used from time to time and drained using their external drains during preflight. Crossfeed is ordinarily not used for completing single-engine flights when an alternate airport is readily at hand, and it is never used during takeoff or landings.

Combustion Heater

Combustion heaters are common on multiengine airplanes. A combustion heater is best described as a small furnace that burns gasoline to produce heated air for occupant comfort and windshield defogging. Most are thermostatically operated and have a separate hour meter to record time in service for maintenance purposes. Automatic over temperature protection is provided by a thermal switch mounted on the unit that cannot be accessed in flight. This requires the pilot or mechanic to actually visually inspect the unit for possible heat damage in order to reset the switch.

When finished with the combustion heater, a cool-down period is required. Most heaters require that outside air be permitted to circulate through the unit for at least 15 seconds in flight or that the ventilation fan can be operated for at least 2 minutes on the ground. Failure to provide an adequate cool down usually trips the thermal switch and renders the heater inoperative until the switch is reset.

Flight Director/Autopilot

Flight director/autopilot (FD/AP) systems are common on the better-equipped multiengine airplanes. The system integrates pitch, roll, heading, altitude, and radio navigation signals in a computer. The outputs, called computed commands, are displayed on a flight command indicator (FCI). The FCI replaces the conventional attitude indicator on the instrument panel. The FCI is occasionally referred to as a flight director indicator (FDI) or as an attitude director indicator (ADI).

The entire flight director/autopilot system is sometimes called an integrated flight control system (IFCS) by some manufacturers. Others may use the term automatic flight control system (AFCS).

The FD/AP system may be employed at the following different levels:

- Off (raw data)
- Flight director (computed commands)
- Autopilot

With the system off, the FCI operates as an ordinary attitude indicator. On most FCIs, the command bars are biased out of view when the FD is off. The pilot maneuvers the airplane as though the system were not installed.

To maneuver the airplane using the FD, the pilot enters the desired modes of operation (heading, altitude, navigation (NAV) intercept, and tracking) on the FD/AP mode controller. The computed flight commands are then displayed to the pilot through either a single-cue or dual-cue system in the FCI. On a single-cue system, the commands are indicated by "V" bars. On a dual-cue system, the commands are displayed on two separate command bars, one for pitch and one for roll. To maneuver the airplane using computed commands, the pilot "flies" the symbolic airplane of the FCI to match the steering cues presented.

On most systems, to engage the autopilot the FD must first be operating. At any time thereafter, the pilot may engage the autopilot through the mode controller. The autopilot then maneuvers the airplane to satisfy the computed commands of the FD.

Like any computer, the FD/AP system only does what it is told. The pilot must ensure that it has been programmed properly for the particular phase of flight desired. The armed and/or engaged modes are usually displayed on the mode controller or separate annunciator lights. When the airplane is being hand-flown, if the FD is not being used at any particular moment, it should be off so that the command bars are pulled from view.

Prior to system engagement, all FD/AP computer and trim checks should be accomplished. Many newer systems cannot be engaged without the completion of a self-test. The pilot must also be very familiar with various methods of disengagement, both normal and emergency. System details, including approvals and limitations, can be found in the supplements section of the AFM/POH. Additionally, many avionics manufacturers can provide informative pilot operating guides upon request.

Yaw Damper

The yaw damper is a servo that moves the rudder in response to inputs from a gyroscope or accelerometer that detects yaw rate. The yaw damper minimizes motion about the vertical axis caused by turbulence. (Yaw dampers on swept wing airplanes provide another, more vital function of damping dutch roll characteristics.) Occupants feel a smoother ride, particularly if seated in the rear of the airplane, when the yaw damper is engaged. The yaw damper should be off for takeoff and landing. There may be additional restrictions against its use during single-engine operation. Most yaw dampers can be engaged independently of the autopilot.

Alternator/Generator

Alternator or generator paralleling circuitry matches the output of each engine's alternator/generator so that the electrical system load is shared equally between them. In the event of an alternator/generator failure, the inoperative unit can be isolated and the entire electrical system powered from the remaining one. Depending upon the electrical capacity of the alternator/generator, the pilot may need to reduce the electrical load (referred to as load shedding) when operating on a single unit. The AFM/POH contains system description and limitations.

Nose Baggage Compartment

Nose baggage compartments are common on multiengine airplanes (and are even found on a few single-engine airplanes). There is nothing strange or exotic about a nose baggage compartment, and the usual guidance concerning observation of load limits applies. Pilots occasionally neglect to secure the latches properly. When improperly secured, the door opens and the contents may be drawn out, usually into the propeller arc and just after takeoff. Even when the nose baggage compartment is empty, airplanes have been lost when the pilot became distracted by the open door. Security of the nose baggage compartment latches and locks is a vital preflight item.

Most airplanes continue to fly with a nose baggage door open. There may be some buffeting from the disturbed airflow, and there is an increase in noise. Pilots should never become so preoccupied with an open door (of any kind) that they fail to fly the airplane.

Inspection of the compartment interior is also an important preflight item. More than one pilot has been surprised to find a supposedly empty compartment packed to capacity or loaded with ballast. The tow bars, engine inlet covers, windshield sun screens, oil containers, spare chocks, and miscellaneous small hand tools that find their way into baggage compartments should be secured to prevent damage from shifting in flight.

Anti-Icing/Deicing

Anti-icing/deicing equipment is frequently installed on multiengine airplanes and consists of a combination of different systems. These may be classified as either anti-icing or deicing, depending upon function. The presence of anti-icing and deicing equipment, even though it may appear elaborate and complete, does not necessarily mean that the airplane is approved for flight in icing conditions. The AFM/POH, placards, and even the manufacturer should be consulted for specific determination of approvals and limitations. Anti-icing equipment is provided to prevent ice from forming on certain protected surfaces. Anti-icing equipment includes heated pitot tubes, heated or non-icing static ports and fuel vents, propeller blades with electrothermal boots or alcohol slingers, windshields with alcohol spray or electrical resistance heating, windshield defoggers, and heated stall warning lift detectors. On many turboprop engines, the "lip" surrounding the air intake is heated either electrically or with bleed air. In the absence of AFM/POH guidance to the contrary, anti-icing equipment should be actuated prior to flight into known or suspected icing conditions.

Deicing equipment is generally limited to pneumatic boots on wing and tail leading edges. Deicing equipment is installed to remove ice that has already formed on protected surfaces. Upon pilot actuation, the boots inflate with air from the pneumatic pumps to break off accumulated ice. After a few seconds of inflation, they are deflated back to their normal position with the assistance of a vacuum. The pilot monitors the buildup of ice and cycles the boots as directed in the AFM/POH. An ice light on the left engine nacelle allows the pilot to monitor wing ice accumulation at night.

Other airframe equipment necessary for flight in icing conditions includes an alternate induction air source and an alternate static system source. Ice tolerant antennas are also installed.

In the event of impact ice accumulating over normal engine air induction sources, carburetor heat (carbureted engines) or alternate air (fuel injected engines) should be selected. Ice buildup on normal induction sources can be detected by a loss of engine rpm with fixed-pitch propellers and a loss of manifold pressure with constant-speed propellers. On some fuel injected engines, an alternate air source is automatically activated with blockage of the normal air source.

An alternate static system provides an alternate source of static air for the pitot-static system in the unlikely event that the primary static source becomes blocked. In non-pressurized airplanes, most alternate static sources are plumbed to the cabin. On pressurized airplanes, they are usually plumbed to a non-pressurized baggage compartment. The pilot must activate the alternate static source by opening a valve or a fitting in the flightdeck. Upon activation, the airspeed indicator, altimeter, and the vertical speed indicator (VSI) is affected and reads somewhat in error. A correction table is frequently provided in the AFM/POH.

Anti-icing/deicing equipment only eliminates ice from the protected surfaces. Significant ice accumulations may form on unprotected areas, even with proper use of anti-ice and deice systems. Flight at high angles of attack (AOA) or even normal climb speeds permit significant ice accumulations on lower wing surfaces, which are unprotected. Many AFM/POHs mandate minimum speeds to be maintained in icing conditions. Degradation of all flight characteristics and large performance losses can be expected with ice accumulations. Pilots should not rely upon the stall warning devices for adequate stall warning with ice accumulations.

Ice accumulates unevenly on the airplane. It adds weight and drag (primarily drag) and decreases thrust and lift. Even wing shape affects ice accumulation; thin airfoil sections are more prone to ice accumulation than thick, highly-cambered sections. For this reason, certain surfaces, such as the horizontal stabilizer, are more prone to icing than the wing. With ice accumulations, landing approaches should be made with a minimum wing flap setting (flap extension increases the AOA of the horizontal stabilizer) and with an added margin of airspeed. Sudden and large configuration and airspeed changes should be avoided.

Unless otherwise recommended in the AFM/POH, the autopilot should not be used in icing conditions. Continuous use of the autopilot masks trim and handling changes that occur with ice accumulation. Without this control feedback, the pilot may not be aware of ice accumulation building to hazardous levels. The autopilot suddenly disconnects when it reaches design limits, and the pilot may find the airplane has assumed unsatisfactory handling characteristics.

The installation of anti-ice/deice equipment on airplanes without AFM/POH approval for flight into icing conditions is to facilitate escape when such conditions are inadvertently encountered. Even with AFM/POH approval, the prudent pilot avoids icing conditions to the maximum extent practicable and avoids extended flight in any icing conditions. No multiengine airplane is approved for flight into severe icing conditions and none are intended for indefinite flight in continuous icing conditions.

Performance and Limitations

Discussion of performance and limitations requires the definition of the following terms.

- Accelerate-stop distance is the runway length required to accelerate to a specified speed (either V_R or V_{LOF}, as specified by the manufacturer), experience an engine failure, and bring the airplane to a complete stop.

- Accelerate-go distance is the horizontal distance required to continue the takeoff and climb to 50 feet, assuming an engine failure at V_R or V_{LOF}, as specified by the manufacturer.

- Climb gradient is a slope most frequently expressed in terms of altitude gain per 100 feet of horizontal distance, whereupon it is stated as a percentage. A 1.5 percent climb gradient is an altitude gain of one and one-half feet per 100 feet of horizontal travel. Climb gradient may also be expressed as a function of altitude gain per nautical mile(NM), or as a ratio of the horizontal distance to the vertical distance (50:1, for example). Unlike rate of climb, climb gradient is affected by wind. Climb gradient is improved with a headwind component and reduced with a tailwind component. *[Figure 12-5]*

- The all-engine service ceiling of multiengine airplanes is the highest altitude at which the airplane can maintain a steady rate of climb of 100 fpm with both engines operating. The airplane has reached its absolute ceiling when climb is no longer possible.

- The single-engine service ceiling is reached when the multiengine airplane can no longer maintain a 50 fpm rate of climb with OEI, and its single-engine absolute ceiling when climb is no longer possible.

The takeoff in a multiengine airplane should be planned in sufficient detail so that the appropriate action is taken in the event of an engine failure. The pilot should be thoroughly familiar with the airplane's performance capabilities and limitations in order to make an informed takeoff decision as part of the preflight planning. That decision should be reviewed as the last item of the "before takeoff" checklist.

In the event of an engine failure shortly after takeoff, the decision is basically one of continuing flight or landing, even off-airport. If single-engine climb performance is adequate for continued flight, and the airplane has been promptly and correctly configured, the climb after takeoff may be continued. If single-engine climb performance is such that climb is unlikely or impossible, a landing has to be made in the most suitable area. To be avoided above all is attempting to continue flight when it is not within the airplane's performance capability to do so. *[Figure 12-6]*

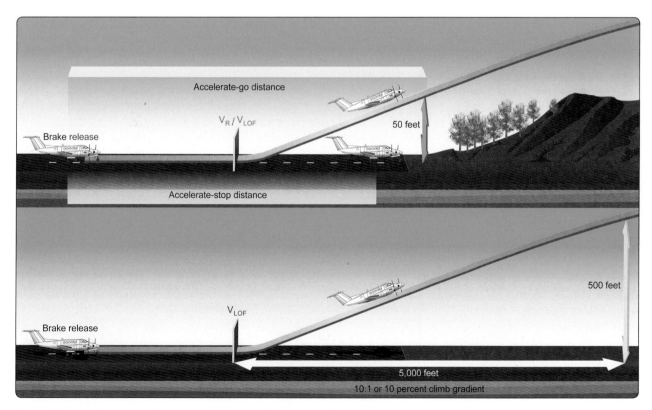

Figure 12-5. *Accelerate-stop distance, accelerate-go distance, and climb gradient.*

Takeoff planning factors include weight and balance, airplane performance (both single and multiengine), runway length, slope and contamination, terrain and obstacles in the area, weather conditions, and pilot proficiency. Most multiengine airplanes have AFM/POH performance charts and the pilot should be highly proficient in their use. Prior to takeoff, the multiengine pilot should ensure that the weight and balance limitations have been observed, the runway length is adequate, and the normal flightpath clears obstacles and terrain. A clear and definite course of action to follow in the event of engine failure is essential.

The regulations do not specifically require that the runway length be equal to or greater than the accelerate-stop distance. Most AFM/POHs publish accelerate-stop distances only as an advisory. It becomes a limitation only when published in the limitations section of the AFM/POH. Experienced multiengine pilots, however, recognize the safety margin of runway lengths in excess of the bare minimum required for normal takeoff. They insist on runway lengths of at least accelerate-stop distance as a matter of safety and good operating practice.

The multiengine pilot must keep in mind that the accelerate-go distance, as long as it is, has only brought the airplane, under ideal circumstances, to a point a mere 50 feet above the takeoff elevation. To achieve even this meager climb, the pilot had to instantaneously recognize and react to an unanticipated engine failure, retract the landing gear, identify and feather the correct engine, all the while maintaining precise airspeed control and bank angle as the airspeed is nursed to V_{YSE}. Assuming flawless airmanship thus far, the airplane has now arrived at a point little more than one wingspan above the terrain, assuming it was absolutely level and without obstructions.

For the purpose of illustration, with a near 150 fpm rate of climb at a 90-knot V_{YSE}, it takes approximately 3 minutes to climb an additional 450 feet to reach 500 feet AGL. In doing so, the airplane has traveled an additional 5 NM beyond the original accelerate-go distance, with a climb gradient of about 1.6 percent. Any turn, such as to return to the airport, seriously degrades the already marginal climb performance of the airplane.

Not all multiengine airplanes have published accelerate-go distances in their AFM/POH and fewer still publish climb gradients. When such information is published, the figures have been determined under ideal flight testing conditions. It is unlikely that this performance is duplicated in service conditions.

The point of the previous discussion is to illustrate the marginal climb performance of a multiengine airplane that suffers an engine failure shortly after takeoff, even under ideal conditions. The prudent multiengine pilot should pick a decision point in the takeoff and climb sequence in advance. If an engine fails before this point the takeoff should be rejected, even if airborne, for a landing on whatever runway or surface lies essentially ahead. If an engine fails after this point, the pilot should promptly execute the appropriate engine failure procedure and continue the climb, assuming the performance capability exists. As a general recommendation, if the landing gear has not been selected up, the takeoff should be rejected, even if airborne.

As a practical matter for planning purposes, the option of continuing the takeoff probably does not exist unless the published single-engine rate-of-climb performance is at least 100 to 200 fpm. Thermal turbulence, wind gusts, engine and propeller wear, or poor technique in airspeed, bank angle, and rudder control can easily negate even a 200 fpm rate of climb.

A pre-takeoff safety brief clearly defines all pre planned emergency actions to all crewmembers. Even if operating the aircraft alone, the pilot should review and be familiar with takeoff emergency considerations. Indecision at the moment an emergency occurs degrades reaction time and the ability to make a proper response.

Figure 12-6. *Area of decision for engine failure after lift-off.*

Weight and Balance

The weight and balance concept is no different than that of a single-engine airplane. The actual execution, however, is almost invariably more complex due to a number of new loading areas, including nose and aft baggage compartments, nacelle lockers, main fuel tanks, auxiliary fuel tanks, nacelle fuel tanks, and numerous seating options in a variety of interior configurations. The flexibility in loading offered by the multiengine airplane places a responsibility on the pilot to address weight and balance prior to each flight.

The terms empty weight, licensed empty weight, standard empty weight, and basic empty weight as they appear on the manufacturer's original weight and balance documents are sometimes confused by pilots.

In 1975, the General Aviation Manufacturers Association (GAMA) adopted a standardized format for AFM/POHs. It was implemented by most manufacturers in model year 1976. Airplanes whose manufacturers conform to the GAMA standards utilize the following terminology for weight and balance:

standard empty weight + optional equipment = basic empty weight

Standard empty weight is the weight of the standard airplane, full hydraulic fluid, unusable fuel, and full oil. Optional equipment includes the weight of all equipment installed beyond standard. Basic empty weight is the standard empty weight plus optional equipment. Note that basic empty weight includes no usable fuel, but full oil.

Airplanes manufactured prior to the GAMA format generally utilize the following terminology for weight and balance, although the exact terms may vary somewhat:

empty weight + unusable fuel = standard empty weight

standard empty weight + optional equipment = licensed empty weight

Empty weight is the weight of the standard airplane, full hydraulic fluid, and undrainable oil. Unusable fuel is the fuel remaining in the airplane not available to the engines. Standard empty weight is the empty weight plus unusable fuel. When optional equipment is added to the standard empty weight, the result is licensed empty weight. Licensed empty weight, therefore, includes the standard airplane, optional equipment, full hydraulic fluid, unusable fuel, and undrainable oil.

The major difference between the two formats (GAMA and the old) is that basic empty weight includes full oil and

licensed empty weight does not. Oil must always be added to any weight and balance utilizing a licensed empty weight.

When the airplane is placed in service, amended weight and balance documents are prepared by appropriately-rated maintenance personnel to reflect changes in installed equipment. The old weight and balance documents are customarily marked "superseded" and retained in the AFM/POH. Maintenance personnel are under no regulatory obligation to utilize the GAMA terminology, so weight and balance documents subsequent to the original may use a variety of terms. Pilots should use care to determine whether or not oil has to be added to the weight and balance calculations or if it is already included in the figures provided.

The multiengine airplane is where most pilots encounter the term "zero fuel weight" for the first time. Not all multiengine airplanes have a zero fuel weight limitation published in their AFM/POH, but many do. Zero fuel weight is simply the maximum allowable weight of the airplane and payload, assuming there is no usable fuel on board. The actual airplane is not devoid of fuel at the time of loading, of course. This is merely a calculation that assumes it was. If a zero fuel weight limitation is published, then all weight in excess of that figure must consist of usable fuel. The purpose of a zero fuel weight is to limit load forces on the wing spars with heavy fuselage loads.

Assume a hypothetical multiengine airplane with the following weights and capacities:

Basic empty weight . . . . 3,200 lb
Zero fuel weight. . . . . 4,400 lb
Maximum takeoff weight. . . . . 5,200 lb
Maximum usable fuel . . . . 180 gal

1. Calculate the useful load:

Maximum takeoff weight. . . . . 5,200 lb
Basic empty weight . . . . –3,200 lb
Useful load. . . . . 2,000 lb

The useful load is the maximum combination of usable fuel, passengers, baggage, and cargo that the airplane is capable of carrying.

2. Calculate the payload:

Zero fuel weight. . . . . .4,400 lb
Basic empty weight . . . . –3,200 lb
Payload. . . . . .1,200 lb

The payload is the maximum combination of passengers, baggage, and cargo that the airplane is capable of carrying. A zero fuel weight, if published, is the limiting weight.

3. Calculate the fuel capacity at maximum payload (1,200 lb):

Maximum takeoff weight. .5,200 lb
Zero fuel weight. . . . .–4,400 lb
Fuel allowed. . . . . .800 lb

Assuming maximum payload, the only weight permitted in excess of the zero fuel weight must consist of usable fuel. In this case, 133.3 gallons (gal).

4. Calculate the payload at maximum fuel capacity (180 gal):

Basic empty weight . . . . .3,200 lb
Maximum usable fuel . . . .+1,080 lb
Weight with max. fuel . . . . .4,280 lb
Maximum takeoff weight. . . . . .5,200 lb
Weight with max. fuel . . . .–4,280 lb
Payload allowed. . . . . .920 lb

Assuming maximum fuel, the payload is the difference between the weight of the fueled airplane and the maximum takeoff weight.

Some multiengine airplanes have a ramp weight, which is in excess of the maximum takeoff weight. The ramp weight is an allowance for fuel that would be burned during taxi and run-up, permitting a takeoff at full maximum takeoff weight. The airplane must weigh no more than maximum takeoff weight at the beginning of the takeoff roll.

A maximum landing weight is a limitation against landing at a weight in excess of the published value. This requires preflight planning of fuel burn to ensure that the airplane weight upon arrival at destination is at or below the maximum landing weight. In the event of an emergency requiring an immediate landing, the pilot should recognize that the structural margins designed into the airplane are not fully available when over landing weight. An overweight landing inspection may be advisable—the service manual or manufacturer should be consulted.

Although the foregoing problems only dealt with weight, the balance portion of weight and balance is equally vital. The flight characteristics of the multiengine airplane vary significantly with shifts of the center of gravity (CG) within the approved envelope.

At forward CG, the airplane is more stable, with a slightly higher stalling speed, a slightly slower cruising speed, and

favorable stall characteristics. At aft CG, the airplane is less stable, with a slightly lower stalling speed, a slightly faster cruising speed, and less desirable stall characteristics. Forward CG limits are usually determined in certification by elevator/stabilator authority in the landing roundout. Aft CG limits are determined by the minimum acceptable longitudinal stability. It is contrary to the airplane's operating limitations and 14 CFR to exceed any weight and balance parameter.

Some multiengine airplanes may require ballast to remain within CG limits under certain loading conditions. Several models require ballast in the aft baggage compartment with only a student and instructor on board to avoid exceeding the forward CG limit. When passengers are seated in the aft-most seats of some models, ballast or baggage may be required in the nose baggage compartment to avoid exceeding the aft CG limit. The pilot must direct the seating of passengers and placement of baggage and cargo to achieve a CG within the approved envelope. Most multiengine airplanes have general loading recommendations in the weight and balance section of the AFM/POH. When ballast is added, it must be securely tied down, and it must not exceed the maximum allowable floor loading.

Some airplanes make use of a special weight and balance plotter. It consists of several movable parts that can be adjusted over a plotting board on which the CG envelope is printed. The reverse side of the typical plotter contains general loading recommendations for the particular airplane. A pencil line plot can be made directly on the CG envelope imprinted on the working side of the plotting board. This plot can easily be erased and recalculated anew for each flight. This plotter is to be used only for the make and model airplane for which it was designed.

Ground Operation

Good habits learned with single-engine airplanes are directly applicable to multiengine airplanes for preflight and engine start. Upon placing the airplane in motion to taxi, the new multiengine pilot notices several differences, however. The most obvious is the increased wingspan and the need for even greater vigilance while taxiing in close quarters. Ground handling may seem somewhat ponderous and the multiengine airplane is not as nimble as the typical two- or four-place single-engine airplane. As always, use care not to ride the brakes by keeping engine power to a minimum. One ground handling advantage of the multiengine airplane over single-engine airplanes is the differential power capability. Turning with an assist from differential power minimizes both the need for brakes during turns and the turning radius.

The pilot should be aware, however, that making a sharp turn assisted by brakes and differential power can cause the airplane to pivot about a stationary inboard wheel and landing gear. This is abuse for which the airplane was not designed and should be guarded against. Unless otherwise directed by the AFM/POH, all ground operations should be conducted with the cowl flaps fully open. The use of strobe lights is normally deferred until taxiing onto the active runway.

Normal and Crosswind Takeoff and Climb

With the Before Takeoff checklist, which includes a pre-takeoff safety brief complete and air traffic control (ATC) clearance received, the airplane should be taxied into position on the runway centerline. If departing from an airport without an operating control tower, a careful check for approaching aircraft should be made along with a radio advisory on the appropriate frequency. Sharp turns onto the runway combined with a rolling takeoff are not a good operating practice and may be prohibited by the AFM/POH due to the possibility of "unporting" a fuel tank pickup. (The takeoff itself may be prohibited by the AFM/POH under any circumstances below certain fuel levels.) The flight controls should be positioned for a crosswind, if present. Exterior lights, such as landing and taxi lights, and wingtip strobes should be illuminated immediately prior to initiating the takeoff roll, day or night. If holding in takeoff position for any length of time, particularly at night, the pilot should activate all exterior lights upon taxiing into position.

Takeoff power should be set as recommended in the AFM/POH. With normally aspirated (non-turbocharged) engines, this is full throttle. Full throttle is also used in most turbocharged engines. There are some turbocharged engines, however, that require the pilot to set a specific power setting, usually just below red line manifold pressure. This yields takeoff power with less than full throttle travel. Turbocharged engines often require special consideration. Throttle motion with turbocharged engines should be exceptionally smooth and deliberate. It is acceptable, and may even be desirable, to hold the airplane in position with brakes as the throttles are advanced. Brake release customarily occurs after significant boost from the turbocharger is established. This prevents wasting runway with slow, partial throttle acceleration as the engine power is increased. If runway length or obstacle clearance is critical, full power should be set before brake release as specified in the performance charts.

As takeoff power is established, initial attention should be divided between tracking the runway centerline and monitoring the engine gauges. Many novice multiengine pilots tend to fixate on the airspeed indicator just as soon as the airplane begins its takeoff roll. Instead, the pilot should confirm that both engines are developing full-rated manifold pressure and rpm, and that as the fuel flows, fuel pressures, exhaust gas temperatures (EGTs), and oil pressures are matched in their normal ranges. A directed and purposeful scan of the engine gauges can be accomplished well before the airplane approaches rotation speed. If a crosswind is present, the aileron displacement in the direction of the crosswind may be reduced as the airplane accelerates. The elevator/stabilator control should be held neutral throughout.

Full rated takeoff power should be used for every takeoff. Partial power takeoffs are not recommended. There is no evidence to suggest that the life of modern reciprocating engines is prolonged by partial power takeoffs. Paradoxically, excessive heat and engine wear can occur with partial power as the fuel metering system fails to deliver the slightly over-rich mixture vital for engine cooling during takeoff.

There are several key airspeeds to be noted during the takeoff and climb sequence in any twin. The first speed to consider is V_{MC}. If an engine fails below V_{MC} while the airplane is on the ground, the takeoff must be rejected. Directional control can only be maintained by promptly closing both throttles and using rudder and brakes as required. If an engine fails below V_{MC} while airborne, directional control is not possible with the remaining engine producing takeoff power. On takeoffs, therefore, the airplane should never be airborne before the airspeed reaches and exceeds V_{MC}. Pilots should use the manufacturer's recommended rotation speed (V_R) or lift-off speed (V_{LOF}). If no such speeds are published, a minimum of V_{MC} plus 5 knots should be used for V_R.

The rotation to a takeoff pitch attitude is performed with smooth control inputs. With a crosswind, the pilot should ensure that the landing gear does not momentarily touch the runway after the airplane has lifted off, as a side drift is present. The rotation may be accomplished more positively and/or at a higher speed under these conditions. However, the pilot should keep in mind that the AFM/POH performance figures for accelerate-stop distance, takeoff ground roll, and distance to clear an obstacle were calculated at the recommended V_R and/or V_{LOF} speed.

After lift-off, the next consideration is to gain altitude as rapidly as possible. To assist the pilot in takeoff and initial climb profile, some AFM/POHs give a "50-foot" or "50-foot barrier" speed to use as a target during rotation, lift-off, and acceleration to V_Y. Prior to takeoff, pilots should review the takeoff distance to 50 feet above ground level (AGL) and the stopping distance from 50 feet AGL and add the distance together. If the runway is no longer than the total value, the odds are very good that if anything fails, it will be an off-runway landing at the least. After leaving the ground, altitude gain is more important than achieving an excess of airspeed. Experience has shown that excessive speed cannot be effectively converted into altitude in the event of an engine failure. Additional altitude increases the time available to recognize and respond to any aircraft abnormality or emergency during the climb segment.

Excessive climb attitudes can be just as dangerous as excessive airspeed. Steep climb attitudes limit forward visibility and impede the pilot's ability to detect and avoid other traffic. The airplane should be allowed to accelerate in a shallow climb to attain V_Y, the best all-engine rate-of-climb speed. V_Y should then be maintained until a safe single-engine maneuvering altitude, considering terrain and obstructions is achieved. Any speed above or below V_Y reduces the performance of the airplane. Even with all engines operating normally, terrain and obstruction clearance during the initial climb after takeoff is an important preflight consideration. Most airliners and most turbine powered airplanes climb out at an attitude that yields best rate of climb (V_Y) usually utilizing a flight management system (FMS).

When to raise the landing gear after takeoff depends on several factors. Normally, the gear should be retracted when there is insufficient runway available for landing and after a positive rate of climb is established as indicated on the altimeter. If an excessive amount of runway is available, it would not be prudent to leave the landing gear down for an extended period of time and sacrifice climb performance and acceleration. Leaving the gear extended after the point at which a landing cannot be accomplished on the runway is a hazard. In some multiengine airplanes, operating in a high-density altitude environment, a positive rate of climb with the landing gear down is not possible. Waiting for a positive rate of climb under these conditions is not practicable. An important point to remember is that raising the landing gear as early as possible after liftoff drastically decreases the drag profile and significantly increases climb performance should an engine failure occur. An equally important point to remember is that leaving the gear down to land on sufficient runway or overrun is a much better option than landing with the gear retracted. A general recommendation is to raise the landing gear not later than V_{YSE} airspeed, and once the gear is up, consider it a GO commitment if climb performance is available. Some AFM/POHs direct the pilot to apply the wheel brakes momentarily after lift-off to stop wheel rotation prior to landing gear retraction. If flaps were extended for takeoff, they should be retracted as recommended in the AFM/POH.

Once a safe, single-engine maneuvering altitude has been reached, typically a minimum of 400–500 feet AGL, the transition to an en route climb speed should be made. This speed is higher than V_Y and is usually maintained to cruising altitude. En route climb speed gives better visibility, increased engine cooling, and a higher groundspeed. Takeoff power can be reduced, if desired, as the transition to en route climb speed is made.

Some airplanes have a climb power setting published in the AFM/POH as a recommendation (or sometimes as a limitation), which should then be set for en route climb. If there is no climb power setting published, it is customary, but not a requirement, to reduce manifold pressure and rpm somewhat for en route climb. The propellers are usually synchronized after the first power reduction and the yaw damper, if installed, engaged. The AFM/POH may also recommend leaning the mixtures during climb. The Climb checklist should be accomplished as traffic and work load allow. *[Figure 12-7]*

Level Off and Cruise

Upon leveling off at cruising altitude, the pilot should allow the airplane to accelerate at climb power until cruising airspeed is achieved, and then cruise power and rpm should be set. To extract the maximum cruise performance from any airplane, the power setting tables provided by the manufacturer should be closely followed. If the cylinder head and oil temperatures are within their normal ranges, the cowl flaps may be closed. When the engine temperatures have stabilized, the mixtures may be leaned per AFM/POH recommendations. The remainder of the Cruise checklist should be completed by this point.

Fuel management in multiengine airplanes is often more complex than in single-engine airplanes. Depending upon system design, the pilot may need to select between main tanks and auxiliary tanks or even employ fuel transfer from one tank to another. In complex fuel systems, limitations are often found restricting the use of some tanks to level flight only or requiring a reserve of fuel in the main tanks for descent and landing. Electric fuel pump operation can vary widely among different models also, particularly during tank switching or fuel transfer. Some fuel pumps are to be on for takeoff and landing; others are to be off. There is simply no substitute for thorough systems and AFM/POH knowledge when operating complex aircraft.

Normal Approach and Landing

Given the higher cruising speed (and frequently altitude) of multiengine airplanes over most single-engine airplanes, the descent must be planned in advance. A hurried, last minute descent with power at or near idle is inefficient and can cause excessive engine cooling. It may also lead to passenger discomfort, particularly if the airplane is unpressurized. As a rule of thumb, if terrain and passenger conditions permit, a maximum of a 500 fpm rate of descent should be planned. Pressurized airplanes can plan for higher descent rates, if desired.

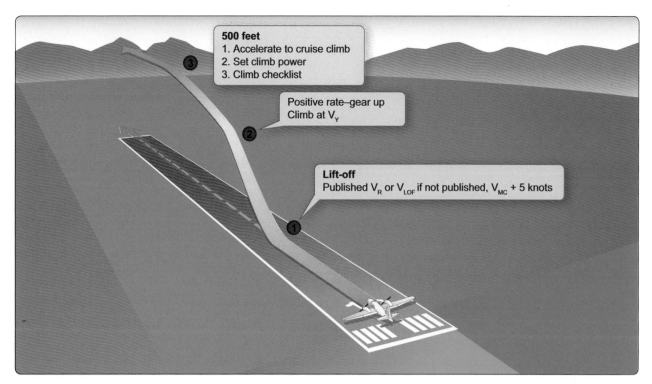

Figure 12-7. *Takeoff and climb profile.*

In a descent, some airplanes require a minimum EGT or may have a minimum power setting or cylinder head temperature to observe. In any case, combinations of very low manifold pressure and high rpm settings are strongly discouraged by engine manufacturers. If higher descent rates are necessary, the pilot should consider extending partial flaps or lowering the landing gear before retarding the power excessively. The Descent checklist should be initiated upon leaving cruising altitude and completed before arrival in the terminal area. Upon arrival in the terminal area, pilots are encouraged to turn on their landing and recognition lights when operating below 10,000 feet, day or night, and especially when operating within 10 miles of any airport or in conditions of reduced visibility.

The traffic pattern and approach are typically flown at somewhat higher indicated airspeeds in a multiengine airplane contrasted to most single-engine airplanes. The pilot may allow for this through an early start on the Before Landing checklist. This provides time for proper planning, spacing, and thinking well ahead of the airplane. Many multiengine airplanes have partial flap extension speeds above V_{FE}, and partial flaps can be deployed prior to traffic pattern entry. Normally, the landing gear should be selected and confirmed down when abeam the intended point of landing as the downwind leg is flown. *[Figure 12-8]*

The FAA recommends a stabilized approach concept. To the greatest extent practical, on final approach and within 500 feet AGL, the airplane should be on speed, in trim, configured for landing, tracking the extended centerline of the runway, and established in a constant angle of descent towards an aim point in the touchdown zone. Absent unusual flight conditions, only minor corrections are required to maintain this approach to the round out and touchdown.

The final approach should be made with power and at a speed recommended by the manufacturer; if a recommended speed is not furnished, the speed should be no slower than the single-engine best rate-of-climb speed (V_{YSE}) until short final with the landing assured, but in no case less than critical engine-out minimum control speed (V_{MC}). Some multiengine pilots prefer to delay full flap extension to short final with the landing assured. This is an acceptable technique with appropriate experience and familiarity with the airplane.

In the round out for landing, residual power is gradually reduced to idle. With the higher wing loading of multiengine airplanes and with the drag from two windmilling propellers, there is minimal float. Full stall landings are generally undesirable in twins. The airplane should be held off as with a high performance single-engine model, allowing touchdown of the main wheels prior to a full stall.

Under favorable wind and runway conditions, the nosewheel can be held off for best aerodynamic braking. Even as the nosewheel is gently lowered to the runway centerline, continued elevator back pressure greatly assists the wheel brakes in stopping the airplane.

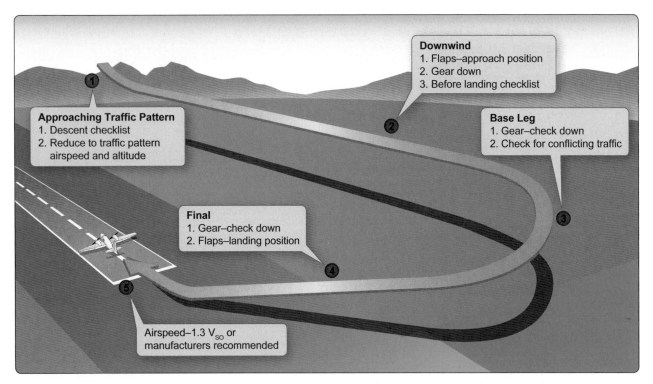

Figure 12-8. *Normal two-engine approach and landing.*

If runway length is critical, or with a strong crosswind, or if the surface is contaminated with water, ice or snow, it is undesirable to rely solely on aerodynamic braking after touchdown. The full weight of the airplane should be placed on the wheels as soon as practicable. The wheel brakes are more effective than aerodynamic braking alone in decelerating the airplane.

Once on the ground, elevator back pressure should be used to place additional weight on the main wheels and to add additional drag. When necessary, wing flap retraction also adds additional weight to the wheels and improves braking effectivity. Flap retraction during the landing rollout is discouraged, however, unless there is a clear, operational need. It should not be accomplished as routine with each landing.

Some multiengine airplanes, particularly those of the cabin class variety, can be flown through the round out and touchdown with a small amount of power. This is an acceptable technique to prevent high sink rates and to cushion the touchdown. The pilot should keep in mind, however, that the primary purpose in landing is to get the airplane down and stopped. This technique should only be attempted when there is a generous margin of runway length. As propeller blast flows directly over the wings, lift as well as thrust is produced. The pilot should taxi clear of the runway as soon as speed and safety permit, and then accomplish the After Landing checklist. Ordinarily, no attempt should be made to retract the wing flaps or perform other checklist duties until the airplane has been brought to a halt when clear of the active runway. Exceptions to this would be the rare operational needs discussed above, to relieve the weight from the wings and place it on the wheels. In these cases, AFM/POH guidance should be followed. The pilot should not indiscriminately reach out for any switch or control on landing rollout. An inadvertent landing gear retraction while meaning to retract the wing flaps may result.

Crosswind Approach and Landing

The multiengine airplane is often easier to land in a crosswind than a single-engine airplane due to its higher approach and landing speed. In any event, the principles are no different between singles and twins. Prior to touchdown, the longitudinal axis must be aligned with the runway centerline to avoid landing gear side loads.

The two primary methods, crab and wing-low, are typically used in conjunction with each other. As soon as the airplane rolls out onto final approach, the crab angle to track the extended runway centerline is established. This is coordinated flight with adjustments to heading to compensate for wind drift either left or right. Prior to touchdown, the transition to a sideslip is made with the upwind wing lowered and opposite rudder applied to prevent a turn. The airplane touches down on the landing gear of the upwind wing first, followed by that of the downwind wing, and then the nose gear. Follow-through with the flight controls involves an increasing application of aileron into the wind until full control deflection is reached.

The point at which the transition from the crab to the sideslip is made is dependent upon pilot familiarity with the airplane and experience. With high skill and experience levels, the transition can be made during the round out just before touchdown. With lesser skill and experience levels, the transition is made at increasing distances from the runway. Some multiengine airplanes (as some single-engine airplanes) have AFM/POH limitations against slips in excess of a certain time period; 30 seconds, for example. This is to prevent engine power loss from fuel starvation as the fuel in the tank of the lowered wing flows towards the wingtip, away from the fuel pickup point. This time limit must be observed if the wing-low method is utilized.

Some multiengine pilots prefer to use differential power to assist in crosswind landings. The asymmetrical thrust produces a yawing moment little different from that produced by the rudder. When the upwind wing is lowered, power on the upwind engine is increased to prevent the airplane from turning. This alternate technique is completely acceptable, but most pilots feel they can react to changing wind conditions quicker with rudder and aileron than throttle movement. This is especially true with turbocharged engines where the throttle response may lag momentarily. The differential power technique should be practiced with an instructor before being attempted alone.

Short-Field Takeoff and Climb

The short-field takeoff and climb differs from the normal takeoff and climb in the airspeeds and initial climb profile. Some AFM/POHs give separate short-field takeoff procedures and performance charts that recommend specific flap settings and airspeeds. Other AFM/POHs do not provide separate short-field procedures. In the absence of such specific procedures, the airplane should be operated only as recommended in the AFM/POH. No operations should be conducted contrary to the recommendations in the AFM/POH.

On short-field takeoffs in general, just after rotation and lift-off, the airplane should be allowed to accelerate to V_X, making the initial climb over obstacles at V_X and transitioning to V_Y as obstacles are cleared. *[Figure 12-9]*

When partial flaps are recommended for short-field takeoffs, many light-twins have a strong tendency to become airborne prior to V_{MC} plus 5 knots. Attempting to prevent premature lift-off with forward elevator pressure results in wheel barrowing. To prevent this, allow the airplane to become airborne, but only a few inches above the runway. The pilot should be prepared to promptly abort the takeoff and land in the event of engine failure on takeoff with landing gear and flaps extended at airspeeds below V_X.

Engine failure on takeoff, particularly with obstructions, is compounded by the low airspeeds and steep climb attitudes utilized in short-field takeoffs. V_X and V_{XSE} are often perilously close to V_{MC}, leaving scant margin for error in the event of engine failure as V_{XSE} is assumed. If flaps were used for takeoff, the engine failure situation becomes even more critical due to the additional drag incurred. If V_X is less than 5 knots higher than V_{MC}, give strong consideration to reducing useful load or using another runway in order to increase the takeoff margins so that a short-field technique is not required.

Short-Field Approach and Landing

The primary elements of a short-field approach and landing do not differ significantly from a normal approach and landing. Many manufacturers do not publish short-field landing techniques or performance charts in the AFM/POH. In the absence of specific short-field approach and landing procedures, the airplane should be operated as recommended in the AFM/POH. No operations should be conducted contrary to the AFM/POH recommendations.

The emphasis in a short-field approach is on configuration (full flaps), a stabilized approach with a constant angle of descent, and precise airspeed control. As part of a short-

Figure 12-9. *Short-field takeoff and climb.*

field approach and landing procedure, some AFM/POHs recommend a slightly slower than normal approach airspeed. If no such slower speed is published, use the AFM/POH-recommended normal approach speed.

Full flaps are used to provide the steepest approach angle. If obstacles are present, the approach should be planned so that no drastic power reductions are required after they are cleared. The power should be smoothly reduced to idle in the round out prior to touchdown. Pilots should keep in mind that the propeller blast blows over the wings providing some lift in addition to thrust. Reducing power significantly, just after obstacle clearance, usually results in a sudden, high sink rate that may lead to a hard landing. After the short-field touchdown, maximum stopping effort is achieved by retracting the wing flaps, adding back pressure to the elevator/stabilator, and applying heavy braking. However, if the runway length permits, the wing flaps should be left in the extended position until the airplane has been stopped clear of the runway. There is always a significant risk of retracting the landing gear instead of the wing flaps when flap retraction is attempted on the landing rollout.

Landing conditions that involve a short-field, high-winds, or strong crosswinds are just about the only situations where flap retraction on the landing rollout should be considered. When there is an operational need to retract the flaps just after touchdown, it must be done deliberately with the flap handle positively identified before it is moved.

Go-Around

When the decision to go around is made, the throttles should be advanced to takeoff power. With adequate airspeed, the airplane should be placed in a climb pitch attitude. These actions, which are accomplished simultaneously, arrest the sink rate and place the airplane in the proper attitude for transition to a climb. The initial target airspeed is V_Y or V_X if obstructions are present. With sufficient airspeed, the flaps should be retracted from full to an intermediate position and

the landing gear retracted when there is a positive rate of climb and no chance of runway contact. The remaining flaps should then be retracted. *[Figure 12-10]*

If the go-around was initiated due to conflicting traffic on the ground or aloft, the pilot should maneuver to the side so as to keep the conflicting traffic in sight. This may involve a shallow bank turn to offset and then parallel the runway/landing area.

If the airplane was in trim for the landing approach when the go-around was commenced, it soon requires a great deal of forward elevator/stabilator pressure as the airplane accelerates away in a climb. The pilot should apply appropriate forward pressure to maintain the desired pitch attitude. Trim should be commenced immediately. The Balked Landing checklist should be reviewed as work load permits.

Flaps should be retracted before the landing gear for two reasons. First, on most airplanes, full flaps produce more drag than the extended landing gear. Secondly, the airplane tends to settle somewhat with flap retraction, and the landing gear should be down in the event of an inadvertent, momentary touchdown.

Many multiengine airplanes have a landing gear retraction speed significantly less than the extension speed. Care should be exercised during the go-around not to exceed the retraction speed. If the pilot desires to return for a landing, it is essential to re-accomplish the entire Before Landing checklist. An interruption to a pilot's habit patterns, such as a go-around, is a classic scenario for a subsequent gear up landing.

The preceding discussion about doing a go-around assumes that the maneuver was initiated from normal approach speeds or faster. If the go-around was initiated from a low airspeed, the initial pitch up to a climb attitude must be tempered with the necessity of maintaining adequate flying speed throughout the maneuver. Examples of where this applies include a go-around initiated from the landing round out or recovery

Figure 12-10. *Go-around procedure.*

from a bad bounce, as well as a go-around initiated due to an inadvertent approach to a stall. The first priority is always to maintain control and obtain adequate flying speed. A few moments of level or near level flight may be required as the airplane accelerates up to climb speed.

Rejected Takeoff

A takeoff can be rejected for the same reasons a takeoff in a single-engine airplane would be rejected. Once the decision to reject a takeoff is made, the pilot should promptly close both throttles and maintain directional control with the rudder, nosewheel steering, and brakes. Aggressive use of rudder, nosewheel steering, and brakes may be required to keep the airplane on the runway. Particularly, if an engine failure is not immediately recognized and accompanied by prompt closure of both throttles. However, the primary objective is not necessarily to stop the airplane in the shortest distance, but to maintain control of the airplane as it decelerates. In some situations, it may be preferable to continue into the overrun area under control, rather than risk directional control loss, landing gear collapse, or tire/brake failure in an attempt to stop the airplane in the shortest possible distance.

Engine Failure After Lift-Off

A takeoff or go-around is the most critical time to suffer an engine failure. The airplane will be slow, close to the ground, and may even have landing gear and flaps extended. Altitude and time is minimal. Until feathered, the propeller of the failed engine is windmilling, producing a great deal of drag and yawing tendency. Airplane climb performance is marginal or even non-existent, and obstructions may lie ahead. An emergency contingency plan and safety brief should be clearly understood well before the takeoff roll commences. An engine failure before a predetermined airspeed or point results in an aborted takeoff. An engine failure after a certain airspeed and point, with the gear up, and climb performance assured result in a continued takeoff. With loss of an engine, it is paramount to maintain airplane control and comply with the manufacturer's recommended emergency procedures. Complete failure of one engine shortly after takeoff can be broadly categorized into one of three following scenarios.

Landing Gear Down

If the engine failure occurs prior to selecting the landing gear to the UP position *[Figure 12-11]*: Keep the nose as straight as possible, close both throttles, allow the nose to maintain airspeed and descend to the runway. Concentrate on a normal landing and do not force the aircraft on the ground. Land on the remaining runway or overrun. Depending upon how quickly the pilot reacts to the sudden yaw, the airplane may run off the side of the runway by the time action is taken. There are really no other practical options. As discussed earlier, the chances of maintaining directional control while retracting the flaps (if extended), landing gear, feathering the propeller, and accelerating are minimal. On some airplanes with a single-engine-driven hydraulic pump, failure of that engine means the only way to raise the landing gear is to allow the engine to windmill or to use a hand pump. This is not a viable alternative during takeoff.

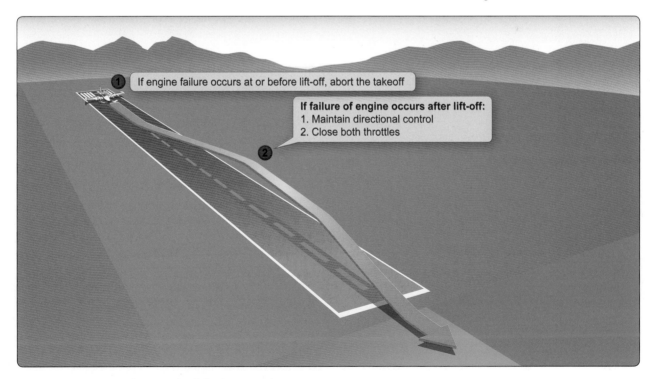

Figure 12-11. *Engine failure on takeoff, landing gear down.*

Figure 12-12. *Engine failure on takeoff, inadequate climb performance.*

Landing Gear Control Selected Up, Single-Engine Climb Performance Inadequate

When operating near or above the single-engine ceiling and an engine failure is experienced shortly after lift-off, a landing must be accomplished on whatever essentially lies ahead. *[Figure 12-12]* There is also the option of continuing ahead, in a descent at V_{YSE} with the remaining engine producing power, as long as the pilot is not tempted to remain airborne beyond the airplane's performance capability. Remaining airborne and bleeding off airspeed in a futile attempt to maintain altitude is almost invariably fatal. Landing under control is paramount. The greatest hazard in a single-engine takeoff is attempting to fly when it is not within the performance capability of the airplane to do so. An accident is inevitable.

Analysis of engine failures on takeoff reveals a very high success rate of off-airport engine inoperative landings when the airplane is landed under control. Analysis also reveals a very high fatality rate in stall spin accidents when the pilot attempts flight beyond the performance capability of the airplane.

As mentioned previously, if the airplane's landing gear retraction mechanism is dependent upon hydraulic pressure from a certain engine-driven pump, failure of that engine can mean a loss of hundreds of feet of altitude as the pilot either windmills the engine to provide hydraulic pressure to raise the gear or raises it manually with a backup pump.

Landing Gear Control Selected Up, Single-Engine Climb Performance Adequate

If the single-engine rate of climb is adequate, the procedures for continued flight should be followed. *[Figure 12-13]* There are four areas of concern: control, configuration, climb, and checklist.

Control

The first consideration following engine failure during takeoff is to maintain control of the airplane. Maintaining directional control with prompt and often aggressive rudder application and STOPPING THE YAW is critical to the safety of flight. Ensure that airspeed stays above V_{MC}. If the yaw cannot be controlled with full rudder applied, reducing thrust on the

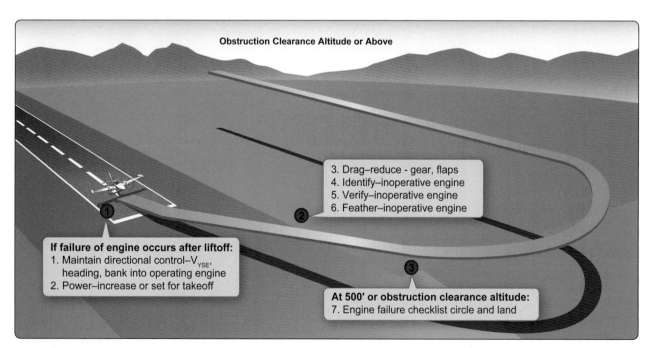

Figure 12-13. *Landing gear up—adequate climb performance.*

operative engine is the only alternative. Attempting to correct the roll with aileron without first applying rudder increases drag and adverse yaw and further degrades directional control. After rudder is applied to stop the yaw, a slight amount of aileron should be used to bank the airplane toward the operative engine. This is the most efficient way to control the aircraft, minimize drag, and gain the most performance. Control forces, particularly on the rudder, may be high. The pitch attitude for V_{YSE} has to be lowered from that of V_Y. At least 5° of bank should be used initially to stop the yaw and maintain directional control. This initial bank input is held only momentarily, just long enough to establish or ensure directional control. Climb performance suffers when bank angles exceed approximately 2 or 3°, but obtaining and maintaining V_{YSE} and directional control are paramount. Trim should be adjusted to lower the control forces.

Configuration

The memory items from the Engine Failure After Takeoff checklist should be promptly executed to configure the airplane for climb. *[Figure 12-14]* The specific procedures to follow are found in the AFM/POH and checklist for the particular airplane. Most direct the pilot to assume V_{YSE}, set takeoff power, retract the flaps and landing gear, identify, verify, and feather the failed engine. (On some airplanes, the landing gear is to be retracted before the flaps.)

The "identify" step is for the pilot to initially identify the failed engine. Confirmation on the engine gauges may or may not be possible, depending upon the failure mode. Identification should be primarily through the control inputs required to maintain straight flight, not the engine gauges. The "verify" step directs the pilot to retard the throttle of the engine thought to have failed. No change in performance

Engine Failure After Takeoff	
Airspeed	Maintain V_{YSE}
Mixtures	**RICH**
Propellers	**HIGH RPM**
Throttles	**FULL POWER**
Flaps	**UP**
Landing gear	**UP**
Identify	**Determine failed engine**
Verify	**Close throttle of failed engine**
Propeller	**FEATHER**
Trim tabs	ADJUST
Failed engine	SECURE
As soon as practical	LAND
Bold-faced items require immediate action and are to be accomplished from memory.	

Figure 12-14. *Typical "engine failure after takeoff" emergency checklist.*

when the suspected throttle is retarded is verification that the correct engine has been identified as failed. The corresponding propeller control should be brought fully aft to feather the engine.

Climb

As soon as directional control is established and the airplane configured for climb, the bank angle should be reduced to that producing best climb performance. Without specific guidance for zero sideslip, a bank of 2° and one-third to one-half ball deflection on the slip/skid indicator is suggested. V_{YSE} is maintained with pitch control. As turning flight reduces climb performance, climb should be made straight ahead or with shallow turns to avoid obstacles to an altitude of at least 400 feet AGL before attempting a return to the airport.

Checklist

Having accomplished the memory items from the Engine Failure After Takeoff checklist, the printed copy should be reviewed as time permits. The Securing Failed Engine checklist should then be accomplished. *[Figure 12-15]* Unless the pilot suspects an engine fire, the remaining items should be accomplished deliberately and without undue haste. Airplane control should never be sacrificed to execute the remaining checklists. The priority items have already been accomplished from memory.

Other than closing the cowl flap of the failed engine, none of these items, if left undone, adversely affects airplane climb performance. There is a distinct possibility of actuating an incorrect switch or control if the procedure is rushed. The pilot should concentrate on flying the airplane and extracting maximum performance. If an ATC facility is available, an emergency should be declared.

The memory items in the Engine Failure After Takeoff checklist may be redundant with the airplane's existing configuration. For example, in the third takeoff scenario, the gear and flaps were assumed to already be retracted, yet the memory items included gear and flaps. This is not an oversight. The purpose of the memory items is to either initiate the appropriate action or to confirm that a condition exists. Action on each item may not be required in all cases. The memory items also apply to more than one circumstance. In an engine failure from a go-around, for example, the landing gear and flaps would likely be extended when the failure occurred.

The three preceding takeoff scenarios all include the landing gear as a key element in the decision to land or continue. With the landing gear selector in the DOWN position, for example, continued takeoff and climb is not recommended. This situation, however, is not justification to retract the

Securing Failed Engine	
Mixture	IDLE CUT OFF
Magnetos	OFF
Alternator	OFF
Cowl flap	CLOSE
Boost pump	OFF
Fuel selector	OFF
Prop sync	OFF
Electrical load	Reduce
Crossfeed	Consider

Figure 12-15. *Typical "securing failed engine" emergency checklist.*

landing gear the moment the airplane lifts off the surface on takeoff as a normal procedure. The landing gear should remain selected down as long as there is usable runway or overrun available to land on. The use of wing flaps for takeoff virtually eliminates the likelihood of a single-engine climb until the flaps are retracted.

There are two time-tested memory aids the pilot may find useful in dealing with engine-out scenarios. The first, "dead foot–dead engine" is used to assist in identifying the failed engine. Depending on the failure mode, the pilot will not be able to consistently identify the failed engine in a timely manner from the engine gauges. In maintaining directional control, however, rudder pressure is exerted on the side (left or right) of the airplane with the operating engine. Thus, the "dead foot" is on the same side as the "dead engine." Variations on this saying include "idle foot–idle engine" and "working foot–working engine."

The second memory aid has to do with climb performance. The phrase "raise the dead" is a reminder that the best climb performance is obtained with a very shallow bank, about 2° toward the operating engine. Therefore, the inoperative, or "dead" engine should be "raised" with a very slight bank.

Not all engine power losses are complete failures. Sometimes the failure mode is such that partial power may be available. If there is a performance loss when the throttle of the affected engine is retarded, the pilot should consider allowing it to run until altitude and airspeed permit safe single-engine flight, if this can be done without compromising safety. Attempts to save a malfunctioning engine can lead to a loss of the entire airplane.

Engine Failure During Flight

Engine failures well above the ground are handled differently than those occurring at lower speeds and altitudes. Cruise airspeed allows better airplane control and altitude, which may permit time for a possible diagnosis and remedy of the failure. Maintaining airplane control, however, is still paramount. Airplanes have been lost at altitude due to apparent fixation on the engine problem to the detriment of flying the airplane.

Not all engine failures or malfunctions are catastrophic in nature (catastrophic meaning a major mechanical failure that damages the engine and precludes further engine operation). Many cases of power loss are related to fuel starvation, where restoration of power may be made with the selection of another tank. An orderly inventory of gauges and switches may reveal the problem. Carburetor heat or alternate air can be selected. The affected engine may run smoothly on just one magneto or at a lower power setting. Altering the mixture may help. If fuel vapor formation is suspected, fuel boost pump operation may be used to eliminate flow and pressure fluctuations.

Although it is a natural desire among pilots to save an ailing engine with a precautionary shutdown, the engine should be left running if there is any doubt as to needing it for further safe flight. Catastrophic failure accompanied by heavy vibration, smoke, blistering paint, or large trails of oil, on the other hand, indicate a critical situation. The affected engine should be feathered and the Securing Failed Engine checklist completed. The pilot should divert to the nearest suitable airport and declare an emergency with ATC for priority handling.

Fuel crossfeed is a method of getting fuel from a tank on one side of the airplane to an operating engine on the other. Crossfeed is used for extended single-engine operation. If a suitable airport is close at hand, there is no need to consider crossfeed. If prolonged flight on a single-engine is inevitable due to airport non-availability, then crossfeed allows use of fuel that would otherwise be unavailable to the operating engine. It also permits the pilot to balance the fuel consumption to avoid an out-of-balance wing heaviness.

The AFM/POH procedures for crossfeed vary widely. Thorough fuel system knowledge is essential if crossfeed is to be conducted. Fuel selector positions and fuel boost pump usage for crossfeed differ greatly among multiengine airplanes. Prior to landing, crossfeed should be terminated and the operating engine returned to its main tank fuel supply.

If the airplane is above its single-engine absolute ceiling at the time of engine failure, it slowly loses altitude. The pilot should maintain V_{YSE} to minimize the rate of altitude loss. This "drift down" rate is greatest immediately following the failure and decreases as the single-engine ceiling is approached. Due to performance variations caused by engine and propeller wear, turbulence, and pilot technique, the airplane may not maintain altitude even at its published single-engine ceiling. Any further rate of sink, however, would likely be modest.

An engine failure in a descent or other low power setting can be deceiving. The dramatic yaw and performance loss is absent. At very low power settings, the pilot may not even be aware of a failure. If a failure is suspected, the pilot should advance both engine mixtures, propellers, and throttles significantly, to the takeoff settings if necessary, to correctly identify the failed engine. The power on the operative engine can always be reduced later.

Engine Inoperative Approach and Landing

The approach and landing with OEI is essentially the same as a two-engine approach and landing. The traffic pattern should be flown at similar altitudes, airspeeds, and key positions as a two-engine approach. The differences are the reduced power available and the fact that the remaining thrust is asymmetrical. A higher-than-normal power setting is necessary on the operative engine.

With adequate airspeed and performance, the landing gear can still be extended on the downwind leg. In which case it should be confirmed DOWN no later than abeam the intended point of landing. Performance permitting, initial extension of wing flaps (typically 10°) and a descent from pattern altitude can also be initiated on the downwind leg. The airspeed should be no slower than V_{YSE}. The direction of the traffic pattern, and therefore the turns, is of no consequence as far as airplane controllability and performance are concerned. It is perfectly acceptable to make turns toward the failed engine.

On the base leg, if performance is adequate, the flaps may be extended to an intermediate setting (typically 25°). If the performance is inadequate, as measured by decay in airspeed or high sink rate, delay further flap extension until closer to the runway. V_{YSE} is still the minimum airspeed to maintain.

On final approach, a normal, 3° glidepath to a landing is desirable. Visual approach slope indicator (VASI) or other vertical path lighting aids should be utilized if available. Slightly steeper approaches may be acceptable. However, a long, flat, low approach should be avoided. Large, sudden power applications or reductions should also be avoided. Maintain V_{YSE} until the landing is assured, then slow to 1.3 V_{SO} or the AFM/POH recommended speed. The final flap setting may be delayed until the landing is assured or the airplane may be landed with partial flaps.

The airplane should remain in trim throughout. The pilot must be prepared, however, for a rudder trim change as the power of the operating engine is reduced to idle in the round out just prior to touchdown. With drag from only one windmilling propeller, the airplane tends to float more than on a two-engine approach. Precise airspeed control therefore is essential, especially when landing on a short, wet, and/or slippery surface.

Some pilots favor resetting the rudder trim to neutral on final and compensating for yaw by holding rudder pressure for the remainder of the approach. This eliminates the rudder trim change close to the ground as the throttle is closed during the round out for landing. This technique eliminates the need for groping for the rudder trim and manipulating it to neutral during final approach, which many pilots find to be highly distracting. AFM/POH recommendations or personal preference should be used.

A single-engine go-around must be avoided. As a practical matter in single-engine approaches, once the airplane is on final approach with landing gear and flaps extended, it is committed to land on the intended runway, on another runway, a taxiway, or grassy infield. The light-twin does not have the performance to climb on one engine with landing gear and flaps extended. Considerable altitude is lost while maintaining V_{YSE} and retracting landing gear and flaps. Losses of 500 feet or more are not unusual. If the landing gear has been lowered with an alternate means of extension, retraction may not be possible, virtually negating any climb capability.

Engine Inoperative Flight Principles

Best single-engine climb performance is obtained at V_{YSE} with maximum available power and minimum drag. After the flaps and landing gear have been retracted and the propeller of the failed engine feathered, a key element in best climb performance is minimizing sideslip.

With a single-engine airplane or a multiengine airplane with both engines operative, sideslip is eliminated when the ball of the turn and bank instrument is centered. This is a condition of zero sideslip, and the airplane is presenting its smallest possible profile to the relative wind. As a result, drag is at its minimum. Pilots know this as coordinated flight.

In a multiengine airplane with an inoperative engine, the centered ball is no longer the indicator of zero sideslip due to asymmetrical thrust. In fact, there is no instrument at all that directly tells the pilot the flight conditions for zero sideslip. In the absence of a yaw string, minimizing sideslip is a matter of placing the airplane at a predetermined bank angle and ball position. The AFM/POH performance charts for single-engine flight were determined at zero sideslip. If this performance is even to be approximated, the zero sideslip technique must be utilized.

There are two different control inputs that can be used to counteract the asymmetrical thrust of a failed engine:

1. Yaw from the rudder

2. The horizontal component of lift that results from bank with the ailerons.

Used individually, neither is correct. Used together in the proper combination, zero sideslip and best climb performance are achieved.

Three different scenarios of airplane control inputs are presented below. Neither of the first two is correct. They are presented to illustrate the reasons for the zero sideslip approach to best climb performance.

1. Engine inoperative flight with wings level and ball centered requires large rudder input towards the operative engine. *[Figure 12-16]* The result is a moderate sideslip towards the inoperative engine. Climb performance is reduced by the moderate sideslip. With wings level, V_{MC} is significantly higher than published as there is no horizontal component of lift available to help the rudder combat asymmetrical thrust.

2. Engine inoperative flight using ailerons alone requires an 8–10° bank angle towards the operative engine. *[Figure 12-17]* This assumes no rudder input. The ball is displaced well towards the operative engine. The result is a large sideslip towards the operative engine. Climb performance is greatly reduced by the large sideslip.

3. Rudder and ailerons used together in the proper combination result in a bank of approximately 2° towards the operative engine. The ball is displaced approximately one-third to one-half towards the operative engine. The result is zero sideslip and maximum climb performance. *[Figure 12-18]* Any attitude other than zero sideslip increases drag, decreasing performance. V_{MC} under these circumstances is higher than published, as less than the 5° bank certification limit is employed.

The precise condition of zero sideslip (bank angle and ball position) varies slightly from model to model and with available power and airspeed. If the airplane is not equipped with counter-rotating propellers, it also varies slightly with the engine failed due to P-factor. The foregoing zero sideslip recommendations apply to reciprocating engine multiengine airplanes flown at V_{YSE} with the inoperative engine feathered. The zero sideslip ball position for straight flight is also the zero sideslip position for turning flight.

Wings level, ball centered, airplane slips toward dead engine. Results: high drag, large control surface deflections required, and rudder and fin in opposition due to sideslip.

Figure 12-16. *Wings level engine-out flight.*

When bank angle is plotted against climb performance for a hypothetical twin, zero sideslip results in the best (however marginal) climb performance or the least rate of descent. Zero bank (all rudder to counteract yaw) degrades climb performance as a result of moderate sideslip. Using bank angle alone (no rudder) severely degrades climb performance as a result of a large sideslip.

The actual bank angle for zero sideslip varies among airplanes from one and one-half to two and one-half degrees. The position of the ball varies from one-third to one-half of a ball width from instrument center.

Figure 12-17. *Excessive bank engine-out flight.*

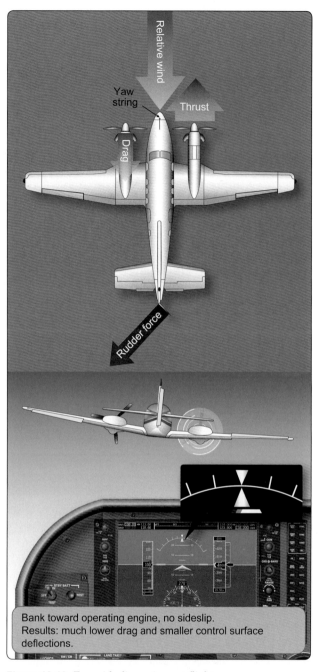

Figure 12-18. *Zero sideslip engine-out flight.*

For any multiengine airplane, zero sideslip can be confirmed through the use of a yaw string. A yaw string is a piece of string or yarn approximately 18 to 36 inches in length taped to the base of the windshield or to the nose near the windshield along the airplane centerline. In two-engine coordinated flight, the relative wind causes the string to align itself with the longitudinal axis of the airplane, and it positions itself straight up the center of the windshield. This is zero sideslip. Experimentation with slips and skids vividly displays the location of the relative wind. Adequate altitude and flying speed must be maintained while accomplishing these maneuvers.

With an engine set to zero thrust (or feathered) and the airplane slowed to V_{YSE}, a climb with maximum power on the remaining engine reveals the precise bank angle and ball deflection required for zero sideslip and best climb performance. Zero sideslip is again indicated by the yaw string when it aligns itself vertically on the windshield. There are very minor changes from this attitude depending upon the engine failed (with non-counter-rotating propellers), power available, airspeed and weight; but without more sensitive testing equipment, these changes are difficult to detect. The only significant difference would be the pitch attitude required to maintain V_{YSE} under different density altitude, power available, and weight conditions.

If a yaw string is attached to the airplane at the time of a V_{MC} demonstration, it is noted that V_{MC} occurs under conditions of sideslip. V_{MC} was not determined under conditions of zero sideslip during aircraft certification and zero sideslip is not part of a V_{MC} demonstration for pilot certification.

To review, there are two different sets of bank angles used in OEI flight.

1. To maintain directional control of a multiengine airplane suffering an engine failure at low speeds (such as climb), momentarily bank at least 5° and a maximum of 10° towards the operative engine as the pitch attitude for V_{YSE} is set. This maneuver should be instinctive to the proficient multiengine pilot and take only 1 to 2 seconds to attain. It is held just long enough to assure directional control as the pitch attitude for V_{YSE} is assumed.

2. To obtain the best climb performance, the airplane must be flown at V_{YSE} and zero sideslip with the failed engine feathered and maximum available power from the operating engine. Zero sideslip is approximately 2° of bank toward the operating engine and a one-third to one-half ball deflection also toward the operating engine. The precise bank angle and ball position varies somewhat with make and model and power available. If above the airplane's single-engine ceiling, this attitude and configuration results in the minimum rate of sink.

In OEI flight at low altitudes and airspeeds such as the initial climb after takeoff, pilots must operate the airplane so as to guard against the three major accident factors: (1) loss of directional control, (2) loss of performance, and (3) loss of flying speed. All have equal potential to be lethal. Loss of flying speed is not a factor, however, when the airplane is operated with due regard for directional control and performance.

Slow Flight

There is nothing unusual about maneuvering during slow flight in a multiengine airplane. Slow flight may be conducted in straight-and-level flight, turns, climbs, or descents. It can also be conducted in the clean configuration, landing configuration, or at any other combination of landing gear and flaps. Slow flight in a multiengine airplane should be conducted so the maneuver can be completed no lower than 3,000 feet AGL or higher if recommended by the manufacturer. In all cases, practicing slow flight should be conducted at an adequate height above the ground for recovery should the airplane inadvertently stall.

Pilots should closely monitor cylinder head and oil temperatures during slow flight. Some high performance multiengine airplanes tend to heat up fairly quickly under some conditions of slow flight, particularly in the landing configuration. Simulated engine failures should not be conducted during slow flight. The airplane will be well below V_{SSE} and very close to V_{MC}. Stability, stall warning, or stall avoidance devices should not be disabled while maneuvering during slow flight.

Stalls

Stall characteristics vary among multiengine airplanes just as they do with single-engine airplanes, and therefore, a pilot must be familiar with them. Yet, the most important stall recovery step in a multiengine airplane is the same as it is in all airplanes: reduce the angle of attack (AOA). For reference, the stall recovery procedure described in Chapter 4 is included in *Figure 12-19*.

Following a reduction in the AOA and the stall warning being eliminated, the wings should be rolled level and power added as needed. Immediate full application of power in a stalled condition has an associated risk due to the possibility of asymmetric thrust. In addition, single-engine stalls or stalls with significantly more power on one engine than the other should not be attempted due to the likelihood of a departure from controlled flight and possible spin entry. Similarly, simulated engine failures should not be performed during stall entry and recovery.

It is recommended that stalls be practiced at an altitude that allows recovery no lower than 3,000 feet AGL for multiengine airplanes, or higher if recommended by the AFM/POH. Losing altitude during recovery from a stall is to be expected.

Power-Off Approach to Stall (Approach and Landing)

A power-off approach to stall is trained and checked to simulate problematic approach and landing scenarios. A power-off approach to stall may be performed with wings level, or from shallow and medium banked turns (20 degrees of bank). To initiate a power-off approach to stall maneuver, the area surrounding the airplane should first be cleared for possible traffic. The airplane should then be slowed and configured for an approach and landing. A stabilized descent should be established (approximately 500 fpm) and trim adjusted. A turn should be initiated at this point, if desired. The pilot should then smoothly increase the AOA to induce a stall warning. Power is reduced further during this phase, and trimming should cease at speeds slower than takeoff.

When the airplane reaches the stall warning (e.g., aural alert, buffet, etc.), the recovery is accomplished by first reducing

Stall Recovery Template		
1. Wing leveler or autopilot	1. Disconnect
2. a) Nose-down pitch control	2. a) Apply until impending stall indications are eliminated
b) Nose-down pitch trim	b) As needed
3. Bank	3. Wings Level
4. Thrust/Power	4. As needed
5. Speed brakes/spoilers	5. Retract
6. Return to the desired flight path	

Figure 12-19. *Stall recovery procedure.*

the AOA until the stall warning is eliminated. The pilot then rolls the wings level with coordinated use of the rudder and smoothly applies power as required. The airplane should be accelerated to V_X (if simulated obstacles are present) or V_Y during recovery and climb. Considerable forward elevator/ stabilator pressure will be required after the stall recovery as the airplane accelerates to V_X or V_Y. Appropriate trim input should be anticipated. The flap setting should be reduced from full to approach, or as recommended by the manufacturer. Then, with a positive rate of climb, the landing gear is selected up. The remaining flaps are then retracted as a positive rate-of-climb continues.

Power-On Approach to Stall (Takeoff and Departure)

A power-on approach to stall is trained and checked to simulate problematic takeoff scenarios. A power-on approach to stall may be performed from straight-and-level flight or from shallow and medium banked turns (20 degrees of bank). To initiate a power-on approach to stall maneuver, the area surrounding the airplane should always be cleared to look for potential traffic. The airplane is slowed to the manufacturer's recommended lift-off speed. The airplane should be configured in the takeoff configuration. Trim should be adjusted for this speed. Engine power is then increased to that recommended in the AFM/POH for the practice of power-on approach to stall. In the absence of a recommended setting, use approximately 65 percent of maximum available power. Begin a turn, if desired, while increasing AOA to induce a stall warning (e.g., aural alert, buffet, etc.). Other specified (reduced) power settings may be used to simulate performance at higher gross weights and density altitudes.

When the airplane reaches the stall warning, the recovery is made first by reducing the AOA until the stall warning is eliminated. The pilot then rolls the wings level with coordinated use of the rudder and applying power as needed. However, if simulating limited power available for high gross weight and density altitude situations, the power during the recovery should be limited to that specified. The landing gear should

be retracted when a positive rate of climb is attained, and flaps retracted, if flaps were set for takeoff. The target airspeed on recovery is V_X if (simulated) obstructions are present, or V_Y. The pilot should anticipate the need for nose-down trim as the airplane accelerates to V_X or V_Y after recovery.

Full Stall

It is not recommended that full stalls be practiced unless a qualified flight instructor is present. A power-off or power-on full stall should only be practiced in a structured lesson with clear learning objectives and cautions discussed. The goals of the training are (a) to provide the pilots the experience of the handling characteristics and dynamic cues (e.g., buffet, roll off) near and at full stall and (b) to reinforce the proper application of the stall recovery procedures. Given the associated risk of asymmetric thrust at high angles of attack and low rudder effectiveness due to low airspeeds, this reinforces the primary step of first lowering the AOA, which allows all control surfaces to become more effective and allows for roll to be better controlled. Thrust should only be used as needed in the recovery.

Accelerated Approach to Stall

Accelerated approach to stall should be performed with a bank of approximately 45°, and in no case at a speed greater than the airplane manufacturer's recommended airspeed or the specified design maneuvering speed (V_A). The entry altitude for this maneuver should be no lower than 5,000 feet AGL.

The entry method for the maneuver is no different than for a single-engine airplane. Once at an appropriate speed, begin increasing the back pressure on the elevator while maintaining a coordinated 45° turn. A good speed reduction rate is approximately 3-5 knots per second. Once a stall warning occurs, recover promptly by reducing the AOA until the stall warning stops. Then roll the wings level with coordinated rudder and add power as necessary to return to the desired flightpath.

Spin Awareness

No multiengine airplane is approved for spins, and their spin recovery characteristics are generally very poor. It is therefore necessary to practice spin avoidance and maintain a high awareness of situations that can result in an inadvertent spin.

In order to spin any airplane, it must first be stalled. At the stall, a yawing moment must be introduced. In a multiengine airplane, the yawing moment may be generated by rudder input or asymmetrical thrust. It follows, then, that spin awareness be at its greatest during V_{MC} demonstrations, stall practice, slow flight, or any condition of high asymmetrical thrust, particularly at low speed/high AOA. Single-engine stalls are not part of any multiengine training curriculum.

No engine failure should ever be introduced below safe, intentional one-engine inoperative speed (V_{SSE}). If no V_{SSE} is published, use V_{YSE}. Other than training situations, the multiengine airplane is only operated below V_{SSE} for mere seconds just after lift-off or during the last few dozen feet of altitude in preparation for landing.

For spin avoidance when practicing engine failures, the flight instructor should pay strict attention to the maintenance of proper airspeed and bank angle as the student executes the appropriate procedure. The instructor should also be particularly alert during stall and slow flight practice. Forward center-of-gravity positions result in favorable stall and spin avoidance characteristics, but do not eliminate the hazard.

When performing a V_{MC} demonstration, the instructor should also be alert for any sign of an impending stall. The student may be highly focused on the directional control aspect of the maneuver to the extent that impending stall indications go unnoticed. If a V_{MC} demonstration cannot be accomplished under existing conditions of density altitude, it may, for training purposes, be done utilizing the rudder blocking technique described in the following section.

As very few twins have ever been spin-tested (none are required to), the recommended spin recovery techniques are based only on the best information available. The departure from controlled flight may be quite abrupt and possibly disorienting. The direction of an upright spin can be confirmed from the turn needle or the symbolic airplane of the turn coordinator, if necessary. Do not rely on the ball position or other instruments.

If a spin is entered, most manufacturers recommend immediately retarding both throttles to idle, applying full rudder opposite the direction of rotation, and applying full forward elevator/stabilator pressure (with ailerons neutral). These actions should be taken as near simultaneously as possible. The controls should then be held in that position until the spin has stopped. At that point adjust rudder pressure, back elevator pressure, and power as necessary to return to the desired flight path. Pilots should be aware that a spin recovery will take considerable altitude therefore it is critical that corrective action be taken immediately.*

* For more information, please see the Addendum to Chapter 12 at the end of the book.

Chapter 13

Transition to Tailwheel Airplanes

Introduction

Due to their design and structure, tailwheel airplanes (tailwheels) exhibit operational and handling characteristics different from those of tricycle-gear airplanes (nosewheels). *[Figure 13-1]* In general, tailwheels are less forgiving of pilot error while in contact with the ground than are nosewheels. This chapter focuses on the operational differences that occur during ground operations, takeoffs, and landings.

Although still termed "conventional-gear airplanes," tailwheel designs are most likely to be encountered today by pilots who have first learned in nosewheels. Therefore, tailwheel operations are approached as they appear to a pilot making a transition from nosewheel designs.

Profile behind
pivot point

Wind

Motion

Point of
wheel pivoting

CG

Normal glide

Start roundout
to landing altitude

Main gear and tailwheel
touch down simultaneously

Hold elevator
full up

Figure 13-1. *The Piper Super Cub on the left is a popular tailwheel airplane. The airplane on the right is a Mooney M20, which is a nosewheel (tricycle gear) airplane.*

Landing Gear

The main landing gear forms the principal support of the airplane on the ground. The tailwheel also supports the airplane, but steering and directional control are its primary functions. With the tailwheel-type airplane, the two main struts are attached to the airplane slightly ahead of the airplane's center of gravity (CG), so that the plane naturally rests in a nose-high attitude on the triangle created by the main gear and the tailwheel. This arrangement is responsible for the three major handling differences between nosewheel and tailwheel airplanes. They center on directional instability, angle of attack (AOA), and crosswind weathervaning tendencies.

Proper usage of the rudder pedals is crucial for directional control while taxiing. Steering with the pedals may be accomplished through the forces of airflow or propeller slipstream acting on the rudder surface or through a mechanical linkage acting through springs to communicate steering inputs to the tailwheel. Initially, the pilot should taxi with the heels of the feet resting on the floor and the balls of the feet on the bottom of the rudder pedals. The feet should be slid up onto the brake pedals only when it is necessary to depress the brakes. This permits the simultaneous application of rudder and brake whenever needed. Some models of tailwheel airplanes are equipped with heel brakes rather than toe brakes. As in nosewheel airplanes, brakes are used to slow and stop the aircraft and to increase turning authority when tailwheel steering inputs prove insufficient. Whenever used, brakes should be applied smoothly and evenly.

Instability

Because of the relative placement of the main gear and the CG, tailwheel aircraft are inherently unstable on the ground. As taxi turns are started, the aircraft begins to pivot on one or the other of the main wheels. From that point, with the CG aft of that pivot point, the forward momentum of the plane acts to continue and even tighten the turn without further steering inputs. In consequence, removal of rudder pressure does not stop a turn that has been started, and it is necessary to apply an opposite input (opposite rudder) to bring the aircraft back to straight-line travel.

If the initial rudder input is maintained after a turn has been started, the turn continues to tighten, an unexpected result for pilots accustomed to a nosewheel. In consequence, it is common for pilots making the transition between the two types to experience difficulty in early taxi attempts. As long as taxi speeds are kept low, however, no serious problems result, and pilots typically adjust quickly to the technique of using rudder pressure to start a turn, then neutralizing the pedals as the turn continues, and finally using an opposite pedal input to stop the turn and regain straight line travel.

Because of this inbuilt instability, the most important lesson that can be taught in tailwheel airplanes is to taxi and make turns at slow speeds.

Angle of Attack

A second strong contrast to nosewheel airplanes, tailwheel aircraft make lift while on the ground any time there is a relative headwind. The amount of lift obviously depends on the wind speed, but even at slow taxi speeds, the wings and ailerons are doing their best to aid in liftoff. This phenomenon requires care and management, especially during the takeoff and landing rolls, and is again unexpected by nosewheel pilots making the transition.

Taxiing

On most tailwheel-type airplanes, directional control while taxiing is facilitated by the use of a steerable tailwheel, which operates along with the rudder. The tailwheel steering mechanism remains engaged when the tailwheel is operated through an arc of about 30° each side of center. Beyond that limit, the tailwheel breaks free and becomes full swiveling. In full swivel mode, the airplane can be pivoted within its own length, if desired. While taxiing, the steerable tailwheel should be used for making normal turns and the pilot's feet kept off the brake pedals to avoid unnecessary wear on the brakes.

When beginning to taxi, the brakes should be tested immediately for proper operation. This is done by first applying power to start the airplane moving slowly forward, then retarding the throttle and simultaneously applying pressure smoothly to both brakes. If braking action is unsatisfactory, the engine should be shut down immediately.

To turn the airplane on the ground, the pilot should apply rudder in the desired direction of turn and use whatever power or brake necessary to control the taxi speed. At very low taxi speeds, directional response is sluggish as surface friction acting on the tailwheel inhibits inputs trough the steering springs. At normal taxi speeds, rudder inputs alone should be sufficient to start and stop most turns. During taxi, the AOA built in to the structure gives control placement added importance when compared to nosewheel models.

When taxiing in a quartering headwind, the upwind wing can easily be lifted by gusting or strong winds unless ailerons are positioned to "kill" lift on that side (stick held into the wind). At the same time, elevator should be held full back to add downward pressure to the tailwheel assembly and improve tailwheel steering response. This is standard control positioning for both nosewheel and tailwheel airplanes, so the difference lies only in the added tailwheel vulnerability created by the fuselage pitch attitude.

When taxiing with a quartering tailwind, this fuselage angle reduces the tendency of the wind to lift either wing. Nevertheless, the basic vulnerability to surface winds common to all tailwheel airplanes makes it essential to be aware of wind direction at all times, so holding the stick away from the cross wind is good practice (left aileron in a right quartering tailwind).

Elevator positioning in tailwinds is a bit more complex. Standard teaching tends to recommend full forward stick in any degree of tailwind, arguing that a tailwind striking the elevator when it is deflected full down increases downward pressure on the tailwheel assembly and increases directional control. Equally important, if the elevator were to remain deflected up, a strong tailwind can get under the control surface and lift the tail with unfortunate consequences for the propeller and engine.

While stick-forward positioning is essential in strong tailwinds, it is not likely to be an appropriate response when winds are light. The propeller wash in even lightly-powered airplanes is usually strong enough to overcome the effects of light tailwinds, producing a net headwind over the tail. This in turn suggests that back stick, not forward, does the most to help with directional control. If in doubt, it is best to sample the wind as you taxi and position the elevator where it will do the most good.

Weathervaning

Tailwheel airplanes have an exaggerated tendency to weathervane, or turn into the wind, when operated on the ground in crosswinds. This tendency is greatest when taxing with a direct crosswind, a factor that makes maintaining directional control more difficult, sometimes requiring use of the brakes when tailwheel steering alone proves inadequate to counteract the weathervane effect.

Visibility

In the normal nose-high attitude, the engine cowling may be high enough to restrict the pilot's vision of the area directly ahead of the airplane while on the ground. Consequently, objects directly ahead are difficult, if not impossible, to see. In aircraft that are completely blind ahead, all taxi movements should be started with a small turn to ensure no other plane or ground vehicle has positioned itself directly under the nose while the pilot's attention was distracted with getting ready to takeoff. In taxiing such an airplane, the pilot should alternately turn the nose from one side to the other (zigzag) or make a series of short S-turns. This should be done slowly, smoothly, positively, and cautiously.

Directional Control

After absorbing all the information presented to this point, the transitioning pilot may conclude that the best approach to maintaining directional control is to limit rudder inputs from fear of overcontrolling. Although intuitive, this is an incorrect assumption: the disadvantages built in to the tailwheel design sometimes require vigorous rudder inputs to maintain or retain directional control. The best approach is to understand the fact that tailwheel aircraft are not damaged from the use of too much rudder, but rather from rudder inputs held for too long.

Normal Takeoff Roll

Wing flaps should be lowered prior to takeoff if recommended by the manufacturer. After taxiing onto the runway, the airplane should be aligned with the intended takeoff direction, and the tailwheel positioned straight or centered. In airplanes equipped with a locking device, the tailwheel should be locked in the centered position. After releasing the brakes, the throttle should be smoothly and continuously advanced to takeoff power. At all times on the takeoff roll, care must be taken to avoid applying brake pressure.

After a brief period of acceleration, positive forward elevator should be applied to smoothly lift the tail. The goal is to achieve a pitch attitude that improves forward visibility and produces a smooth transition to climbing flight as the aircraft continues to accelerate. If the attitude chosen is excessively steep, weight transfers rapidly to the wings, making crosswind control more difficult. If the attitude is too flat, crosswind control is also diminished, a counter-intuitive result that is discussed in the Crosswind section of this chapter.

It is important to note that nose-down pitch movement produces left yaw, the result of gyroscopic precession created by the propeller. The amount of force created by this precession is directly related to the rate the propeller axis is tilted when the tail is raised, so it is best to avoid an abrupt pitch change. Whether smooth or abrupt, the need to react to this yaw with rudder inputs emphasizes the increased directional demands common to tailwheel airplanes, a demand likely to be unanticipated by pilots transitioning from nosewheel models.

As speed is gained on the runway, the added authority of the elevator naturally continues to pitch the nose forward. During this stage, the pilot should concentrate on maintaining a constant-pitch attitude by gradually reducing elevator deflection. At the same time, directional control must be maintained with smooth, prompt, positive rudder corrections. All this activity emphasizes the point that tailwheel planes start to "fly" long before leaving the runway surface.

Liftoff

When the appropriate pitch attitude is maintained throughout the takeoff roll, liftoff occurs when the AOA and airspeed combine to produce the necessary lift without any additional "rotation" input. The ideal takeoff attitude requires only minimum pitch adjustments shortly after the airplane lifts off to attain the desired climb speed.

All modern tailwheel aircraft can be lifted off in the three-point attitude. That is, the AOA with all three wheels on the ground does not exceed the critical AOA, and the wings will not be stalled. While instructive, this technique results in an unusually high pitch attitude and an AOA excessively close to stall, both inadvisable circumstances when flying only inches from the ground.

As the airplane leaves the ground, the pilot must continue to maintain straight flight and hold the proper pitch attitude. During takeoffs in strong, gusty winds, it is advisable to add an extra margin of speed before the airplane is allowed to leave the ground. A takeoff at the normal takeoff speed may result in a lack of positive control, or a stall, when the airplane encounters a sudden lull in strong, gusty wind or other turbulent air currents. In this case, the pilot should hold the airplane on the ground longer to attain more speed, then make a smooth, positive rotation to leave the ground.

Crosswind Takeoff

It is important to establish and maintain proper crosswind corrections prior to lift-off; that is, application of aileron deflection into the wind to keep the upwind wing from rising and rudder deflection as needed to prevent weathervaning.

Takeoffs made into strong crosswinds are the reason for maintaining a positive AOA (tail-low attitude) while accelerating on the runway. Because the wings are making lift during the takeoff roll, a strong upwind aileron deflection can bank the airplane into the wind and provide positive crosswind correction before the aircraft lifts from the runway. The remainder of the takeoff roll is then made on the upwind main wheel. As the aircraft leaves the runway, the wings can be leveled as appropriate drift correction (crab) is established.

Short-Field Takeoff

With the exception of flap settings and initial climb speed as recommended by the manufacturer, there is little difference between the techniques described above for normal takeoffs. After liftoff, the pitch attitude should be adjusted as required for obstacle clearance.

Soft-Field Takeoff

Wing flaps may be lowered prior to starting the takeoff (if recommended by the manufacturer) to provide additional lift and transfer the airplane's weight from the wheels to the wings as early as possible. The airplane should be taxied onto the takeoff surface without stopping on a soft surface. Stopping on a soft surface, such as mud or snow, might bog the airplane down. The airplane should be kept in continuous motion with sufficient power while lining up for the takeoff roll.

As the airplane is aligned with the proposed takeoff path, takeoff power is applied smoothly and as rapidly as the powerplant will accept without faltering. The tail should be kept very low to maintain the inherent positive AOA and to avoid any tendency of the airplane to nose over as a result of soft spots, tall grass, or deep snow.

When the airplane is held at a nose-high attitude throughout the takeoff run, the wings progressively relieve the wheels of more and more of the airplane's weight, thereby minimizing the drag caused by surface irregularities or adhesion. Once airborne, the airplane should be allowed to accelerate to climb speed in ground effect.

Landing

The difference between nosewheel and tailwheel airplanes becomes apparent when discussing the touchdown and the period of deceleration to taxi speed. In the nosewheel design, touchdown is followed quite naturally by a reduction in pitch attitude to bring the nosewheel tire into contact with the runway. This pitch change reduces AOA, removes almost all wing lift, and rapidly transfers aircraft weight to the tires.

In tailwheel designs, this reduction of AOA and weight transfer are not practical and, as noted in the section on Takeoffs, it is rare to encounter tailwheel planes designed so that the wings are beyond critical AOA in the three-point attitude. In consequence, the airplane continues to "fly" in the three-point attitude after touchdown, requiring careful attention to heading, roll, and pitch for an extended period.

Touchdown

Tailwheel airplanes are less forgiving of crosswind landing errors than nosewheel models. It is important that touchdown occurs with the airplane's longitudinal axis parallel to the direction the airplane is moving along the runway. *[Figure 13-2]* Failure to accomplish this imposes Side loads on the landing gear which leads to directional instability. To avoid side stresses and directional problems, the pilot should not allow the airplane to touch down while in a crab or while drifting.

There are two significantly different techniques used to manage tailwheel aircraft touchdowns: three-point and wheel landings. In the first, the airplane is held off the surface of the runway until the attitude needed to remain aloft matches the geometry of the landing gear. When touchdown occurs at this point, the main gear and the tailwheel make contact at the same time. In the second technique (wheel landings),

the airplane is allowed to touch down earlier in the process in a lower pitch attitude, so that the main gear touch while the tail remains off the runway.

Three-Point Landing

As with all landings, success begins with an orderly arrival: airspeed, alignment, and configuration well in hand crossing the threshold. Round out (level-off) should be made with the main wheels about one foot off the surface. From that point forward, the technique is essentially the same that is used in nosewheels: a gentle increase in AOA to maintain flight while slowing. In a tailwheel aircraft, however, the goal is to attain a much steeper fuselage angle than that commonly used in nosewheel models; one that touches the tailwheels at the same time as the mainwheels.

With the tailwheel on the surface, a further increase in pitch attitude is impossible, so the plane remains on the runway, albeit tenuously. With deceleration, weight shifts increasingly from wings to wheels, with the final result that the plane once again becomes a ground vehicle after shedding most of its speed.

There are two potential errors in attempting a three-point landing. In the first, the mainwheels are allowed to make runway contact a little early with the tail still in the air. With the CG aft of the mainwheels, the tail naturally drops when the mainwheels touch, AOA increases, and the plane becomes airborne again. This "skip" is easily managed by re-flaring and again trying to hold the plane off until reaching the three-point attitude.

In the second error, the plane is held off the ground a bit too long so that the in-flight pitch attitude is steeper than the three-point attitude. When touchdown is made in this attitude,

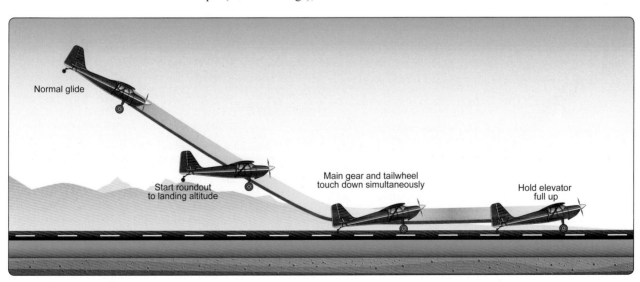

Figure 13-2. *Tailwheel touchdown.*

the tail makes contact first. Provided this happens from no more than a foot off the surface, the result is undramatic: the tail touches, the plane pitches forward slightly onto the mainwheels, and rollout proceeds normally.

In every case, once the tailwheel makes contact, the elevator control should be eased fully back to press the tailwheel on the runway. Without this elevator input, the AOA of the horizontal stabilizer develops enough lift to lighten pressure on the tailwheel and render it useless as a directional control with possibly unwelcome consequences. This after-landing elevator input is quite foreign to nosewheel pilots and must be stressed during transition training.

NOTE: Before the tailwheel is on the ground, application of full back elevator during the flare lowers the tail, increases the AOA, and quite naturally puts the plane in climbing flight.

Wheel Landing

In some wind conditions, the need to retain control authority may make it desirable to make contact with the runway at a higher airspeed than that associated with the three-point attitude. This necessitates landing in a flatter pitch attitude on the mainwheels only, with the tailwheel still off the surface. *[Figure 13-3]* As noted, if the tail is off the ground, it tends to drop and put the plane airborne, so a soft touchdown and a slight relaxation of back elevator just after the wheels touch are key ingredients to a successful wheel landing.

If the touchdown is made at too high a rate of descent, the tail is forced down by its own weight, resulting in a sudden increase in lift. If the pilot now pushes forward in an attempt to again make contact with the surface, a potentially dangerous pilot-induced oscillation may develop. It is far better to respond to a bounced wheel landing attempt by initiating a go-around or converting to a three-point landing if conditions permit.

Once the mainwheels are on the surface, the tail should be permitted to drop on its own accord until it too makes ground contact. At this point, the elevator should be brought to the full aft position and deceleration should be allowed to proceed as in a three-point landing.

NOTE: The only difference between three-point and wheel landings is the timing of the touchdown (early and later). There is no difference between the approach angles and airspeeds in the two techniques.

Crosswinds

As noted, it is highly desirable to eliminate crab and drift at touchdown. By far the best approach to crosswind management is a side-slip or wing-low touchdown. Landing in this attitude, only one mainwheel makes initial contact, either in concert with the tailwheel in three-point landings or by itself in wheel landings.

After-Landing Roll

The landing process must never be considered complete until the airplane decelerates to the normal taxi speed during the landing roll or has been brought to a complete stop when clear of the landing area. The pilot must be alert for directional control difficulties immediately upon and after touchdown, and the elevator control should be held back as far as possible and as firmly as possible until the airplane stops. This provides more positive control with tailwheel steering, tends to shorten the after-landing roll, and prevents bouncing and skipping.

Any difference between the direction the airplane is traveling and the direction it is headed (drift or crab) produces a moment about the pivot point of the wheels, and the airplane tends to swerve. Loss of directional control may lead to an aggravated, uncontrolled, tight turn on the ground, or a ground loop. The

Figure 13-3. *Wheel landing.*

combination of inertia acting on the CG and ground friction of the main wheels during the ground loop may cause the airplane to tip enough for the outside wingtip to contact the ground and may even impose a sideward force that could collapse one landing gear leg. *[Figure 13-4]* In general, this combination of events is eliminated by landing straight and avoiding turns at higher than normal running speed.

To use the brakes, the pilot should slide the toes or feet up from the rudder pedals to the brake pedals (or apply heel pressure in airplanes equipped with heel brakes). If rudder pressure is being held at the time braking action is needed, that pressure should not be released as the feet or toes are being slid up to the brake pedals because control may be lost before brakes can be applied. During the ground roll, the airplane's direction of movement may be changed by carefully applying pressure on one brake or uneven pressures on each brake in the desired direction. Caution must be exercised when applying brakes to avoid overcontrolling.

If a wing starts to rise, aileron control should be applied toward that wing to lower it. The amount required depends on speed because as the forward speed of the airplane decreases, the ailerons become less effective.

If available runway permits, the speed of the airplane should be allowed to dissipate in a normal manner by the friction and drag of the wheels on the ground. Brakes may be used if needed to help slow the airplane. After the airplane has been slowed sufficiently and has been turned onto a taxiway or clear of the landing area, it should be brought to a complete stop. Only after this is done should the pilot retract the flaps and perform other checklist items.

Crosswind After-Landing Roll

Particularly during the after-landing roll, special attention must be given to maintaining directional control by the use of rudder and tailwheel steering while keeping the upwind wing from rising by the use of aileron. Characteristically, an airplane has a greater profile or side area behind the main landing gear than forward of it. With the main wheels acting as a pivot point and the greater surface area exposed to the crosswind behind that pivot point, the airplane tends to turn or weathervane into the wind. *[Figure 13-5]* This weathervaning tendency is more prevalent in the tailwheel-type because the airplane's surface area behind the main landing gear is greater than in nosewheel-type airplanes.

Pilots should be familiar with the crosswind component of each airplane they fly and avoid operations in wind conditions that exceed the capability of the airplane, as well as their own limitations. While the airplane is decelerating during the after-landing roll, more aileron must be applied to keep the upwind wing from rising. Since the airplane is slowing down, there is less airflow around the ailerons and they become less effective. At the same time, the relative wind is becoming more of a crosswind and exerting a greater lifting force on the upwind wing. Consequently, when the airplane is coming to a stop, the aileron control must be held fully toward the wind.

Short-Field Landing

Upon touchdown, the airplane should be firmly held in a three-point attitude. This provides aerodynamic braking by the wings. Immediately upon touchdown and closing the throttle, the brakes should be applied evenly and firmly to minimize the after-landing roll. The airplane should be stopped within the shortest possible distance consistent with safety.

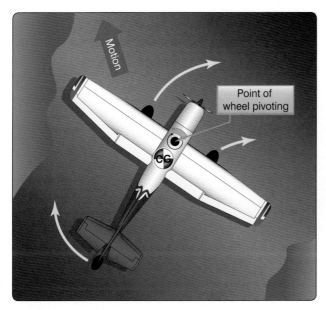

Figure 13-4. *Effect of CG on directional control.*

Figure 13-5. *Weathervaning tendency.*

Soft-Field Landing

The tailwheel should touchdown simultaneously with or just before the main wheels and should then be held down by maintaining firm back-elevator pressure throughout the landing roll. This minimizes any tendency for the airplane to nose over and provides aerodynamic braking. The use of brakes on a soft field is not needed because the soft or rough surface itself provides sufficient reduction in the airplane's forward speed. Often, it is found that upon landing on a very soft field, the pilot needs to increase power to keep the airplane moving and from becoming stuck in the soft surface.

Ground Loop

A ground loop is an uncontrolled turn during ground operations that may occur during taxi, takeoff, or during the after-landing roll. Ground loops start with a swerve that is allowed to continue for too long. The swerve may be the result of side-load on landing, a taxi turn started with too much groundspeed, overcorrection, or even an uneven ground surface or a soft spot that retards one main wheel of the airplane.

Due to the inbuilt instability of the tailwheel design, the forces that lead to a ground loop accumulate as the angle between the fuselage and inertia, acting from the CG, increase. If allowed to develop, these forces may become great enough to tip the airplane to the outside of the turn until one wing strikes the ground.

To counteract the possibility of an uncontrolled turn, the pilot should counter any swerve with firm rudder input. In stronger swerves, differential braking is essential as tailwheel steering proves inadequate. It is important to note, however, that as corrections begin to become apparent, rudder and braking inputs need to be removed promptly to avoid starting yet another departure in the opposite direction.

Chapter Summary

This chapter focuses on the operational differences between tailwheel and nosewheel airplanes that occur during ground operations, takeoffs, and landings. The chapter covers specific topics, such as landing gear, taxiing, visibility, liftoff, and landing. Comparisons are given as to how each react during the takeoff and landing, as well as situations that should be avoided. Pilots who use proper rudder control techniques should be able to transition to tailwheel airplanes without too much difficulty.

Chapter 14

Transition to Turbopropeller-Powered Airplanes

Normal "Forward" Pitch

Pull Up
Idle
Beta
Reverse
Feather
Fit Idle
Low Idle
Fuel Cut Off

Power   Prop   Condition

Feather "Maximum Forward Pitch"

Reverse
Feather
Fit Idle
Low Idle
Fuel Cut Off

Introduction

The turbopropeller-powered airplane flies and handles just like any other airplane of comparable size and weight. The aerodynamics are the same. The major differences between flying a turboprop and other non-turbine-powered airplanes are found in the handling of the airplane's powerplant and its associated systems. The powerplant is different and requires operating procedures that are unique to gas turbine engines. But so, too, are other systems, such as electrical, hydraulics, environmental, flight control, rain and ice protection, and avionics. The turbopropeller-powered airplane also has the advantage of being equipped with a constant speed, full feathering and reversing propeller—something normally not found on piston-powered airplanes.

Air inlet | Compression | Combustion chambers | Turbine | Exhaust

Gas Turbine Engine

Both piston (reciprocating) engines and gas turbine engines are internal combustion engines. They have a similar cycle of operation that consists of induction, compression, combustion, expansion, and exhaust. In a piston engine, each of these events is a separate distinct occurrence in each cylinder. Also in a piston engine, an ignition event must occur during each cycle in each cylinder. Unlike reciprocating engines, in gas turbine engines these phases of power occur simultaneously and continuously instead of successively one cycle at a time. Additionally, ignition occurs during the starting cycle and is continuous thereafter. The basic gas turbine engine contains four sections: intake, compression, combustion, and exhaust. *[Figure 14-1]*

To start the engine, the compressor section is rotated by an electrical starter on small engines or an air-driven starter on large engines. As compressor rates per minute (rpm) accelerates, air is brought in through the inlet duct, compressed to a high pressure, and delivered to the combustion section (combustion chambers). Fuel is then injected by a fuel controller through spray nozzles and ignited by igniter plugs. (Not all of the compressed air is used to support combustion. Some of the compressed air bypasses the burner section and circulates within the engine to provide internal cooling, enhanced thrust, and noise abatement. In turbojet engines, by-pass airflow may be augmented by the action of a fan located at the engine's intake.) The fuel/air mixture in the combustion chamber is then burned in a continuous combustion process and produces a very high temperature, typically around 4,000° Fahrenheit (F), which heats the entire air mass to 1,600 – 2,400 °F. The mixture of hot air and gases expands and is directed to the turbine blades forcing the turbine section to rotate, which in turn drives the compressor by means of a direct shaft, a concentric shaft, or a combination of both. After powering the turbine section,

the high velocity excess exhaust exits the tail pipe or exhaust section. (The exhaust section of a turbojet engine may also incorporate a system of moving doors to redirect airflow for the purpose of slowing an airplane down after landing or back-powering it away from a gate. They are referred to as thrust reversers). Once the turbine section is powered by gases from the burner section, the starter is disengaged, and the igniters are turned off. Combustion continues until the engine is shut down by turning off the fuel supply.

NOTE: Because compression produces heat, some pneumatic aircraft systems tap into the source of hot (480 °F) compressed air from the engine compressor (bleed air) and use it for engine anti-ice, airfoil anti-ice, aircraft pressurization, and other ancillary systems after further conditioning its internal pressure and temperature.

High-pressure exhaust gases can be used to provide jet thrust as in a turbojet engine. Or, the gases can be directed through an additional turbine to drive a propeller through reduction gearing, as in a turbopropeller (turboprop) engine.

Turboprop Engines

The turbojet engine excels the reciprocating engine in top speed and altitude performance. On the other hand, the turbojet engine has limited takeoff and initial climb performance as compared to that of a reciprocating engine. In the matter of takeoff and initial climb performance, the reciprocating engine is superior to the turbojet engine. Turbojet engines are most efficient at high speeds and high altitudes, while propellers are most efficient at slow and medium speeds (less than 400 miles per hour (mph)). Propellers also improve takeoff and climb performance. The development of the turboprop engine was an attempt to combine in one engine the best characteristics of both the turbojet and propeller-driven reciprocating engine.

Figure 14-1. *Basic components of a gas turbine engine.*

The turboprop engine offers several advantages over other types of engines, such as:

- Lightweight

- Mechanical reliability due to relatively few moving parts

- Simplicity of operation

- Minimum vibration

- High power per unit of weight

- Use of propeller for takeoff and landing

Turboprop engines are most efficient at speeds between 250 and 400 mph and altitudes between 18,000 and 30,000 feet. They also perform well at the slow speeds required for takeoff and landing and are fuel efficient. The minimum specific fuel consumption of the turboprop engine is normally available in the altitude range of 25,000 feet up to the tropopause.

The power output of a piston engine is measured in horsepower and is determined primarily by rpm and manifold pressure. The power of a turboprop engine, however, is measured in shaft horsepower (shp). Shaft horsepower is determined by the rpm and the torque (twisting moment) applied to the propeller shaft. Since turboprop engines are gas turbine engines, some jet thrust is produced by exhaust leaving the engine. This thrust is added to the shaft horsepower to determine the total engine power or equivalent shaft horsepower (eshp). Jet thrust usually accounts for less than 10 percent of the total engine power.

Although the turboprop engine is more complicated and heavier than a turbojet engine of equivalent size and power, it delivers more thrust at low subsonic airspeeds. However, the advantages decrease as flight speed increases. In normal cruising speed ranges, the propulsive efficiency (output divided by input) of a turboprop decreases as speed increases.

The propeller of a typical turboprop engine is responsible for roughly 90 percent of the total thrust under sea level conditions on a standard day. The excellent performance of a turboprop during takeoff and climb is the result of the ability of the propeller to accelerate a large mass of air while the airplane is moving at a relatively low ground and flight speed. "Turboprop," however, should not be confused with "turbo supercharged" or similar terminology. All turbine engines have a similarity to normally aspirated (non-supercharged) reciprocating engines in that maximum available power decreases almost as a direct function of increased altitude.

Although power decreases as the airplane climbs to higher altitudes, engine efficiency in terms of specific fuel consumption (expressed as pounds of fuel consumed per horsepower per hour) is increased. Decreased specific fuel consumption plus the increased true airspeed at higher altitudes is a definite advantage of a turboprop engine.

All turbine engines, turboprop or turbojet, are defined by limiting temperatures, rotational speeds, and (in the case of turboprops) torque. Depending on the installation, the primary parameter for power setting might be temperature, torque, fuel flow, or rpm (either propeller rpm, gas generator (compressor) rpm, or both). In cold weather conditions, torque limits can be exceeded while temperature limits are still within acceptable range. While in hot weather conditions, temperature limits may be exceeded without exceeding torque limits. In any weather, the maximum power setting of a turbine engine is usually obtained with the throttles positioned somewhat aft of the full forward position. The transitioning pilot must understand the importance of knowing and observing limits on turbine engines. An over temperature or over torque condition that lasts for more than a few seconds can literally destroy internal engine components.

Turboprop Engine Types
Fixed Shaft
One type of turboprop engine is the fixed shaft constant speed type, such as the Garrett TPE331. *[Figure 14-2]* In this type engine, ambient air is directed to the compressor section through the engine inlet. An acceleration/diffusion process in the two stage compressor increases air pressure and directs it rearward to a combustor. The combustor is made up of a combustion chamber, a transition liner, and a turbine plenum. Atomized fuel is added to the air in the combustion chamber. Air also surrounds the combustion chamber to provide for cooling and insulation of the combustor.

The gas mixture is initially ignited by high-energy igniter plugs, and the expanding combustion gases flow to the turbine. The energy of the hot, high-velocity gases is converted to torque on the main shaft by the turbine rotors. The reduction gear converts the high rpm—low torque of the main shaft to low rpm—high torque to drive the accessories and the propeller. The spent gases leaving the turbine are directed to the atmosphere by the exhaust pipe.

Only about 10 percent of the air that passes through the engine is actually used in the combustion process. Up to approximately 20 percent of the compressed air may be bled off for the purpose of heating, cooling, cabin pressurization, and pneumatic systems. Over half the engine power is devoted to driving the compressor, and it is the compressor that can potentially produce very high drag in the case of a failed, windmilling engine.

Figure 14-2. *Fixed shaft turboprop engine.*

In the fixed shaft constant-speed engine, the engine rpm may be varied within a narrow range of 96 percent to 100 percent. During ground operation, the rpm may be reduced to 70 percent. In flight, the engine operates at a constant speed that is maintained by the governing section of the propeller. Power changes are made by increasing fuel flow and propeller blade angle rather than engine speed. An increase in fuel flow causes an increase in temperature and a corresponding increase in energy available to the turbine. The turbine absorbs more energy and transmits it to the propeller in the form of torque. The increased torque forces the propeller blade angle to be increased to maintain the constant speed. Turbine temperature is a very important factor to be considered in power production. It is directly related to fuel flow and thus to the power produced. It must be limited because of strength and durability of the material in the combustion and turbine section. The control system schedules fuel flow to produce specific temperatures and to limit those temperatures so that the temperature tolerances of the combustion and turbine sections are not exceeded. The engine is designed to operate for its entire life at 100 percent. All of its components, such as compressors and turbines, are most efficient when operated at or near the rpm design point.

Powerplant (engine and propeller) control is achieved by means of a power lever and a condition lever for each engine. *[Figure 14-3]* There is no mixture control and/or rpm lever as found on piston-engine airplanes.

On the fixed shaft constant-speed turboprop engine, the power lever is advanced or retarded to increase or decrease forward thrust. The power lever is also used to provide reverse thrust. The condition lever sets the desired engine rpm within a narrow range between that appropriate for ground operations and flight.

Powerplant instrumentation in a fixed shaft turboprop engine typically consists of the following basic indicators. *[Figure 14-4]*

- Torque or horsepower
- Interturbine temperature (ITT)
- Fuel flow
- RPM

Figure 14-3. *Powerplant controls—fixed shaft turboprop engine.*

14-4

Figure 14-4. *Powerplant instrumentation—fixed shaft turboprop engine.*

Torque developed by the turbine section is measured by a torque sensor. The torque is then reflected on the instrument panel horsepower gauge calibrated in horsepower times 100. ITT is a measurement of the combustion gas temperature between the first and second stages of the turbine section. The gauge is calibrated in degrees Celsius (°C). Propeller rpm is reflected on a tachometer as a percentage of maximum rpm. Normally, a vernier indicator on the gauge dial indicates rpm in 1 percent graduations as well. The fuel flow indicator indicates fuel flow rate in pounds per hour.

Propeller feathering in a fixed shaft constant-speed turboprop engine is normally accomplished with the condition lever. An engine failure in this type engine, however, results in a serious drag condition due to the large power requirements of the compressor being absorbed by the propeller. This could create a serious airplane control problem in twin-engine airplanes unless the failure is recognized immediately and the affected propeller feathered. For this reason, the fixed shaft turboprop engine is equipped with negative torque sensing (NTS).

NTS is a condition wherein propeller torque drives the engine, and the propeller is automatically driven to high pitch to reduce drag. The function of the negative torque sensing system is to limit the torque the engine can extract from the propeller during windmilling and thereby prevent large drag forces on the airplane. The NTS system causes a movement of the propeller blades automatically toward their feathered position should the engine suddenly lose power while in flight. The NTS system is an emergency backup system in the event of sudden engine failure. It is not a substitution for the feathering device controlled by the condition lever.

Split Shaft/ Free Turbine Engine

In a free power-turbine engine, such as the Pratt & Whitney PT-6 engine, the propeller is driven by a separate turbine through reduction gearing. The propeller is not on the same shaft as the basic engine turbine and compressor. *[Figure 14-5]* Unlike the fixed shaft engine, in the split shaft engine the propeller can be feathered in flight or on the ground with the basic engine still running. The free power-turbine design allows the pilot to select a desired propeller governing rpm, regardless of basic engine rpm.

A typical free power-turbine engine has two independent counter-rotating turbines. One turbine drives the compressor, while the other drives the propeller through a reduction gearbox. The compressor in the basic engine consists of three axial flow compressor stages combined with a single centrifugal compressor stage. The axial and centrifugal stages are assembled on the same shaft and operate as a single unit.

Inlet air enters the engine via a circular plenum near the rear of the engine and flows forward through the successive compressor stages. The flow is directed outward by the

Figure 14-5. *Split shaft/free turbine engine.*

centrifugal compressor stage through radial diffusers before entering the combustion chamber, where the flow direction is actually reversed. The gases produced by combustion are once again reversed to expand forward through each turbine stage. After leaving the turbines, the gases are collected in a peripheral exhaust scroll and are discharged to the atmosphere through two exhaust ports near the front of the engine.

A pneumatic fuel control system schedules fuel flow to maintain the power set by the gas generator power lever. Except in the beta range, propeller speed within the governing range remains constant at any selected propeller control lever position through the action of a propeller governor.

The accessory drive at the aft end of the engine provides power to drive fuel pumps, fuel control, oil pumps, a starter/generator, and a tachometer transmitter. At this point, the speed of the drive (N_1) is the true speed of the compressor side of the engine, approximately 37,500 rpm.

Powerplant (engine and propeller) operation is achieved by three sets of controls for each engine: the power lever, propeller lever, and condition lever. *[Figure 14-6]* The power lever serves to control engine power in the range from idle through takeoff power. Forward or aft motion of the power lever increases or decreases gas generator rpm (N_1) and thereby increases or decreases engine power. The propeller lever is operated conventionally and controls the constant-speed propellers through the primary governor. The propeller rpm range is normally from 1,500 to 1,900. The condition lever controls the flow of fuel to the engine. Like the mixture lever in a piston-powered airplane, the condition

lever is located at the far right of the power quadrant. But the condition lever on a turboprop engine is really just an on/off valve for delivering fuel. There are HIGH IDLE and LOW IDLE positions for ground operations, but condition levers have no metering function. Leaning is not required in turbine engines; this function is performed automatically by a dedicated fuel control unit.

Engine instruments in a split shaft/free turbine engine typically consist of the following basic indicators. *[Figure 14-7]*

- ITT indicator
- Torquemeter
- Propeller tachometer
- N_1 (gas generator) tachometer
- Fuel flow indicator
- Oil temperature/pressure indicator

The ITT indicator gives an instantaneous reading of engine gas temperature between the compressor turbine and the power turbines. The torquemeter responds to power lever movement and gives an indication in foot-pounds (ft/lb) of the torque being applied to the propeller. Because in the free turbine engine the propeller is not attached physically to the shaft of the gas turbine engine, two tachometers are justified—one for the propeller and one for the gas generator. The propeller tachometer is read directly in revolutions per minute. The N_1 or gas generator is read in percent of rpm. In the Pratt & Whitney PT-6 engine, it is based on a figure of 37,000 rpm at 100 percent. Maximum continuous gas generator is limited to 38,100 rpm or 101.5 percent N_1.

Figure 14-6. *Powerplant controls—split shaft/free turbine engine.*

The ITT indicator and torquemeter are used to set takeoff power. Climb and cruise power are established with the torquemeter and propeller tachometer while observing ITT limits. Gas generator (N_1) operation is monitored by the gas generator tachometer. Proper observation and interpretation of these instruments provide an indication of engine performance and condition.

Reverse Thrust and Beta Range Operations

The thrust that a propeller provides is a function of the angle of attack (AOA) at which the air strikes the blades, and the

Figure 14-7. *Engine instruments—split shaft/free turbine engine.*

speed at which this occurs. The AOA varies with the pitch angle of the propeller.

So called "flat pitch" is the blade position offering minimum resistance to rotation and no net thrust for moving the airplane. Forward pitch produces forward thrust—higher pitch angles being required at higher airplane speeds.

The "feathered" position is the highest pitch angle obtainable. *[Figure 14-8]* The feathered position produces no forward thrust. The propeller is generally placed in feather only in case of in-flight engine failure to minimize drag and prevent the air from using the propeller as a turbine.

In the "reverse" pitch position, the engine/propeller turns in the same direction as in the normal (forward) pitch position, but the propeller blade angle is positioned to the other side of flat pitch. *[Figure 14-8]* In reverse pitch, air is pushed away from the airplane rather than being drawn over it. Reverse pitch results in braking action, rather than forward thrust of the airplane. It is used for backing away from obstacles when taxiing, controlling taxi speed, or to aid in bringing the airplane to a stop during the landing roll. Reverse pitch does not mean reverse rotation of the engine. The engine delivers power just the same, no matter which side of flat pitch the propeller blades are positioned.

Figure 14-8. *Propeller pitch angle characteristics.*

With a turboprop engine, in order to obtain enough power for flight, the power lever is placed somewhere between flight idle (in some engines referred to as "high idle") and maximum. The power lever directs signals to a fuel control unit to manually select fuel. The propeller governor selects the propeller pitch needed to keep the propeller/engine on speed. This is referred to as the propeller governing or "alpha" mode of operation. When positioned aft of flight idle, however, the power lever directly controls propeller blade angle. This is known as the "beta" range of operation.

The beta range of operation consists of power lever positions from flight idle to maximum reverse. Beginning at power lever positions just aft of flight idle, propeller blade pitch angles become progressively flatter with aft movement of the power lever until they go beyond maximum flat pitch and into negative pitch, resulting in reverse thrust. While in a fixed shaft/constant-speed engine, the engine speed remains largely unchanged as the propeller blade angles achieve their negative values. On the split shaft PT-6 engine, as the negative 5° position is reached, further aft movement of the power lever also results in a progressive increase in engine (N_1) rpm until a maximum value of about negative 11° of blade angle and 85 percent N_1 are achieved.

Operating in the beta range and/or with reverse thrust requires specific techniques and procedures depending on the particular airplane make and model. There are also specific engine parameters and limitations for operations within this area that must be adhered to. It is essential that a pilot transitioning to turboprop airplanes become knowledgeable and proficient in these areas, which are unique to turbine-engine powered airplanes.

Turboprop Airplane Electrical Systems

The typical turboprop airplane electrical system is a 28-volt direct current (DC) system, which receives power from one or more batteries and a starter/generator for each engine. The batteries may either be of the lead-acid type commonly used on piston-powered airplanes, or they may be of the nickel-cadmium (NiCad) type. The NiCad battery differs from the lead-acid type in that its output remains at relatively high power levels for longer periods of time. When the NiCad battery is depleted, however, its voltage drops off very suddenly. When this occurs, its ability to turn the compressor for engine start is greatly diminished, and the possibility of engine damage due to a hot start increases. Therefore, it is essential to check the battery's condition before every engine start. Compared to lead-acid batteries, high-performance NiCad batteries can be recharged very quickly. But the faster the battery is recharged, the more heat it produces. Therefore, NiCad battery-equipped airplanes are fitted with battery overheat annunciator lights signifying maximum safe and critical temperature thresholds.

The DC generators used in turboprop airplanes double as starter motors and are called "starter/generators." The starter/generator uses electrical power to produce mechanical torque to start the engine and then uses the engine's mechanical torque to produce electrical power after the engine is running. Some of the DC power produced is changed to 28 volt 400 cycle alternating current (AC) power for certain avionic, lighting, and indicator synchronization functions. This is accomplished by an electrical component called an inverter.

The distribution of DC and AC power throughout the system is accomplished through the use of power distribution buses. These "buses" as they are called are actually common terminals from which individual electrical circuits get their power. *[Figure 14-9]*

Buses are usually named for what they power (avionics bus, for example) or for where they get their power (right generator bus, battery bus). The distribution of DC and AC power is often divided into functional groups (buses) that give priority to certain equipment during normal and emergency operations. Main buses serve most of the airplane's electrical equipment. Essential buses feed power to equipment having top priority. *[Figure 14-10]*

Multiengine turboprop airplanes normally have several power sources—a battery and at least one generator per engine. The electrical systems are usually designed so that any bus can be energized by any of the power sources. For example, a typical system might have a right and left generator buses powered normally by the right and left engine-driven generators. These buses are connected by a normally open switch, which isolates them from each other. If one generator fails, power is lost to its bus, but power can be restored to that bus by closing a bus tie switch. Closing this switch connects the buses and allows the operating generator to power both.

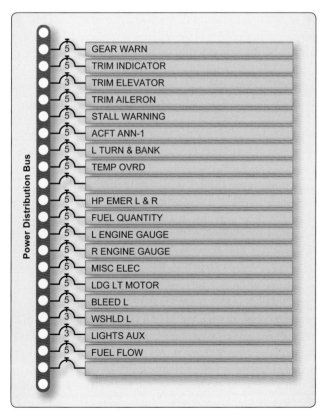

Figure 14-9. *Typical individual power distribution bus.*

Power distribution buses are protected from short circuits and other malfunctions by a type of fuse called a current limiter. In the case of excessive current supplied by any power source, the current limiter opens the circuit and thereby isolates that power source and allows the affected bus to become separated from the system. The other buses continue to operate normally. Individual electrical components are connected to the buses through circuit breakers. A circuit breaker is a device that opens an electrical circuit when an excess amount of current flows.

Operational Considerations

As previously stated, a turboprop airplane flies just like any other piston engine airplane of comparable size and weight. It is the operation of the engines and airplane systems that makes the turboprop airplane different from its piston engine counterpart. Pilot errors in engine and/or systems operation are the most common cause of aircraft damage or mishap. The time of maximum vulnerability to pilot error in any gas turbine engine is during the engine start sequence.

Turbine engines are extremely heat sensitive. They cannot tolerate an over temperature condition for more than a very few seconds without serious damage being done. Engine temperatures get hotter during starting than at any other time. Thus, turbine engines have minimum rotational speeds for introducing fuel into the combustion chambers during startup. Vigilant monitoring of temperature and acceleration on the part of the pilot remain crucial until the engine is running at a stable speed. Successful engine starting depends on assuring the correct minimum battery voltage before initiating start or employing a ground power unit (GPU) of adequate output.

After fuel is introduced to the combustion chamber during the start sequence, "light-off" and its associated heat rise occur very quickly. Engine temperatures may approach the maximum in a matter of 2 or 3 seconds before the engine stabilizes and temperatures fall into the normal operating range. During this time, the pilot must watch for any tendency of the temperatures to exceed limitations and be prepared to cut off fuel to the engine.

An engine tendency to exceed maximum starting temperature limits is termed a hot start. The temperature rise may be preceded by unusually high initial fuel flow, which may be the first indication the pilot has that the engine start is not proceeding normally. Serious engine damage occurs if the hot start is allowed to continue.

A condition where the engine is accelerating more slowly than normal is termed a hung start or false start. During a hung start/false start, the engine may stabilize at an engine

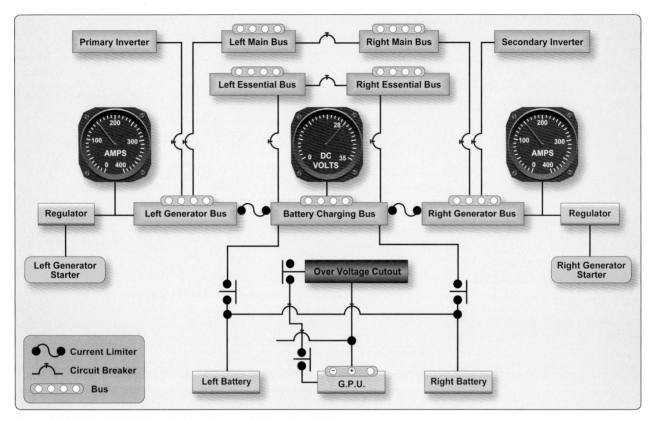

Figure 14-10. *Simplified schematic of turboprop airplane electrical system.*

rpm that is not high enough for the engine to continue to run without help from the starter. This is usually the result of low battery power or the starter not turning the engine fast enough for it to start properly.

Takeoffs in turboprop airplanes are not made by automatically pushing the power lever full forward to the stops. Depending on conditions, takeoff power may be limited by either torque or by engine temperature. Normally, the power lever position on takeoff is somewhat aft of full forward.

Takeoff and departure in a turboprop airplane (especially a twin-engine cabin-class airplane) should be accomplished in accordance with a standard takeoff and departure "profile" developed for the particular make and model. *[Figure 14-11]* The takeoff and departure profile should be in accordance with the airplane manufacturer's recommended procedures as outlined in the Federal Aviation Administration (FAA)-approved Airplane Flight Manual and/or the Pilot's Operating Handbook (AFM/POH). The increased complexity of turboprop airplanes makes the standardization of procedures a necessity for safe and efficient operation. The transitioning pilot should review the profile procedures before each takeoff to form a mental picture of the takeoff and departure process.

For any given high horsepower operation, the pilot can expect that the engine temperature will climb as altitude increases

at a constant power. On a warm or hot day, maximum temperature limits may be reached at a rather low altitude, making it impossible to maintain high horsepower to higher altitudes. Also, the engine's compressor section has to work harder with decreased air density. Power capability is reduced by high-density altitude and power use may have to be modulated to keep engine temperature within limits.

In a turboprop airplane, the pilot can close the throttles(s) at any time without concern for cooling the engine too rapidly. Consequently, rapid descents with the propellers in low pitch can be dramatically steep. Like takeoffs and departures, approach and landing should be accomplished in accordance with a standard approach and landing profile. *[Figure 14-12]*

A stabilized approach is an essential part of the approach and landing process. In a stabilized approach, the airplane, depending on design and type, is placed in a stabilized descent on a glidepath ranging from 2.5 to 3.5°. The speed is stabilized at some reference from the AFM/POH—usually 1.25 to 1.30 times the stall speed in approach configuration. The descent rate is stabilized from 500 fpm to 700 fpm until the landing flare.

Landing some turboprop airplanes (as well as some piston twins) can result in a hard, premature touchdown if the engines are idled too soon. This is because large propellers spinning rapidly in low pitch create considerable drag. In such airplanes,

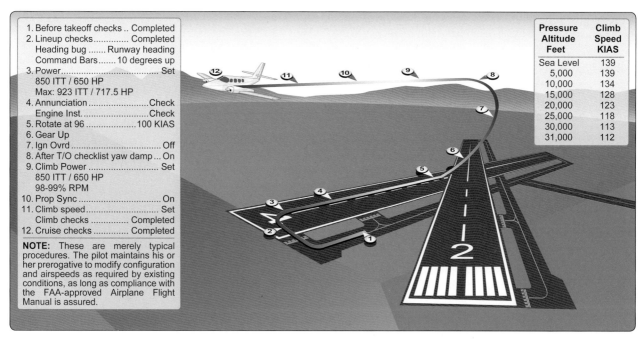

Figure 14-11. *Example of a typical turboprop airplane takeoff and departure profile.*

it may be preferable to maintain power throughout the landing flare and touchdown. Once firmly on the ground, propeller beta range operation dramatically reduces the need for braking in comparison to piston airplanes of similar weights.

Training Considerations

The medium and high altitudes at which turboprop airplanes are flown provide an entirely different environment in terms of regulatory requirements, airspace structure, physiological requirements, and even meteorology. The pilot transitioning to turboprop airplanes, particularly those who are not familiar with operations in the high/medium altitude environment, should approach turboprop transition training with this in mind. Thorough ground training should cover all aspects of high/medium altitude flight, including the flight environment, weather, flight planning and navigation, physiological aspects of high-altitude flight, oxygen and pressurization system operation, and high-altitude emergencies.

Figure 14-12. *Example of a typical turboprop airplane arrival and landing profile.*

Flight training should prepare the pilot to demonstrate a comprehensive knowledge of airplane performance, systems, emergency procedures, and operating limitations, along with a high degree of proficiency in performing all flight maneuvers and in-flight emergency procedures. The training outline below covers the minimum information needed by pilots to operate safely at high altitudes.

Ground Training

1. High-Altitude Flight Environment

   a. Airspace and Reduced Vertical Separation Minimum (RVSM) Operations

   b. Title 14 Code of Federal Regulations (14 CFR) part 91, section 91.211, Requirements for Use of Supplemental Oxygen

2. Weather

   a. Atmosphere

   b. Winds and clear air turbulence

   c. Icing

3. Flight Planning and Navigation

   a. Flight planning

   b. Weather charts

   c. Navigation

   d. Navigation aids (NAVAIDs)

   e. High Altitude Redesign (HAR)

   f. RNAV/Required Navigation Performance (RNP) and Receiver Autonomous Integrity Monitoring (RAIM) prediction

4. Physiological Training

   a. Respiration

   b. Hypoxia

   c. Effects of prolonged oxygen use

   d. Decompression sickness

   e. Vision

   f. Altitude chamber (optional)

5. High-Altitude Systems and Components

   a. Oxygen and oxygen equipment

   b. Pressurization systems

   c. High-altitude components

6. Aerodynamics and Performance Factors

   a. Acceleration and deceleration

   b. Gravity (G)-forces

   c. MACH Tuck and MACH Critical (turbojet airplanes)

   d. Swept wing concept

7. Emergencies

   a. Decompression

   b. Donning of oxygen masks

   c. Failure of oxygen mask or complete loss of oxygen supply/system

   d. In-flight fire

   e. Flight into severe turbulence or thunderstorms

   f. Compressor stalls

Flight Training

1. Preflight Briefing

2. Preflight Planning

   a. Weather briefing and considerations

   b. Course plotting

   c. Airplane Flight Manual (AFM)

   d. Flight plan

3. Preflight Inspection

   a. Functional test of oxygen system, including the verification of supply and pressure, regulator operation, oxygen flow, mask fit, and pilot and air traffic control (ATC) communication using mask microphones

4. Engine Start Procedures, Runup, Takeoff, and Initial Climb

5. Climb to High Altitude and Normal Cruise Operations While Operating Above 25,000 Feet Mean Sea Level (MSL)

6. Emergencies

   a. Simulated rapid decompression, including the immediate donning of oxygen masks

   b. Emergency descent

7. Planned Descents

8. Shutdown Procedures

9. Postflight Discussion

Chapter Summary

Transitioning from a non-turbopropeller airplane to a turbopropeller-powered airplane is discussed in this chapter. The major differences are introduced specifically handling, powerplant, and the associated systems. Turbopropeller electrical systems and operational considerations are explained to include starting procedures and high temperature considerations. Training considerations are also discussed and a sample training syllabus is given to show the topics that a pilot should become proficient in when transitioning to a turbopropeller-powered airplane.

Chapter 15

Transition to Jet-Powered Airplanes

Introduction

This chapter contains an overview of jet powered airplane operations. The information contained in this chapter is meant to be a useful preparation for, and a supplement to, formal and structured jet airplane qualification training. The intent of this chapter is to provide information on the major differences a pilot will encounter when transitioning to jet powered airplanes. In order to achieve this in a logical manner, the major differences between jet powered airplanes and piston powered airplanes have been approached by addressing two distinct areas: differences in technology, or how the airplane itself differs; and differences in pilot technique, or how the pilot addresses the technological differences through the application of different techniques. For airplane-specific information, a pilot should refer to the FAA-approved Airplane Flight Manual for that airplane.

Jet Engine Basics

A jet engine is a gas turbine engine. A jet engine develops thrust by accelerating a relatively small mass of air to very high velocity, as opposed to a propeller, which develops thrust by accelerating a much larger mass of air to a much slower velocity.

Piston and gas turbine engines are internal combustion engines and have a similar basic cycle of operation; that is, induction, compression, combustion, expansion, and exhaust. Air is taken in and compressed, and fuel is injected and burned. The hot gases then expand and supply a surplus of power over that required for compression and are finally exhausted. In both piston and jet engines, the efficiency of the cycle is improved by increasing the volume of air taken in and the compression ratio.

Part of the expansion of the burned gases takes place in the turbine section of the jet engine providing the necessary power to drive the compressor, while the remainder of the expansion takes place in the nozzle of the tail pipe in order to accelerate the gas to a high velocity jet thereby producing thrust. *[Figure 15-1]*

In theory, the jet engine is simpler and more directly converts thermal energy (the burning and expansion of gases) into mechanical energy (thrust). The piston or reciprocating engine, with all of its moving parts, must convert the thermal energy into mechanical energy and then finally into thrust by rotating a propeller.

One of the advantages of the jet engine over the piston engine is the jet engine's capability of producing much greater amounts of thrust horsepower at the high altitudes and high speeds. In fact, turbojet engine efficiency increases with altitude and speed.

Although the propeller-driven airplane is not nearly as efficient as the jet, particularly at the higher altitudes and cruising speeds required in modern aviation, one of the few advantages the propeller-driven airplane has over the jet is that maximum thrust is available almost at the start of the takeoff roll. Initial thrust output of the jet engine on takeoff is relatively lower and does not reach peak efficiency until the higher speeds. The fanjet or turbofan engine was developed to help compensate for this problem and is, in effect, a compromise between the pure jet engine (turbojet) and the propeller engine.

Like other gas turbine engines, the heart of the turbofan engine is the gas generator—the part of the engine that produces the hot, high-velocity gases. Similar to turboprops, turbofans have a low-pressure turbine section that uses most of the energy produced by the gas generator. The low pressure turbine is mounted on a concentric shaft that passes through the hollow shaft of the gas generator, connecting it to a ducted fan at the front of the engine. *[Figure 15-2]*

Air enters the engine, passes through the fan, and splits into two separate paths. Some of it flows around—bypasses the engine core, hence its name, bypass air. The air drawn into the engine for the gas generator is the core airflow. The amount of air that bypasses the core compared to the amount drawn into the gas generator determines a turbofan's bypass ratio. Turbofans efficiently convert fuel into thrust because they produce low-pressure energy spread over a large fan disk area. While a turbojet engine uses the entire gas generator's output to produce thrust in the form of a high-velocity exhaust gas jet, cool, low-velocity bypass air produces between 30 percent and 70 percent of the thrust produced by a turbofan engine.

The fan-jet concept increases the total thrust of the jet engine, particularly at the lower speeds and altitudes. Although efficiency at the higher altitudes is lost (turbofan engines are subject to a large lapse in thrust with increasing altitude), the

Figure 15-1. *Basic turbojet engine.*

Figure 15-2. *Turbofan engine.*

turbofan engine increases acceleration, decreases the takeoff roll, improves initial climb performance, and often has the effect of decreasing specific fuel consumption. Specific fuel consumption is a ratio of the fuel used by an engine and the amount of thrust it produces.

Operating the Jet Engine

In a jet engine, thrust is determined by the amount of fuel injected into the combustion chamber. The power controls on most turbojet-and turbofan-powered airplanes consist of just one thrust lever for each engine, because most engine control functions are automatic. The thrust lever is linked to a fuel control and/or electronic engine computer that meters fuel flow based upon revolutions per minute (rpm), internal temperatures, ambient conditions, and other factors. *[Figure 15-3]*

In a jet engine, each major rotating section usually has a separate gauge devoted to monitoring its speed of rotation. Depending on the make and model, a jet engine may have an N_1 gauge that monitors the low-pressure compressor section and/or fan speed in turbofan engines. The gas generator section may be monitored by an N_2 gauge, while triple spool engines may have an N_3 gauge as well. Each engine section rotates at many thousands of rpm. Their gauges therefore are calibrated in percent of rpm rather than actual rpm, for ease of display and interpretation. *[Figure 15-4]*

The temperature of turbine gases must be closely monitored by the pilot. As in any gas turbine engine, exceeding

temperature limits, even for a very few seconds, may result in serious heat damage to turbine blades and other components. Depending on the make and model, gas temperatures can be measured at a number of different locations within the engine. The associated engine gauges therefore have different names according to their location. For instance:

- Exhaust Gas Temperature (EGT)—the temperature of the exhaust gases as they enter the tail pipe after passing through the turbine.

Figure 15-3. *Jet engine power controls.*

Figure 15-4. *Jet engine RPM gauges.*

- Turbine Inlet Temperature (TIT)—the temperature of the gases from the combustion section of the engine as they enter the first stage of the turbine. The TIT is the highest temperature inside a gas turbine engine and is one of the limiting factors of the amount of power the engine can produce. TIT, however, is difficult to measure. Therefore, EGT, which relates to TIT, is normally the parameter measured.

- Interstage Turbine Temperature (ITT)—the temperature of the gases between the high-pressure and low-pressure turbine wheels.

- Turbine Outlet Temperature (TOT)—like EGT, turbine outlet temperature is taken aft of the turbine wheel(s).

Jet Engine Ignition

Most jet engine ignition systems consist of two igniter plugs, which are used during the ground or air starting of the engine. Once the start is completed, this ignition either automatically goes off or is turned off, and from this point on, the combustion in the engine is a continuous process.

Continuous Ignition

An engine is sensitive to the flow characteristics of the air that enters the intake of the engine nacelle. So long as the flow of air is substantially normal, the engine continues to run smoothly. However, particularly with rear-mounted engines that are sometimes in a position to be affected by disturbed airflow from the wings, there are some abnormal flight situations that could cause a compressor stall or flameout of the engine. These abnormal flight conditions would usually be associated with abrupt pitch changes such as might be encountered in severe turbulence or a stall.

In order to avoid the possibility of engine flameout from the above conditions, or from other conditions that might cause ingestion problems, such as heavy rain, ice, or possible bird strike, most jet engines are equipped with a continuous ignition system. This system can be turned on and used continuously whenever the need arises. In many jets, as an added precaution, this system is normally used during takeoffs and landings. Many jets are also equipped with an automatic ignition system that operates both igniters whenever the airplane stall warning or stick shaker is activated.

Fuel Heaters

Because of the high altitudes and extremely cold outside air temperatures in which the jet flies, it is possible to supercool the jet fuel to the point that the small particles of water suspended in the fuel can turn to ice crystals and clog the fuel filters leading to the engine. For this reason, jet engines are normally equipped with fuel heaters. The fuel heater may be of the automatic type that constantly maintains the fuel temperature above freezing, or they may be manually controlled by the pilot.

Setting Power

On some jet airplanes, thrust is indicated by an engine pressure ratio (EPR) gauge. EPR can be thought of as being equivalent to the manifold pressure on the piston engine. EPR is the difference between turbine discharge pressure and engine inlet pressure. It is an indication of what the engine has done with the raw air scooped in. For instance, an EPR setting of 2.24 means that the discharge pressure relative to the inlet pressure is 2.24:1. On these airplanes, the EPR gauge is the primary reference used to establish power settings. *[Figure 15-5]*

Figure 15-5. *EPR gauge.*

Fan speed (N_1) is the primary indication of thrust on most turbofan engines. Fuel flow provides a secondary thrust indication, and cross-checking for proper fuel flow can help in spotting a faulty N_1 gauge. Turbofans also have a gas generator turbine tachometer (N_2). They are used mainly for engine starting and some system functions.

In setting power, it is usually the primary power reference (EPR or N_1) that is most critical and is the gauge that first limits the forward movement of the thrust levers. However, there are occasions where the limits of either rpm or temperature can be exceeded. The rule is: movement of the thrust levers must be stopped and power set at whichever the limits of EPR, rpm, or temperature is reached first.

Thrust To Thrust Lever Relationship

In a piston-engine, propeller-driven airplane, thrust is proportional to rpm, manifold pressure, and propeller blade angle, with manifold pressure being the most dominant factor. At a constant rpm, thrust is proportional to throttle lever position. In a jet engine, however, thrust is quite disproportional to thrust lever position. This is an important difference that the pilot transitioning into jet-powered airplanes must become accustomed to.

On a jet engine, thrust is proportional to rpm (mass flow) and temperature (fuel/air ratio). These are matched and a further variation of thrust results from the compressor efficiency at varying rpm. The jet engine is most efficient at high rpm, where the engine is designed to be operated most of the time.

As rpm increases, mass flow, temperature, and efficiency also increase. Therefore, much more thrust is produced per increment of throttle movement near the top of the range than near the bottom.

One thing that seems different to the piston pilot transitioning into jet-powered airplanes is the rather large amount of thrust lever movement between the flight idle position and full power as compared to the small amount of movement of the throttle in the piston engine. For instance, an inch of throttle movement on a piston may be worth 400 horsepower wherever the throttle may be. On a jet, an inch of thrust lever movement at a low rpm may be worth only 200 pounds of thrust, but at a high rpm that same inch of movement might amount to closer to 2,000 pounds of thrust. Because of this, in a situation where significantly more thrust is needed and the jet engine is at low rpm, it does not do much good to merely "inch the thrust lever forward." Substantial thrust lever movement is in order. This is not to say that rough or abrupt thrust lever action is standard operating procedure. If the power setting is already high, it may take only a small amount of movement. However, there are two characteristics of the jet engine that work against the normal habits of the piston-engine pilot. One is the variation of thrust with rpm, and the other is the relatively slow acceleration of the jet engine.

Variation of Thrust with RPM

Whereas piston engines normally operate in the range of 40 percent to 70 percent of available rpm, jets operate most efficiently in the 85 percent to 100 percent range, with a flight idle rpm of 50 percent to 60 percent. The range from 90 percent to 100 percent in jets may produce as much thrust as the total available at 70 percent. *[Figure 15-6]*

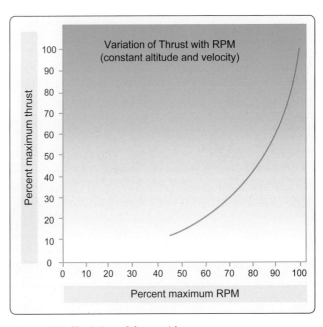

Figure 15-6. *Variation of thrust with rpm.*

Slow Acceleration of the Jet Engine

In a propeller-driven airplane, the constant speed propeller keeps the engine turning at a constant rpm within the governing range, and power is changed by varying the manifold pressure. Acceleration of the piston from idle to full power is relatively rapid, somewhere on the order of 3 to 4 seconds. The acceleration on the different jet engines can vary considerably, but it is usually much slower.

Efficiency in a jet engine is highest at high rpm where the compressor is working closest to its optimum conditions. At low rpm, the operating cycle is generally inefficient. If the engine is operating at normal approach rpm and there is a sudden requirement for increased thrust, the jet engine responds immediately and full thrust can be achieved in about 2 seconds. However, at a low rpm, sudden full-power application tends to over fuel the engine resulting in possible compressor surge, excessive turbine temperatures, compressor stall and/or flameout. To prevent this, various limiters, such as compressor bleed valves, are contained in the system and serve to restrict the engine until it is at an rpm at which it can respond to a rapid acceleration demand without distress. This critical rpm is most noticeable when the engine is at idle rpm, and the thrust lever is rapidly advanced to a high-power position. Engine acceleration is initially very slow, but can change to very fast after about 78 percent rpm is reached. *[Figure 15-7]*

Even though engine acceleration is nearly instantaneous after about 78 percent rpm, total time to accelerate from idle rpm to full power may take as much as 8 seconds. For this reason, most jets are operated at a relatively high rpm

during the final approach to landing or at any other time that immediate power may be needed.

Jet Engine Efficienc

Maximum operating altitudes for general aviation turbojet airplanes now reach 51,000 feet. The efficiency of the jet engine at high-altitudes is the primary reason for operating in the high-altitude environment. The specific fuel consumption of jet engines decreases as the outside air temperature decreases for constant engine rpm and true airspeed (TAS). Thus, by flying at a high altitude, the pilot is able to operate at flight levels where fuel economy is best and with the most advantageous cruise speed. For efficiency, jet airplanes are typically operated at high altitudes where cruise is usually very close to rpm or EGT limits. At high altitudes, little excess thrust may be available for maneuvering. Therefore, it is often impossible for the jet airplane to climb and turn simultaneously, and all maneuvering must be accomplished within the limits of available thrust and without sacrificing stability and controllability.

Absence of Propeller Effect

The absence of a propeller has a significant effect on the operation of jet-powered airplanes that the transitioning pilot must become accustomed to. The effect is due to the absence of lift from the propeller slipstream and the absence of propeller drag.

Absence of Propeller Slipstream

A propeller produces thrust by accelerating a large mass of air rearwards, and (especially with wing-mounted engines) this air passes over a comparatively large percentage of the wing area. On a propeller-driven airplane, the lift that the wing develops is the sum of the lift generated by the wing area not in the wake of the propeller (as a result of airplane speed) and the lift generated by the wing area influenced by the propeller slipstream. By increasing or decreasing the speed of the slipstream air, it is possible to increase or decrease the total lift on the wing without changing airspeed.

For example, a propeller-driven airplane that is allowed to become too low and too slow on an approach is very responsive to a quick blast of power to salvage the situation. In addition to increasing lift at a constant airspeed, stalling speed is reduced with power on. A jet engine, on the other hand, also produces thrust by accelerating a mass of air rearward, but this air does not pass over the wings. Therefore, there is no lift bonus at increased power at constant airspeed and no significant lowering of power-on stall speed.

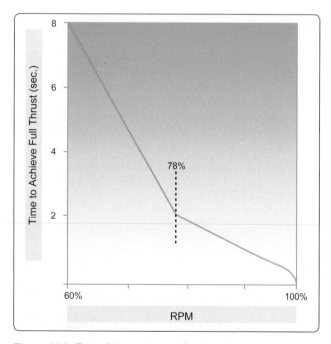

Figure 15-7. *Typical jet engine acceleration times.*

In not having propellers, the jet-powered airplane is minus two assets:

- It is not possible to produce increased lift instantly by simply increasing power.

- It is not possible to lower stall speed by simply increasing power. The 10-knot margin (roughly the difference between power-off and power-on stall speed on a propeller-driven airplane for a given configuration) is lost.

Add the poor acceleration response of the jet engine, and it becomes apparent that there are three ways in which the jet pilot is worse off than the propeller pilot. For these reasons, there is a marked difference between the approach qualities of a piston-engine airplane and a jet. In a piston-engine airplane, there is some room for error. Speed is not too critical and a burst of power salvages an increasing sink rate. In a jet, however, there is little room for error.

If an increasing sink rate develops in a jet, the pilot must remember two points in the proper sequence:

1. Increased lift can be gained only by accelerating airflow over the wings, and this can be accomplished only by accelerating the entire airplane.

2. The airplane can be accelerated, assuming altitude loss cannot be afforded, only by a rapid increase in thrust, and here, the slow acceleration of the jet engine (possibly up to 8 seconds) becomes a factor.

Salvaging an increasing sink rate on an approach in a jet can be a very difficult maneuver. The lack of ability to produce instant lift in the jet, along with the slow acceleration of the engine, necessitates a "stabilized approach" to a landing where full landing configuration, constant airspeed, controlled rate of descent, and relatively high power settings are maintained until over the threshold of the runway. This allows for almost immediate response from the engine in making minor changes in the approach speed or rate of descent and makes it possible to initiate an immediate go-around or missed approach if necessary.

Absence of Propeller Drag

When the throttles are closed on a piston-powered airplane, the propellers create a vast amount of drag, and airspeed is immediately decreased or altitude lost. The effect of reducing power to idle on the jet engine, however, produces no such drag effect. In fact, at an idle power setting, the jet engine still produces forward thrust. The main advantage is that the jet pilot is no longer faced with a potential drag penalty of a runaway propeller or a reversed propeller. A disadvantage, however, is the "freewheeling" effect forward thrust at idle has on the jet. While this occasionally can be used to advantage

(such as in a long descent), it is a handicap when it is necessary to lose speed quickly, such as when entering a terminal area or when in a landing flare. The lack of propeller drag, along with the aerodynamically clean airframe of the jet, are new to most pilots, and slowing the airplane down is one of the initial problems encountered by pilots transitioning into jets.

Speed Margins

The typical piston-powered airplane had to deal with two maximum operating speeds:

- V_{NO}—maximum structural cruising speed, represented on the airspeed indicator by the upper limit of the green arc. It is, however, permissible to exceed V_{NO} and operate in the caution range (yellow arc) in certain flight conditions.

- V_{NE}—never-exceed speed, represented by a red line on the airspeed indicator.

These speed margins in the piston airplanes were never of much concern during normal operations because the high drag factors and relatively low cruise power settings kept speeds well below these maximum limits.

Maximum speeds in jet airplanes are expressed differently and always define the maximum operating speed of the airplane, which is comparable to the V_{NE} of the piston airplane. These maximum speeds in a jet airplane are referred to as:

- V_{MO}—maximum operating speed expressed in terms of knots.

- M_{MO}—maximum operating speed expressed in terms of a decimal of Mach speed (speed of sound).

To observe both limits V_{MO} and M_{MO}, the pilot of a jet airplane needs both an airspeed indicator and a Machmeter, each with appropriate red lines. In some general aviation jet airplanes, these are combined into a single instrument that contains a pair of concentric indicators: one for the indicated airspeed and the other for indicated Mach number. Each is provided with an appropriate red line. *[Figure 15-8]*

It looks much like a conventional airspeed indicator but has a "barber pole" that automatically moves so as to display the applicable speed limit at all times.

Because of the higher available thrust and very low drag design, the jet airplane can very easily exceed its speed margin even in cruising flight and, in fact, in some airplanes in a shallow climb. The handling qualities in a jet can change drastically when the maximum operating speeds are exceeded.

Figure 15-8. *Jet airspeed indicator.*

High-speed airplanes designed for subsonic flight are limited to some Mach number below the speed of sound to avoid the formation of shock waves that begin to develop as the airplane nears Mach 1.0. These shock waves (and the adverse effects associated with them) can occur when the airplane speed is substantially below Mach 1.0. The Mach speed at which some portion of the airflow over the wing first equals Mach 1.0 is termed the critical Mach number (Mcr). This is also the speed at which a shock wave first appears on the airplane.

There is no particular problem associated with the acceleration of the airflow up to Mach Crit, the point where Mach 1.0 is encountered; however, a shock wave is formed at the point where the airflow suddenly returns to subsonic flow. This shock wave becomes more severe and moves aft on the wing as speed of the wing is increased and eventually flow separation occurs behind the well-developed shock wave. *[Figure 15-9]*

If allowed to progress well beyond the M_{MO} for the airplane, this separation of air behind the shock wave can result in severe buffeting and possible loss of control or "upset." Because of the changing center of lift of the wing resulting from the movement of the shock wave, the pilot experiences pitch change tendencies as the airplane moves through the transonic speeds up to and exceeding M_{MO}. *[Figure 15-10]*

As the graph in *Figure 15-10* illustrates, initially as speed is increased up to Mach .72, the wing develops an increasing amount of lift requiring a nose-down force or trim to maintain level flight. With increased speed and the aft movement of the shock wave, the wing's center of pressure also moves aft causing the start of a nose-down tendency or "tuck." By Mach .9, the nose-down forces are well developed to a point where a total of 70 pounds of back pressure are required to

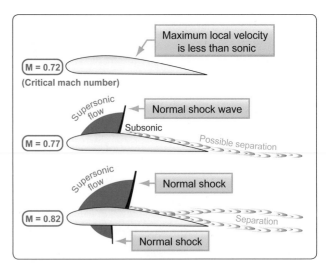

Figure 15-9. *Transonic flow patterns.*

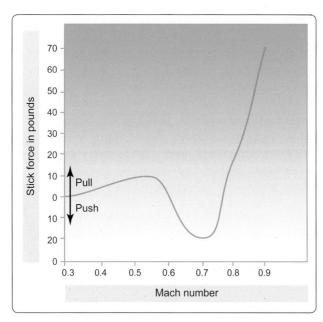

Figure 15-10. *Example of Stick Forces versus Mach Number in a typical jet airplane.*

hold the nose up. If allowed to progress unchecked, Mach tuck may eventually occur. Although Mach tuck develops gradually, if it is allowed to progress significantly, the center of pressure can move so far rearward that there is no longer enough elevator authority available to counteract it, and the airplane could enter a steep, sometimes unrecoverable, dive.

An alert pilot would have observed the high airspeed indications, experienced the onset of buffeting, and responded to aural warning devices long before encountering the extreme stick forces shown. However, in the event that corrective action is not taken and the nose is allowed to drop, increasing airspeed even further, the situation could rapidly become dangerous. As the Mach speed increases beyond the airplane's M_{MO}, the effects of flow separation and turbulence behind the shock wave become more severe. Eventually, the most powerful forces causing Mach tuck are a result of the buffeting and lack of effective downwash on the horizontal stabilizer because of the disturbed airflow over the wing. This is the primary reason for the development of the T-tail configuration on some jet airplanes, which places the horizontal stabilizer as far as practical from the turbulence of the wings. Also, because of the critical aspects of high-altitude/high-Mach flight, most jet airplanes capable of operating in the Mach speed ranges are designed with some form of trim and autopilot Mach compensating device (stick puller) to alert the pilot to inadvertent excursions beyond its certificated M_{MO}.

Recovery From Overspeed Conditions

A pilot must be aware of all the conditions that could lead to exceeding the airplane's maximum operating speeds.

Good attitude instrument flying skills and good power control are essential.

The pilot should be aware of the symptoms that will be experienced in the particular airplane as the V_{MO} or M_{MO} is being approached. These may include:

- Nose-down tendency and need for back pressure or trim.
- Mild buffeting as airflow separation begins to occur after critical Mach speed.
- Activation of an overspeed warning or high speed envelope protection.

The pilot's response to an overspeed condition should be to immediately slow the airplane by reducing the power to flight idle. It will also help to smoothly and easily raise the pitch attitude to help dissipate speed. The use of speed brakes can also aid in slowing the airplane. If, however, the nose-down stick forces have progressed to the extent that they are excessive, some speed brakes will tend to further aggravate the nose-down tendency. Under most conditions, this additional pitch down force is easily controllable, and since speed brakes can normally be used at any speed, they are a very real asset. If the first two options are not successful in slowing the airplane, a last resort option would be to extend the landing gear, if possible. This creates enormous drag and possibly some nose up pitch. This would be considered an emergency maneuver. The pilot transitioning into jet airplanes must be familiar with the manufacturers' recommended procedures for dealing with overspeed conditions contained in the FAA-approved Airplane Flight Manual for the particular make and model airplane.

Mach Buffet Boundaries

Thus far, only the Mach buffet that results from excessive speed has been addressed. The transitioning pilot, however, should be aware that Mach buffet is a function of the speed of the airflow over the wing— not necessarily the airspeed of the airplane. Anytime that too great a lift demand is made on the wing, whether from too fast an airspeed or from too high an angle of attack (AOA) near the M_{MO}, the "high speed buffet" will occur. However, there are also occasions when the buffet can be experienced at much slower speeds known as "low speed Mach buffet."

The most likely situations that could cause the low speed buffet would be when an airplane is flown at too slow of a speed for its weight and altitude causing a high AOA. This very high AOA would have the same effect of increasing airflow over the upper surface of the wing to the point that all of the same effects of the shock waves and buffet would occur as in the high speed buffet situation.

The AOA of the wing has the greatest effect on inducing the Mach buffet, or pre-stall buffet, at either the high or low speed boundaries for the airplane. The conditions that increase the AOA, hence the speed of the airflow over the wing and chances of Mach buffet are:

- High altitudes—The higher the airplane flies, the thinner the air and the greater the AOA required to produce the lift needed to maintain level flight.

- Heavy weights—The heavier the airplane, the greater the lift required of the wing, and all other things being equal, the greater the AOA.

- "G" loading—An increase in the "G" loading of the wing results in the same situation as increasing the weight of the airplane. It makes no difference whether the increase in "G" forces is caused by a turn, rough control usage, or turbulence. The effect of increasing the wing's AOA is the same.

An airplane's indicated airspeed decreases in relation to true airspeed as altitude increases. As the indicated airspeed decreases with altitude, it progressively merges with the low speed buffet boundary where pre-stall buffet occurs for the airplane at a load factor of 1.0 G. The point where the high speed Mach indicated airspeed and low speed buffet boundary indicated airspeed merge is the airplane's absolute or aerodynamic ceiling. This is where if an airplane flew any slower it would exceed its stalling AOA and experience low speed buffet. Additionally, if it flew any faster it would exceed M_{MO}, potentially leading to high speed buffet. This critical area of the airplane's flight envelope is known as "coffin corner." All airplanes are equipped with some form of stall warning system. Crews must be aware of systems installed on their airplanes (stick pushers, stick shakers, audio alarms, etc.) and their intended function. In a high altitude environment, airplane buffet is sometimes the initial indicator of problems.

Mach buffet occurs as a result of supersonic airflow on the wing. Stall buffet occurs at angles of attack that produce airflow disturbances (burbling) over the upper surface of the wing which decreases lift. As density altitude increases, the AOA that is required to produce an airflow disturbance over the top of the wing is reduced until the density altitude is reached where Mach buffet and stall buffet converge (coffin corner). When this phenomenon is encountered, serious consequences may result causing loss of airplane control.

Increasing either gross weight or load factor (G factor) will increase the low speed buffet and decrease Mach buffet speeds. A typical jet airplane flying at 51,000 feet altitude at 1.0 G may encounter Mach buffet slightly above the airplane's M_{MO} (0.82 Mach) and low speed buffet at 0.60 Mach. However, only 1.4 G (an increase of only 0.4 G) may bring on buffet at the optimum speed of 0.73 Mach and any change in airspeed, bank angle, or gust loading may reduce this straight-and-level flight 1.4 G protection to no protection at all. Consequently, a maximum cruising flight altitude must be selected which will allow sufficient buffet margin for necessary maneuvering and for gust conditions likely to be encountered. Therefore, it is important for pilots to be familiar with the use of charts showing cruise maneuver and buffet limits. *[Figure 15-11]*

The transitioning pilot must bear in mind that the maneuverability of the jet airplane is particularly critical, especially at the high altitudes. Some jet airplanes have a narrow span between the high and low speed buffets. One airspeed that the pilot should have firmly fixed in memory is the manufacturer's recommended gust penetration speed for the particular make and model airplane. This speed is normally the speed that would give the greatest margin between the high and low speed buffets, and may be considerably higher than design maneuvering speed (V_A). This means that, unlike piston airplanes, there are times when a jet airplane should be flown in excess of V_A during encounters with turbulence. Pilots operating airplanes at high speeds must be adequately trained to operate them safely. This training cannot be complete until pilots are thoroughly educated in the critical aspects of the aerodynamic factors pertinent to Mach flight at high altitudes.

Low Speed Flight

The jet airplane wing, designed primarily for high speed flight, has relatively poor low speed characteristics. As opposed to the normal piston powered airplane, the jet wing has less area relative to the airplane's weight, a lower aspect ratio (long chord/short span), and thin airfoil shape—all of which amount to the need for speed to generate enough lift. The sweptwing is additionally penalized at low speeds because its effective lift is proportional to airflow speed that is perpendicular to the leading edge. This airflow speed is always less than the airspeed of the airplane itself. In other words, the airflow on the sweptwing has the effect of persuading the wing into believing that it is flying slower than it actually is.

The first real consequence of poor lift at low speeds is a high stall speed. The second consequence of poor lift at low speeds is the manner in which lift and drag vary at those low speeds. As a jet airplane is slowed toward its minimum drag speed (V_{MD} or L/D_{MAX}), total drag increases at a much greater rate than the changes in lift, resulting in a sinking flightpath. If the pilot attempts to increase lift by increasing the AOA, airspeed will be further reduced resulting in a further increase

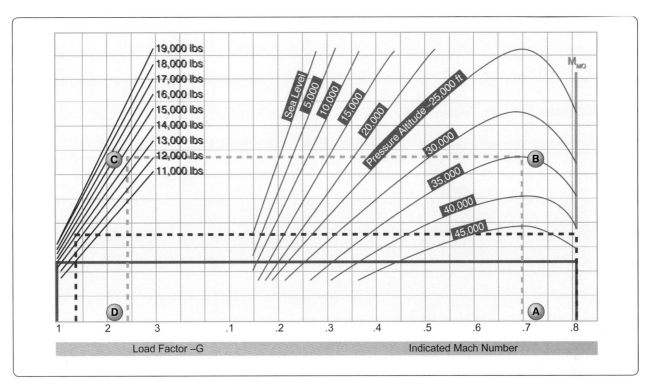

Figure 15-11. *Mach buffet boundary chart.*

in drag and sink rate as the airplane slides up the back side of the power-required curve. The sink rate can be arrested in one of two ways:

- Pitch attitude can be substantially reduced to reduce the AOA and allow the airplane to accelerate to a speed above V_{MD}, where steady flight conditions can be reestablished. This procedure, however, will invariably result in a substantial loss of altitude.

- Thrust can be increased to accelerate the airplane to a speed above V_{MD} to reestablish steady flight conditions. The amount of thrust must be sufficient to accelerate the airplane and regain altitude lost. Also, if the airplane has slid a long way up the back side of the power required (drag) curve, drag will be very high and a very large amount of thrust will be required.

In a typical piston engine airplane, V_{MD} in the clean configuration is normally at a speed of about 1.3 VS. *[Figure 15-12]* Flight below V_{MD} on a piston engine airplane is well identified and predictable. In contrast, in a jet airplane flight in the area of V_{MD} (typically 1.5 – 1.6 V_S) does not normally produce any noticeable changes in flying qualities other than a lack of speed stability—a condition where a decrease in speed leads to an increase in drag which leads to a further decrease in speed and hence a speed divergence. A pilot who is not cognizant of a developing speed divergence may find a serious sink rate developing at a constant power setting, and a pitch attitude that appears to be normal. The

fact that drag increases more rapidly than lift, causing a sinking flightpath, is one of the most important aspects of jet airplane flying qualities.

Stalls

The stalling characteristics of the sweptwing jet airplane can vary considerably from those of the normal straight wing airplane. The greatest difference that will be noticeable to the pilot is the lift developed vs. angle of attack. An increase in angle of attack of the straight wing produces a substantial and constantly increasing lift vector up to its maximum coefficient of lift, and soon thereafter flow separation (stall) occurs with a rapid deterioration of lift.

By contrast, the sweptwing produces a much more gradual buildup of lift with a less well-defined maximum coefficient. This less-defined peak also means that a swept wing may not have as dramatic loss of lift at angles of attack beyond its maximum lift coefficient. However, these high-lift conditions are accompanied by high drag, which results in a high rate of descent. *[Figure 15-13]*

The differences in the stall characteristics between a conventional straight wing/low tailplane (non T-tail) airplane and a sweptwing T-tail airplane center around two main areas.

- The basic pitching tendency of the airplane at the stall.
- Tail effectiveness in stall recovery.

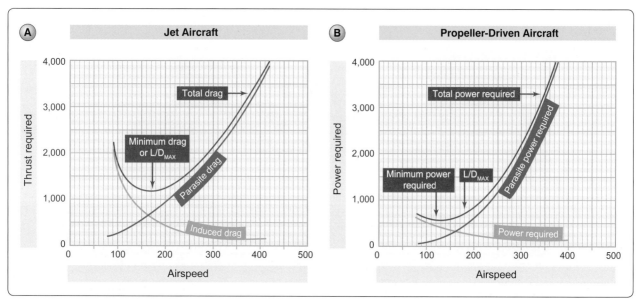

Figure 15-12. *Thrust and power required curves (jet aircraft vs. propeller-driven aircraft).*

On a conventional straight wing/low tailplane airplane, the weight of the airplane acts downwards forward of the lift acting upwards, producing a need for a balancing force acting downwards from the tailplane. As speed is reduced by gentle up elevator deflection, the static stability of the airplane causes a nose-down tendency. This is countered by further up elevator to keep the nose coming up and the speed decreasing. As the pitch attitude increases, the low set tail is immersed in the wing wake, which is slightly turbulent, low energy air. The accompanying aerodynamic buffeting serves as a warning of impending stall. The reduced effectiveness of the tail prevents the pilot from forcing the airplane into a deeper stall. *[Figure 15-14]* The conventional straight wing airplane conforms to the familiar nose-down pitching tendency at the stall and gives the entire airplane a fairly pronounced nose-down pitch. At the moment of stall, the wing wake passes more or less straight rearward and passes above the tail. The tail is now immersed in high energy air where it experiences a sharp increase in positive AOA causing upward lift. This lift then assists the nose-down pitch and decrease in wing AOA essential to stall recovery.

In a sweptwing jet with a T-tail and rear fuselage mounted engines, the two qualities that are different from its straight wing low tailplane counterpart are the pitching tendency

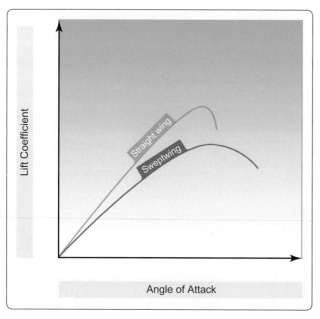

Figure 15-13. *Stall versus angle of attack—sweptwing versus straight wing.*

Figure 15-14. *Stall progression—typical straight wing airplane.*

of the airplane as the stall develops and the loss of tail effectiveness at the stall. The handling qualities down to the stall are much the same as the straight wing airplane except that the high, T-tail remains clear of the wing wake and provides little or no warning in the form of a pre-stall buffet. Also, the tail is fully effective during the speed reduction towards the stall, and remains effective even after the wing has begun to stall. This enables the pilot to drive the wing into a deeper stall at a much greater AOA.

At the stall, two distinct things happen. After the stall, the sweptwing T-tail airplane tends to pitch up rather than down, and the T-tail can become immersed in the wing wake, which is low energy turbulent air. This greatly reduces tail effectiveness and the airplane's ability to counter the nose-up pitch. Also, if the AOA increases further, the disturbed, relatively slow air behind the wing may sweep across the tail at such a large angle that the tail itself stalls. If this occurs, the pilot loses all pitch control and will be unable to lower the nose. The pitch up just after the stall is worsened by large reduction in lift and a large increase in drag, which causes a rapidly increasing descent path, thus compounding the rate of increase of the wing's AOA. *[Figure 15-15]*

A slight pitch up tendency after the stall is a characteristic of a swept or tapered wings. With these types of wings, there is a tendency for the wing to develop a spanwise airflow towards the wingtip when the wing is at high angles of attack. This leads to a tendency for separation of airflow, and the subsequent stall, to occur at the wingtips first. *[Figure 15-16]*

In an unmodified swept wing, the tips first stall, results in a shift of the center of lift of the wing in a forward direction relative to the center of gravity of the airplane, causing a tendency for the nose to pitch up. A disadvantage of a tip first stall is that it can involve the ailerons and erode roll control. To satisfy certification criteria, airplane manufacturers may have to tailor the airfoil characteristics of a wing as it proceeds from the root to the tip so that a pilot can still maintain wings level flight with normal use of the controls. Still, more aileron will be required near stall to correct roll excursion than in normal flight, as the effectiveness of the ailerons will be reduced and feel mushy. This change in feel can be an important recognition cue that the airplane may be stalled.

As previously stated, when flying at a speed near VMD, an increase in AOA causes drag to increase faster than lift and the airplane begins to sink. It is essential to understand that this increasing sinking tendency, at a constant pitch attitude, results in a rapid increase in AOA as the flightpath becomes deflected downwards. *[Figure 15-17]* Furthermore, once the stall has developed and a large amount of lift has been lost, the airplane will begin to sink rapidly and this will be accompanied by a corresponding rapid increase in AOA. This is the beginning of what is termed a deep stall.

As an airplane enters a deep stall, increasing drag reduces forward speed to well below normal stall speed. The sink rate may increase to many thousands of feet per minute. It must be emphasized that this situation can occur without an excessively nose-high pitch attitude. On some airplanes, it can occur at an apparently normal pitch attitude, and it is this quality that can mislead the pilot because it appears similar to the beginning of a normal stall recovery. It can also occur at a negative pitch attitude, that is, with the nose pointing towards the ground. In such situations, it seems counterintuitive to apply the correct recovery action, which is to push forward on the pitch control to reduce the AOA, as this action will also cause the nose to point even further towards the ground. But, that is the right thing to do.

Deep stalls may be unrecoverable. Fortunately, they are easily avoided as long as published limitations are observed. On those airplanes susceptible to deep stalls (not all swept or tapered wing airplanes are), sophisticated stall warning systems such as stick shakers are standard equipment. A stick pusher, as its name implies, acts to automatically reduce the airplane's AOA before the airplane reaches a dangerous stall condition, or it may aid in recovering the airplane from a stall if an airplane's natural aerodynamic characteristics do so weakly.

Pilots undergoing training in jet airplanes are taught to recover at the first sign of an impending stall instead of going beyond those initial cues and into a full stall. Normally, this

Figure 15-15. *Stall progression—sweptwing airplane.*

Prestall

Stalled

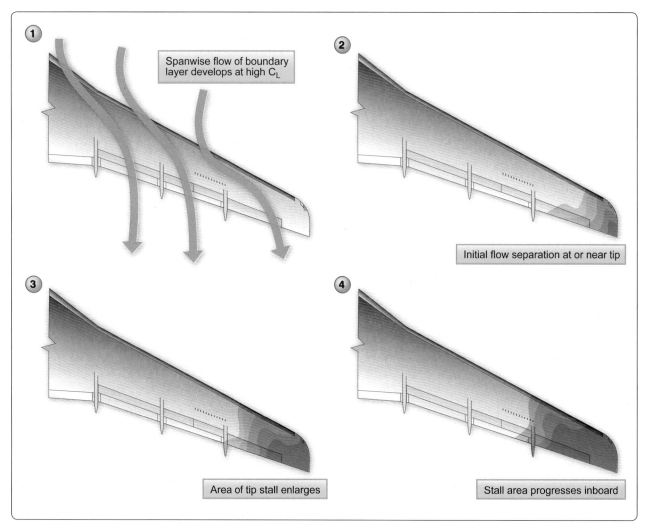

Figure 15-16. *Sweptwing stall characteristics.*

is indicated by aural stall warning devices or activation of the airplane's stick shaker. Stick shakers normally activate around 107 percent of the actual stall speed. In response to a stall warning, the proper action is for the pilot to apply a nose-down input until the stall warning stops (pitch trim may be necessary). Then, the wings are rolled level, followed by adjusting thrust to return to normal flight. The elapsed time will be small between these actions, particularly at low altitude where a significant available thrust exists. It is important to understand that reducing AOA eliminates the stall, but applying thrust will allow the descent to be stopped once the wing is flying again.

At high altitudes the stall recovery technique is the same. A pilot will need to reduce the AOA by lowering the nose until the stall warning stops. However, after the AOA has been reduced to where the wing is again developing efficient lift, the airplane will still likely need to accelerate to a desired airspeed. At high altitudes where the available thrust is significantly less than at lower altitudes, the only way to achieve that

acceleration is to pitch the nose downwards and use gravity. As such, several thousand feet or more of altitude loss may be needed to recover completely. The above discussion covers most airplanes; however, the stall recovery procedures for a particular make and model airplane may differ slightly, as recommended by the manufacturer, and are contained in the FAA-approved Airplane Flight Manual for that airplane.

Drag Devices

To the pilot transitioning into jet airplanes, going faster is seldom a problem. It is getting the airplane to slow down that seems to cause the most difficulty. This is because of the extremely clean aerodynamic design and fast momentum of the jet airplane and because the jet lacks the propeller drag effects that the pilot has been accustomed to. Additionally, even with the power reduced to flight idle, the jet engine still produces thrust, and deceleration of the jet airplane is a slow process. Jet airplanes have a glide performance that is double that of piston-powered airplanes, and jet pilots often cannot

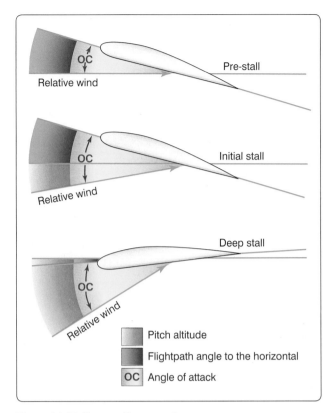

Figure 15-17. *Deep stall progression.*

Figure 15-18. *Spoilers.*

comply with an ATC request to go down and slow down at the same time. Therefore, jet airplanes are equipped with drag devices, such as spoilers and speed brakes.

The primary purpose of spoilers is to spoil lift. The most common type of spoiler consists of one or more rectangular plates that lie flush with the upper surface of each wing. They are installed approximately parallel to the lateral axis of the airplane and are hinged along the leading edges. When deployed, spoilers deflect up against the relative wind, which interferes with the flow of air about the wing. *[Figure 15-18]* This both spoils lift and increases drag. Spoilers are usually installed forward of the flaps but not in front of the ailerons so as not to interfere with roll control.

Deploying spoilers results in a substantial sink rate with little decay in airspeed. Some airplanes exhibit a nose-up pitch tendency when the spoilers are deployed, which the pilot must anticipate.

When spoilers are deployed on landing, most of the wing's lift is destroyed. This action transfers the airplane's weight to the landing gear so that the wheel brakes are more effective. Another beneficial effect of deploying spoilers on landing is that they create considerable drag, adding to the overall aerodynamic braking. The real value of spoilers on landing, however, is creating the best circumstances for using wheel brakes.

The primary purpose of speed brakes is to produce drag. Speed brakes are found in many sizes, shapes, and locations on different airplanes, but they all have the same purpose—to assist in rapid deceleration. The speed brake consists of a hydraulically-operated board that, when deployed, extends into the airstream. Deploying speed brakes results in a rapid decrease in airspeed. Typically, speed brakes can be deployed at any time during flight in order to help control airspeed, but they are most often used only when a rapid deceleration must be accomplished to slow down to landing gear and flap speeds. There is usually a certain amount of noise and buffeting associated with the use of speed brakes, along with an obvious penalty in fuel consumption. Procedures for the use of spoilers and/or speed brakes in various situations are contained in the FAA-approved AFM for the particular airplane.

Thrust Reversers

Jet airplanes have high kinetic energy during the landing roll because of weight and speed. This energy is difficult to dissipate because a jet airplane has low drag with the nose wheel on the ground, and the engines continue to produce forward thrust with the power levers at idle. While wheel brakes normally can cope, there is an obvious need for another speed retarding method. This need is satisfied by the drag provided by reverse thrust.

A thrust reverser is a device fitted in the engine exhaust system that effectively reverses the flow of the exhaust gases. The flow does not reverse through 180°; however, the final path of the exhaust gases is about 45° from straight ahead. This, together with the losses in the reverse flow paths, results in a net efficiency of about 50 percent. It produces even less if the engine rpm is less than maximum in reverse.

Normally, a jet engine has one of two types of thrust reversers: a target reverser or a cascade reverser. *[Figure 15-19]* Target

Figure 15-19. *Thrust reversers.*

reversers are simple clamshell doors that swivel from the stowed position at the engine tailpipe to block all of the outflow and redirect some component of the thrust forward.

Cascade reversers are more complex. They are normally found on turbofan engines and are often designed to reverse only the fan air portion. Blocking doors in the shroud obstructs forward fan thrust and redirects it through cascade vanes for some reverse component. Cascades are generally less effective than target reversers, particularly those that reverse only fan air, because they do not affect the engine core, which continues to produce forward thrust.

On most installations, reverse thrust is obtained with the thrust lever at idle by pulling up the reverse lever to a detent. Doing so positions the reversing mechanisms for operation but leaves the engine at idle rpm. Further upward and backward movement of the reverse lever increases engine power. Reverse is cancelled by closing the reverse lever to the idle reverse position, then dropping it fully back to the forward idle position. This last movement operates the reverser back to the forward thrust position.

Reverse thrust is much more effective at high airplane speed than at low airplane speeds for two reasons: the net amount of reverse thrust increases with speed; and the power produced is higher at higher speeds because of the increased rate of doing work. In other words, the kinetic energy of the airplane is being destroyed at a higher rate at the higher speeds. To get maximum efficiency from reverse thrust, therefore, it should be used as soon as is prudent after touchdown.

When considering the proper time to apply reverse thrust after touchdown, the pilot should remember that some airplanes tend to pitch nose up when reverse is selected on landing and this effect, particularly when combined with the nose-up pitch effect from the spoilers, can cause the airplane to leave the ground again momentarily. On these types, the airplane must be firmly on the ground with the nose wheel down before reverse is selected. Other types of airplanes have no change in pitch, and reverse idle may be selected after the main gear is down and before the nose wheel is down. Specific procedures for reverse thrust operation for a particular airplane/engine combination are contained in the FAA-approved AFM for that airplane.

There is a significant difference between reverse pitch on a propeller and reverse thrust on a jet. Idle reverse on a propeller produces about 60 percent of the reverse thrust available at full power reverse and is therefore very effective at this setting when full reverse is not needed. On a jet engine, however, selecting idle reverse produces very little actual reverse thrust. In a jet airplane, the pilot must not only select reverse as soon as reasonable, but then must open up to full power reverse as soon as possible. Within AFM limitations, full power reverse should be held until the pilot is certain the landing roll is contained within the distance available.

Inadvertent deployment of thrust reversers while airborne is a very serious emergency situation. Therefore, thrust reverser systems are designed with this prospect in mind. The systems normally contain several lock systems: one to keep reversers from operating in the air, another to prevent operation with the thrust levers out of the idle detent, and/or an "auto-stow" circuit to command reverser stowage any time thrust reverser deployment would be inappropriate, such as during takeoff and while airborne. It is essential that pilots understand not only the normal procedures and limitations of thrust reverser use, but also the procedures for coping with uncommanded reverse. Those emergencies demand immediate and accurate response.

Pilot Sensations in Jet Flying

There are usually three general sensations that the pilot transitioning into jets will immediately become aware of. These are: response differences, increased control sensitivity, and a much increased tempo of flight.

In many flight conditions, airspeed changes can occur more slowly than in a propeller airplane. This arises from different effects. At high altitudes, the ability to accelerate lessens due to the reduction in available thrust. Another effect is the long spool-up time required from low throttle settings. Some aircraft can take on the order of 8–10 seconds to develop full thrust when starting from an idle condition. Finally, the clean aerodynamic design of a jet can result in smaller than expected decelerations when thrust is reduced to idle.

The lack of propeller effect is also responsible for the lower drag increment at the reduced power settings and results in other changes that the pilot will have to become accustomed to. These include the lack of effective slipstream over the lifting surfaces and control surfaces, and lack of propeller torque effect.

The aft mounted engines will cause a different reaction to power application and may result in a slightly nosedown pitching tendency with the application of power. On the other hand, power reduction will not cause pitch changes to the same extent the pilot is used to in a propeller airplane. Although neither of these characteristics are radical enough to cause transitioning pilots much of a problem, they must be compensated for.

Power settings required to attain a given performance are almost impossible to memorize in the jets, and the pilot who feels the necessity for having an array of power settings for all occasions will initially feel at a loss. The only way to answer the question of "how much power is needed?" is by saying, "whatever is required to get the job done." The primary reason that power settings vary so much is because of the great changes in weight as fuel is consumed during the flight. Therefore, the pilot will have to learn to use power as needed to achieve the desired performance.

In time, the pilot will find that the only reference to power instruments will be that required to keep from exceeding limits of maximum power settings or to synchronize rpm.

Proper power management is one of the initial problem areas encountered by the pilot transitioning into jet airplanes. Although smooth power applications are still the rule, the pilot will be aware that a greater physical movement of the power levers is required as compared to throttle movement in the piston engines. The pilot will also have to learn to anticipate and lead the power changes more than in the past and must keep in mind that the last 30 percent of engine rpm represents the majority of the engine thrust, and below that the application of power has very little effect. In slowing the airplane, power reduction must be made sooner because there is no longer any propeller drag and the pilot should anticipate the need for drag devices.

Control sensitivity will differ between various airplanes, but in all cases, the pilot will find that they are more sensitive to any change in control displacement, particularly pitch control, than are the conventional propeller airplanes. Because of the higher speeds flown, the control surfaces are more effective and a variation of just a few degrees in pitch attitude in a jet can result in over twice the rate of altitude change that would be experienced in a slower airplane. The sensitive pitch control in jet airplanes is one of the first flight differences that the pilot will notice. Invariably the pilot will have a tendency to overcontrol pitch during initial training flights. The importance of accurate and smooth control cannot be overemphasized, however, and it is one of the first techniques the transitioning pilot must master.

The pilot of a sweptwing jet airplane will soon become adjusted to the fact that it is necessary and normal to fly at higher angles of attack. It is not unusual to have about 5° of nose-up pitch on an approach to a landing. During an approach to a stall at constant altitude, the nose-up angle may be as high as 15° to 20°. The higher deck angles (pitch angle relative to the ground) on takeoff, which may be as high as 15°, will also take some getting used to, although this is not the actual AOA relative to the airflow over the wing.

The greater variation of pitch attitudes flown in a jet airplane are a result of the greater thrust available and the flight characteristics of the low aspect ratio and sweptwing. Flight at the higher pitch attitudes requires a greater reliance on the flight instruments for airplane control since there is not much in the way of a useful horizon or other outside reference to be seen. Because of the high rates of climb and descent, high airspeeds, high altitudes and variety of attitudes flown, the jet airplane can only be precisely flown by applying proficient instrument flight techniques. Proficiency in attitude instrument flying, therefore, is essential to successful transition to jet airplane flying.

Most jet airplanes are equipped with a thumb operated pitch trim button on the control wheel which the pilot must become familiar with as soon as possible. The jet airplane will differ regarding pitch tendencies with the lowering of flaps, landing gear, and drag devices. With experience, the jet airplane pilot will learn to anticipate the amount of pitch change required for a particular operation. The usual method of operating the trim button is to apply several small, intermittent applications of trim in the direction desired rather than holding the trim button for longer periods of time which can lead to overcontrolling.

Jet Airplane Takeoff and Climb

The following information is generic in nature and, since most civilian jet airplanes require a minimum flight crew of two pilots, assumes a two pilot crew. If any of the following information conflicts with FAA-approved AFM procedures for a particular airplane, the AFM procedures take precedence. Also, if any of the following procedures differ from the FAA-approved procedures developed for use by a specific air operator and/or for use in an FAA-approved training center or pilot school curriculum, the FAA-approved procedures for that operator and/or training center/pilot school take precedence.

All FAA certificated jet airplanes are certificated under Title 14 of the Code of Federal Regulations (14 CFR) part 25, which contains the airworthiness standards for transport category airplanes. The FAA-certificated jet airplane is a highly sophisticated machine with proven levels of performance and guaranteed safety margins. The jet airplane's performance and safety margins can only be realized, however, if the airplane is operated in strict compliance with the procedures and limitations contained in the FAA-approved AFM for the particular airplane. Furthermore, in accordance with 14 CFR part 91, section 91.213, a turbine powered airplane may not be operated with inoperable instruments or equipment installed unless an approved Minimum Equipment List (MEL) exists for that aircraft, and the aircraft is operated under all applicable conditions and limitations contained in the MEL.

Minimum Equipment List and Configuration Deviation List

The MEL serves as a reference guide for dispatchers and pilots to determine whether takeoff of an aircraft with inoperative instruments or equipment is authorized under the provisions of applicable regulatory requirements.

The operator's MEL must be modeled after the FAA's Master MEL for each type of aircraft and must be approved by the Administrator before its implementation. The MEL includes a "General Section," comprised of definitions, general policies, as well as operational procedures for flight crews and maintenance personnel. Each aircraft component addressed in the MEL is listed in an alphabetical index for quick reference. A table of contents further divides the manual in different chapters, each numbered for its corresponding aircraft system designation (i.e., the electrical system, also designated as system number 24, would be found in chapter 24 of the MEL).

Maintenance may be deferred only on those aircraft systems and components cataloged in the approved MEL. If a malfunctioning or missing item is not specifically listed in the MEL inventory, takeoff is not authorized until the item is adequately repaired or replaced. In cases where repairs may temporarily be deferred, operation or dispatch of an aircraft whose systems have been impaired is often subject to limitations or other conditional requirements explicitly articulated in the MEL. Such conditional requirements may be of an operational nature, a mechanical nature, or both. Operational conditions generally include one or more of the following:

- Limited use of aircraft systems

- Downgraded instrument flight rule (IFR) landing minima

- Fuel increases due to additional burn, required automatic power unit (APU) usage or potential fuel imbalance situations

- Precautionary checks to be performed by the crew prior to departure, or special techniques to be applied while in flight

- Weight penalties affecting takeoff, cruise, or landing performance (runway limit, climb limit, usable landing distance reduction, and V_{REF}, takeoff V-speeds, N_1/EPR adjustments)

- Specific flight restrictions involving:

  - Authorized areas of operation (clearly defined geographical regions)

  - Type of operations (international, extended operations (ETOPS))

  - Altitude and airspace (reduced vertical separation minimums (RVSM)

  - Minimum navigation performance specifications (MNPS)

  - Speed (knots indicated airspeed (KIAS) or Mach)

  - Routing options (extended overwater, reduced navigation capability, High Altitude Redesign navigation)

  - Environmental conditions (icing, thunderstorms, wind shear, daylight, visual meteorological conditions (VMC), turbulence index, cross-wind component)

  - Airport selection (runway surface, length, contamination, and availability of aircraft maintenance, Airport rescue and firefighting (ARFF) and ATC services)

Listed below are some examples of both operational and mechanical situations that may be encountered:

- A defective Ground Proximity Warning System (GPWS) would require alternate procedures to be developed by the operator to mitigate the loss of the GPWS and would likely only allow continued operation for two days.

- An inoperative air condition (A/C) pack might restrict a Super 80 or a Boeing 737 to a maximum operating altitude of flight level (FL)250, whereas as a Boeing 757 is only restricted to FL350.

- An inoperative Auxiliary Power Unit (APU) will not affect the performance or flying characteristics of an aircraft, but it does prompt the operator to verify that ground air and electrical power is available for that particular type of aircraft at the designated destination and alternate airports.

- A faulty fuel pump in the center tank may lower the Maximum Zero Fuel Weight (MZFW) by the amount of center tank fuel, as that fuel would otherwise be trapped and unusable should the remaining fuel pump fail while in flight. At the same time, the unavailability of center tank fuel unmistakably decreases the aircraft range while perhaps excluding it from operating too far off-shore.

- An inoperable generator (IDG) may require the continuous operation of the APU as an alternate source of electrical power throughout the entire flight (and thus more fuel) as it is tasked with assuming the function of the defunct generator.

- A failure of the Heads-up Display (HUD) or the auto-pilot may restrict the airplane to higher approach minima (taking it out of Category II or Category III authorizations)

Mechanical conditions outlined in the MEL may require precautionary pre-flight checks, partial repairs prior to departure, or the isolation of selected elements of the deficient aircraft system (or related interacting systems), as well as the securing of other system components to avoid further degradation of its operation in flight. The MEL may contain either a step-by-step description of required partial maintenance actions or a list of numerical references to the Maintenance Procedures Manual (MPM) where each corrective procedure is explained in detail. When procedures must be performed to ensure the aircraft can be safely operated, they are categorized as either Operations Procedures or Maintenance Procedures. The MEL will denote which by indicating an "O" or an "M" as appropriate.

If operational and mechanical conditions can be met, a placard is issued and an entry made in the aircraft MEL Deferral Record to authorize the operation for a limited time before more permanent repairs can be accomplished. The placard is affixed by maintenance personnel or the flight crew as appropriate onto the instrument or control mechanism that otherwise governs the operation of the defective device.

In order to use the MEL properly, it is important to clearly understand its purpose and the timing of its applicability. Because it is designed to provide guidance in determining whether a flight can be safely initiated with aircraft equipment that is deficient, inoperative, or missing, the MEL is only relevant while the aircraft is still on the ground awaiting departure or takeoff. It is essentially a dispatching reference tool used in support of all applicable Federal Aviation Regulations. If dispatchers are not required by the Operator's certificate, flight crews still need to refer to the MEL before dispatching themselves and ensure that the flight is planned and conducted within the operating limits set forth in the MEL. However, once the aircraft is airborne, any mechanical failure should be addressed using the appropriate checklists and approved AFM, not the MEL. Although nothing could technically keep a pilot from referring to the MEL for background information and documentation to support his decisions, his actions must be based strictly on instructions provided by the AFM (i.e., Abnormal or Emergency sections).

A Configuration Deviation List (CDL) is used in the same manner as a MEL but it differs in that it addresses missing external parts of the aircraft rather than failing internal systems and their constituent parts. They typically include elements, such as service doors, power receptacle doors, slat track doors, landing gear doors, APU ram air doors, flaps fairings, nose wheel spray deflectors, position light lens covers, slat segment seals, static dischargers, etc. Each CDL item has a corresponding AFM number that identifies successively the system number, sub-system number, and item number. Flight limitations derived from open CDL items typically involve some kind of weight penalty and/or fuel tax due to increased drag and a net performance decrement, although some environmental restrictions may also be of concern in a few isolated cases. For example, a missing nose wheel spray deflector (Super 80 aircraft) requires dry runways for both takeoff and landing.

Each page of the MEL/CDL is divided into 6 columns. From left to right, these columns normally display the following information:

- Functional description/identification of the inoperative or missing aircraft equipment item

- Normal complement of equipment (number installed)

- Minimum equipment required for departure (number of items)

- Conditions required for flight/dispatch including maintenance action required (M) by mechanics or other authorized maintenance personnel and operational procedures or restrictions (O) to be observed by the flight crew

V-Speeds

The following are speeds that affect the jet airplane's takeoff performance. The jet airplane pilot must be thoroughly familiar with each of these speeds and how they are used in the planning of the takeoff.

- V_S—stalling speed or minimum steady flight speed at which the airplane is controllable.

- V_1—critical engine failure speed or takeoff decision speed. It is the speed at which the pilot is to continue the takeoff in the event of an engine failure or other serious emergency. At speeds less than V1, it is considered safer to stop the aircraft within the accelerate-stop distance. It is also the minimum speed in the takeoff, following a failure of the critical engine at V_{EF}, at which the pilot can continue the takeoff and achieve the required height above the takeoff surface within the takeoff distance.

- V_{EF}—speed at which the critical engine is assumed to fail during takeoff. This speed is used during aircraft certification.

- V_R—rotation speed, or speed at which the rotation of the airplane is initiated to takeoff attitude. This speed cannot be less than V_1 or less than $1.05 \times V_{MCA}$ (minimum control speed in the air). On a single-engine takeoff, it must also allow for the acceleration to V_2 at the 35-foot height at the end of the runway.

- V_{LOF}—lift-off speed, or speed at which the airplane first becomes airborne. This is an engineering term used when the airplane is certificated and must meet certain requirements. If it is not listed in the AFM, it is within requirements and does not have to be taken into consideration by the pilot.

- V_2—takeoff safety speed means a referenced airspeed obtained after lift-off at which the required one-engine-inoperative climb performance can be achieved.

Pre-Takeoff Procedures

Takeoff data, including V_1/V_R and V_2 speeds, takeoff power settings, and required field length should be computed prior to each takeoff and recorded on a takeoff data card. This data is based on airplane weight, runway length available, runway gradient, field temperature, field barometric pressure, wind, icing conditions, and runway condition. Both pilots should separately compute the takeoff data and cross-check in the cockpit with the takeoff data card.

A captain's briefing is an essential part of crew resource management (CRM) procedures and should be accomplished just prior to takeoff. *[Figure 15-20]* The captain's briefing is an opportunity to review crew coordination procedures for

Captain's Briefing

I will advance the thrust levers.

Follow me through on the thrust levers.

Monitor all instruments and warning lights on the takeoff roll and call out any discrepancies or malfunctions observed prior to V_1, and I will abort the takeoff. Stand by to arm thrust reversers on my command.

Give me a visual and oral signal for the following:
- 80 knots, and I will disengage nosewheel steering.
- V_1, and I will move my hand from thrust to yoke.
- V_R, and I will rotate.

In the event of engine failure at or after V_1, I will continue the takeoff roll to V_R, rotate and establish V_2 climb speed. I will identify the inoperative engine, and we will both verify. I will accomplish the shutdown, or have you do it on my command.

I will expect you to stand by on the appropriate emergency checklist.

I will give you a visual and oral signal for gear retraction and for power settings after the takeoff.

Our VFR emergency procedure is to............................
Our IFR emergency procedure is to.............................

Figure 15-20. *Sample captain's briefing.*

takeoff, which is always the most critical portion of a flight. The takeoff and climb-out should be accomplished in accordance with a standard takeoff and departure profile developed for the particular make and model airplane. *[Figure 15-21]*

Takeoff Roll

The entire runway length should be available for takeoff, especially if the pre-calculated takeoff performance shows the airplane to be limited by runway length or obstacles. After taxing into position at the end of the runway, the airplane should be aligned in the center of the runway allowing equal distance on either side. The brakes should be held while the thrust levers are brought to a power setting specified in the AFM and the engines allowed to stabilize. The engine instruments should be checked for proper operation before the brakes are released or the power increased further. This procedure assures symmetrical thrust during the takeoff roll and aids in prevention of overshooting the desired takeoff thrust setting. The brakes should then be released and, during the start of the takeoff roll, the thrust levers smoothly advanced to the pre-computed takeoff power setting. All final takeoff thrust adjustments should be made prior to reaching 60 knots. The final engine power adjustments are normally made by the pilot not flying. Once the thrust levers are set for takeoff

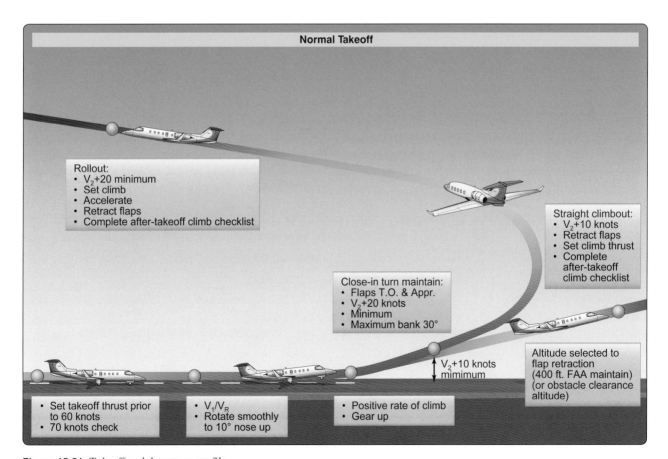

Figure 15-21. *Takeoff and departure profile.*

power, they should not be readjusted after 60 knots. Retarding a thrust lever would only be necessary in case an engine exceeds any limitation, such as ITT, fan, or turbine rpm.

If sufficient runway length is available, a "rolling" takeoff may be made without stopping at the end of the runway. Using this procedure, as the airplane rolls onto the runway, the thrust levers should be smoothly advanced to the recommended intermediate power setting and the engines allowed to stabilize, and then proceed as in the static takeoff outlined above. Rolling takeoffs can also be made from the end of the runway by advancing the thrust levers from idle as the brakes are released.

During the takeoff roll, the pilot flying should concentrate on directional control of the airplane. This is made somewhat easier because there is no torque produced yawing in a jet as there is in a propeller-driven airplane. The airplane must be maintained exactly on centerline with the wings level. This automatically aids the pilot when contending with an engine failure. If a crosswind exists, the wings should be kept level by displacing the control wheel into the crosswind. During the takeoff roll, the primary responsibility of the pilot not flying is to closely monitor the aircraft systems and to call out the proper V speeds as directed in the captain's briefing.

Slight forward pressure should be held on the control column to keep the nose wheel rolling firmly on the runway. If nose-wheel steering is being utilized, the pilot flying should monitor the nose-wheel steering to about 80 knots (or V_{MCG} for the particular airplane) while the pilot not flying applies the forward pressure. After reaching V_{MCG}, the pilot flying should bring his or her left hand up to the control wheel. The pilot's other hand should be on the thrust levers until at least V_1 speed is attained. Although the pilot not flying maintains a check on the engine instruments throughout the takeoff roll, the pilot flying (pilot in command) makes the decision to continue or reject a takeoff for any reason. A decision to reject a takeoff requires immediate retarding of thrust levers.

The pilot not flying should call out V_1. After passing V_1 speed on the takeoff roll, it is no longer mandatory for the pilot flying to keep a hand on the thrust levers. The point for abort has passed, and both hands may be placed on the control wheel. As the airspeed approaches V_R, the control column should be moved to a neutral position. As the pre-computed V_R speed is attained, the pilot not flying should make the appropriate callout, and the pilot flying should smoothly rotate the airplane to the appropriate takeoff pitch attitude.

Rejected Takeoff

Every takeoff could potentially result in a rejected takeoff (RTO) for a variety of reasons: engine failure, fire or smoke,

unsuspected equipment on the runway, bird strike, blown tires, direct instructions from the governing ATC authority, or recognition of a significant abnormality (split airspeed indications, activation of a warning horn, etc.).

Ill-advised rejected takeoff decisions by flight crews and improper pilot technique during the execution of a rejected takeoff contribute to a majority of takeoff-related commercial aviation accidents worldwide. Statistically, although only 2 percent of rejected takeoffs are in this category, high-speed aborts above 120 knots account for the vast majority of RTO overrun accidents. Four out of five rejected takeoffs occur at speeds below 80 knots and generally come to a safe and successful conclusion.

The kinetic energy of any aircraft (and thus the deceleration power required to stop it) increases with aircraft weight and the square of the aircraft speed. Therefore, an increase in weight has a lesser impact on kinetic energy than a proportional increase in groundspeed. A 10 percent increase in takeoff weight produces roughly a 10 percent increase in kinetic energy, while a 10 percent increase in speed results in a 21 percent increase in kinetic energy. Hence, it should be stressed during pilot training that time (delayed decision or reaction) equals higher speed (to the tune of at least 4 knots per second for most jets), and higher speed equals longer stopping distance. A couple of seconds can be the difference between running out of runway and coming to a safe halt. Because weight ceases to be a variable once the doors are closed, the throttles are pushed forward and the airplane is launching down the runway, all focus should be on timely recognition and speed control.

The decision to abort takeoff should not be attempted beyond the calculated V_1, unless there is reason to suspect that the airplane's ability to fly has been impaired or is threatened to cease shortly after takeoff (for example on-board fire, smoke, or identifiable toxic fumes). If a serious failure or malfunction occurs beyond takeoff decision speed (V_1), but the airplane's ability to fly is not in question, takeoff must generally continue.

It is paramount to remember that FAA-approved takeoff data for any aircraft is based on aircraft performance demonstrated in ideal conditions, using a clean, dry runway, and maximum braking (reverse thrust is not used to compute stopping distance). In reality, stopping performance can be further degraded by an array of factors as diversified as:

- Runway friction (grooved/non-grooved)

- Mechanical runway contaminants (rubber, oily residue, debris)

- Natural contaminants (standing water, snow, slush, ice, dust)
- Wind direction and velocity
- Air density
- Flaps configuration
- Bleed air configuration
- Underinflated or failing tires
- Penalizing MEL or CDL items
- Deficient wheel brakes or RTO auto-brakes
- Inoperative anti-skid
- Pilot technique and individual proficiency

Because performance conditions used to determine V_1 do not necessarily consider all variables of takeoff performance, operators and aircraft manufacturers generally agree that the term "takeoff decision speed" is ambiguous at best. By definition, it would suggest that the decision to abort or continue can be made upon reaching the calculated V_1, and invariably result in a safe takeoff or RTO maneuver if initiated at that point in time. In fact, taking into account the pilots' response time, the Go/No Go decision must be made before V_1 so that deceleration can begin no later than V_1. If braking has not begun by V_1, the decision to continue the takeoff is made by default. Delaying the RTO maneuver by just one second beyond V_1 increases the speed 4 to 6 knots on average. Knowing that crews require 3 to 7 seconds to identify an impending RTO and execute the maneuver, it stands to reason that a decision should be made prior to V_1 in order to ensure a successful outcome of the rejected takeoff. This prompted the FAA to expand on the regulatory definition of V_1 and to introduce a couple of new terms through the publication of Advisory Circular (AC) 120-62, "Takeoff Safety Training Aid."

The expanded definition of V_1 is as follows:

a) V_1. The speed selected for each takeoff, based upon approved performance data and specified conditions, which represents:

   (1) The maximum speed by which a rejected takeoff must be initiated to assure that a safe stop can be completed within the remaining runway, or runway and stopway;

   (2) The minimum speed which assures that a takeoff can be safely completed within the remaining runway, or runway and clearway, after failure of the most critical engine the designated speed; and

   (3) The single speed which permits a successful stop or continued takeoff when operating at the minimum allowable field length for a particular weight.

b) Minimum V_1. The minimum permissible V_1 speed for the reference conditions from which the takeoff can be safely completed from a given runway, or runway and clearway, after the critical engine had failed at the designated speed.

c) Maximum V_1. That maximum possible V_1 speed for the reference conditions at which a rejected takeoff can be initiated and the airplane stopped within the remaining runway, or runway and stopway.

d) Reduced V_1. A V_1 less than maximum V_1 or the normal V_1, but more than the minimum V_1, selected to reduce the RTO stopping distance required.

The main purpose for using a reduced V_1 is to properly adjust the RTO stopping distance in light of the degraded stopping capability associated with wet or contaminated runways, while adding approximately 2 seconds of recognition time for the crew.

Most aircraft manufacturers recommend that operators identify a "low-speed" regime (i.e., 80 knots and below) and a "high-speed" regime (i.e., 100 knots and above) of the takeoff run. In the "low speed" regime, pilots should abort takeoff for any malfunction or abnormality (actual or suspected). In the "high speed" regime, takeoff should only be rejected because of catastrophic malfunctions or life-threatening situations. Pilots must weigh the threat against the risk of overshooting the runway during a RTO maneuver. Standard Operating Procedures (SOPs) should be tailored to include a speed callout during the transition from low-speed to high-speed regime, the timing of which serves to remind pilots of the impending critical window of decision-making, to provide them with a last opportunity to crosscheck their instruments, to verify their airspeed, and to confirm that adequate takeoff thrust is set, while at the same time performing a pilot incapacitation check through the "challenge and response" ritual. Ideally, two callouts would enhance a crew's preparedness during takeoff operations. A first callout at the high end of the "low-speed" regime would announce the beginning of the transition from "low speed" to "high-speed," alerting the crew that they have entered a short phase of extreme vigilance where the "Go/No Go" must imminently be decided. A second callout made at the beginning of the "high-speed" regime would signify the end of the transition, thus the end of the decision-making. Short of some catastrophic failure, the crew is then committed to continue the takeoff.

Proper use of brakes should be emphasized in training, as they have the most stopping power during a rejected takeoff. However, experience has shown that the initial tendency of a flight crew is to use normal after-landing braking during a rejected takeoff. Delaying the intervention of the primary deceleration force during a RTO maneuver, when every second counts, could be costly in terms of required stopping distance. Instead of braking after the throttles are retarded and the spoilers are deployed (normal landing), pilots must apply maximum braking immediately while simultaneously retarding the throttles, with spoilers extension and thrust reversers deployment following in short sequence. Differential braking applied to maintain directional control also diminishes the effectiveness of the brakes. And finally, not only does a blown tire eliminate any kind of braking action on that particular tire, but it could also lead to the failure of adjacent tires, and thus further impairing the airplane's ability to stop.

In order to better assist flight crews in making a split second Go/No Go decision during a high speed takeoff run, and subsequently avoid an otherwise unnecessary but risky high speed RTO, some commercial aircraft manufacturers have gone as far as inhibiting aural or visual malfunction warnings of non-critical equipment beyond a preset speed. The purpose is to prevent an overreaction by the crew and a tendency to select a risky high-speed RTO maneuver over a safer takeoff with a non-critical malfunction. Indeed, the successful outcome of a rejected takeoff, one that concludes without damage or injury, does not necessarily point to the best decision-making by the flight crew.

In summary, a rejected takeoff should be perceived as an emergency. RTO safety could be vastly improved by:

- Developing SOPs aiming to advance the expanded FAA definitions of takeoff decision speed and their practical application, including the use of progressive callouts to identify transition from low-to high-speed regime.

- Promoting situational awareness and better recognition of emergency versus abnormal situations through enhanced CRM training.

- Encouraging crews to carefully consider variables that may seriously affect or even compromise available aircraft performance data.

- Expanding practical training in the proper use of brakes, throttles, spoilers, and reverse thrust during RTO demonstrations.

- Encouraging aircraft manufacturers to eliminate non-critical malfunction warnings during the takeoff roll at preset speeds.

Rotation and Lift-Off

Rotation and lift-off in a jet airplane should be considered a maneuver unto itself. It requires planning, precision, and a fine control touch. The objective is to initiate the rotation to takeoff pitch attitude exactly at V_R so that the airplane accelerates through V_{LOF} and attains V_2 speed at 35 feet AGL. Rotation to the proper takeoff attitude too soon may extend the takeoff roll or cause an early lift-off, which results in a lower rate of climb and the predicted flightpath will not be followed. A late rotation, on the other hand, results in a longer takeoff roll, exceeding V_2 speed, and a takeoff and climb path below the predicted path.

Each airplane has its own specific takeoff pitch attitude that remains constant regardless of weight. The takeoff pitch attitude in a jet airplane is normally between 10° and 15° nose up. The rotation to takeoff pitch attitude should be made smoothly but deliberately and at a constant rate. Depending on the particular airplane, the pilot should plan on a rate of pitch attitude increase of approximately 2.5° to 3° per second.

In training, it is common for the pilot to overshoot V_R and then overshoot V_2 because the pilot not flying calls for rotation at or just past V_R. The reaction of the pilot flying is to visually verify V_R and then rotate. The airplane then leaves the ground at or above V_2. The excess airspeed may be of little concern on a normal takeoff, but a delayed rotation can be critical when runway length or obstacle clearance is limited. It should be remembered that on some airplanes, the all-engine takeoff can be more limiting than the engine-out takeoff in terms of obstacle clearance in the initial part of the climb-out. This is because of the rapidly increasing airspeed causing the achieved flightpath to fall below the engine out scheduled flightpath unless care is taken to fly the correct speeds. The transitioning pilot should remember that rotation at the right speed and rate to the right attitude gets the airplane off the ground at the right speed and within the right distance.

Initial Climb

Once the proper pitch attitude is attained, it must be maintained. The initial climb after lift-off is done at this constant pitch attitude. Takeoff power is maintained and the airspeed allowed to accelerate. Landing gear retraction should be accomplished after a positive rate of climb has been established and confirmed. Remember that in some airplanes gear retraction may temporarily increase the airplane drag while landing gear doors open. Premature gear retraction may cause the airplane to settle back towards the runway surface. Remember also that because of ground effect, the vertical speed indicator and the altimeter may not show a positive climb until the airplane is 35 to 50 feet above the runway.

The climb pitch attitude should continue to be held and the airplane allowed to accelerate to flap retraction speed. However, the flaps should not be retracted until obstruction clearance altitude or 400 feet AGL has been passed. Ground effect and landing gear drag reduction results in rapid acceleration during this phase of the takeoff and climb. Airspeed, altitude, climb rate, attitude, and heading must be monitored carefully. When the airplane settles down to a steady climb, longitudinal stick forces can be trimmed out. If a turn must be made during this phase of flight, no more than 15° to 20° of bank should be used. Because of spiral instability and, because at this point an accurate trim state on rudder and ailerons has not yet been achieved, the bank angle should be carefully monitored throughout the turn. If a power reduction must be made, pitch attitude should be reduced simultaneously and the airplane monitored carefully so as to preclude entry into an inadvertent descent. When the airplane has attained a steady climb at the appropriate en route climb speed, it can be trimmed about all axes and the autopilot engaged.

Jet Airplane Approach and Landing

Landing Requirements

The FAA landing field length requirements for jet airplanes are specified in 14 CFR part 25. It defines the minimum field length (and therefore minimum margins) that can be scheduled. The regulation describes the landing profile as the horizontal distance required to land and come to a complete stop on a dry surface runway from a point 50 feet above the runway threshold, through the flare and touchdown, using the maximum stopping capability of the aircraft. The unfactored or certified landing distance is determined during aircraft certification. As such, it may be different from the actual landing distance because certification regulations do not take into account all factors that could potentially affect landing distance. The unfactored landing distance is the baseline landing distance on a dry, level runway at standard temperatures without using thrust reversers, auto brakes, or auto-land systems. In order to meet regulatory requirements however, a safety margin of 67 percent is added to the unfactored dry landing distance in the FAA-approved AFM, after applicable adjustments are made for environmental and aircraft conditions (MEL/CDL penalties). This corrected length is then referred to as the factored dry-landing distance or the minimum dry-landing field length. *[Figure 15-22]*

For minimum wet-landing field length, the factored dry-landing distance is increased by an additional 15 percent. Thus, the minimum dry runway field length is 1.67 times the actual minimum air and ground distance needed, and the wet runway minimum landing field length is 1.92 times the minimum dry air and ground distance needed.

Certified landing field length requirements are computed for the stop made with speed brakes deployed and maximum wheel braking. Reverse thrust is not used in establishing the certified landing distances; however, reversers should definitely be used in service.

Landing Speeds

As in the takeoff planning, there are certain speeds that must be taken into consideration when landing a jet airplane. The speeds are as follows:

- V_{SO}—stall speed in the landing configuration
- V_{REF}—1.3 times the stall speed in the landing configuration

Figure 15-22. *FAR landing field length required.*

- Approach climb—the speed that guarantees adequate performance in a go-around situation with an inoperative engine. The airplane's weight must be limited so that a twin-engine airplane has a 2.1 percent climb gradient capability. (The approach climb gradient requirements for 3 and 4 engine airplanes are 2.4 percent and 2.7 percent, respectively.) These criteria are based on an airplane configured with approach flaps, landing gear up, and takeoff thrust available from the operative engine(s).

- Landing climb—the speed that guarantees adequate performance in arresting the descent and making a go-around from the final stages of landing with the airplane in the full landing configuration and maximum takeoff power available on all engines.

The appropriate speeds should be pre-computed prior to every landing and posted where they are visible to both pilots. The V_{REF} speed, or threshold speed, is used as a reference speed throughout the traffic pattern. For example:

- Downwind leg—V_{REF} plus 20 knots

- Base leg—V_{REF} plus 10 knots

- Final approach—V_{REF} plus 5 knots

- 50 feet over threshold—V_{REF}

The approach and landing sequence in a jet airplane should be accomplished in accordance with an approach and landing profile developed for the particular airplane. *[Figure 15-23]*

Significant Difference

A safe approach in any type of airplane culminates in a particular position, speed, and height over the runway threshold. That final flight condition is the target window at which the entire approach aims. Propeller-powered airplanes are able to approach that target from wider angles, greater speed differentials, and a larger variety of glidepath angles. Jet airplanes are not as responsive to power and course corrections, so the final approach must be more stable, more deliberate, and more constant in order to reach the window accurately.

The transitioning pilot must understand that, in spite of their impressive performance capabilities, there are six ways in which a jet airplane is worse than a piston-engine airplane in making an approach and in correcting errors on the approach.

- The absence of the propeller slipstream in producing immediate extra lift at constant airspeed. There is no such thing as salvaging a misjudged glidepath with a sudden burst of immediately available power. Added lift can only be achieved by accelerating the airframe. Not only must the pilot wait for added power but, even when the engines do respond, added lift is only available when the airframe has responded with speed.

- The absence of the propeller slipstream in significantly lowering the power-on stall speed. There is virtually no difference between power-on and power-off stall speed. It is not possible in a jet airplane to jam the thrust levers forward to avoid a stall.

Figure 15-23. *Typical approach and landing profile.*

- Poor acceleration response in a jet engine from low rpm. This characteristic requires that the approach be flown in a high drag/high power configuration so that sufficient power is available quickly if needed.

- The increased momentum of the jet airplane making sudden changes in the flightpath impossible. Jet airplanes are consistently heavier than comparable sized propeller airplanes. The jet airplane, therefore, requires more indicated airspeed during the final approach due to a wing design that is optimized for higher speeds. These two factors combine to produce higher momentum for the jet airplane. Since force is required to overcome momentum for speed changes or course corrections, the jet is far less responsive than the propeller airplane and requires careful planning and stable conditions throughout the approach.

- The lack of good speed stability being an inducement to a low-speed condition. The drag curve for many jet airplanes is much flatter than for propeller airplanes, so speed changes do not produce nearly as much drag change. Further, jet thrust remains nearly constant with small speed changes. The result is far less speed stability. When the speed does increase or decrease, there is little tendency for the jet airplane to re-acquire the original speed. The pilot, therefore, must remain alert to the necessity of making speed adjustments, and then make them aggressively in order to remain on speed.

- Drag increasing faster than lift producing a high sink rate at low speeds. Jet airplane wings typically have a large increase in drag in the approach configuration. When a sink rate does develop, the only immediate remedy is to increase pitch attitude (AOA). Because drag increases faster than lift, that pitch change rapidly contributes to an even greater sink rate unless a significant amount of power is aggressively applied.

These flying characteristics of jet airplanes make a stabilized approach an absolute necessity.

Stabilized Approach

The performance charts and the limitations contained in the FAA-approved AFM are predicated on momentum values that result from programmed speeds and weights. Runway length limitations assume an exact 50-foot threshold height at an exact speed of 1.3 times V_{SO}. That "window" is critical and is a prime reason for the stabilized approach. Performance figures also assume that once through the target threshold window, the airplane touches down in a target touchdown zone approximately 1,000 feet down the runway, after which maximum stopping capability is used.

The five basic elements to the stabilized approach are listed below.

- The airplane should be in the landing configuration early in the approach. The landing gear should be down, landing flaps selected, trim set, and fuel balanced. Ensuring that these tasks are completed helps keep the number of variables to a minimum during the final approach.

- The airplane should be on profile before descending below 1,000 feet. Configuration, trim, speed, and glidepath should be at or near the optimum parameters early in the approach to avoid distractions and conflicts as the airplane nears the threshold window. An optimum glidepath angle of 2.5° to 3° should be established and maintained.

- Indicated airspeed should be within 10 knots of the target airspeed. There are strong relationships between trim, speed, and power in most jet airplanes, and it is important to stabilize the speed in order to minimize those other variables.

- The optimum descent rate should be 500 to 700 fpm. The descent rate should not be allowed to exceed 1,000 fpm at any time during the approach.

- The engine speed should be at an rpm that allows best response when and if a rapid power increase is needed.

Every approach should be evaluated at 500 feet. In a typical jet airplane, this is approximately 1 minute from touchdown. If the approach is not stabilized at that height, a go-around should be initiated. *[Figure 15-24]*

Approach Speed

On final approach, the airspeed is controlled with power. Any speed diversion from V_{REF} on final approach must be detected immediately and corrected. With experience, the pilot is able to detect the very first tendency of an increasing or decreasing airspeed trend, which normally can be corrected with a small adjustment in thrust. It is imperative the pilot does not allow the airspeed to decrease below the target approach speed or a high sink rate can develop. Remember that with an increasing sink rate, an apparently normal pitch attitude is no guarantee of a normal AOA value. If an increasing sink rate is detected, it must be countered by increasing the AOA and simultaneously increasing thrust to counter the extra drag. The degree of correction required depends on how much the sink rate needs to be reduced. For small amounts, smooth and gentle, almost anticipatory corrections is sufficient. For large sink rates, drastic corrective measures may be required that, even if successful, would destabilize the approach.

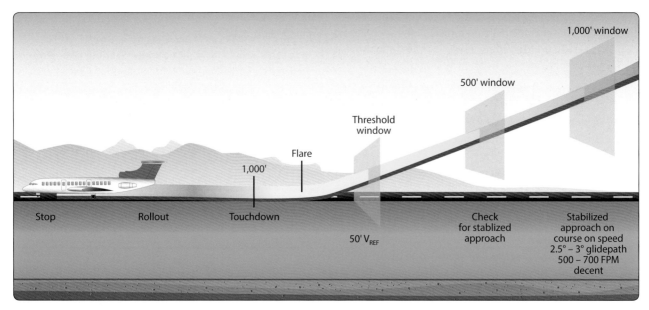

Figure 15-24. *Stabilized approach.*

A common error in the performance of approaches in jet airplanes is excess approach speed. Excess approach speed carried through the threshold window and onto the runway increases the minimum stopping distance required by 20–30 feet per knot of excess speed for a dry runway and 40–50 feet for a wet runway. Worse yet, the excess speed increases the chances of an extended flare, which increases the distance to touchdown by approximately 250 feet for each excess knot in speed.

Proper speed control on final approach is of primary importance. The pilot must anticipate the need for speed adjustment so that only small adjustments are required. It is essential that the airplane arrive at the approach threshold window exactly on speed.

Glidepath Control

On final approach at a constant airspeed, the glidepath angle and rate of descent is controlled with pitch attitude and elevator. The optimum glidepath angle is 2.5° to 3° whether or not an electronic glidepath reference is being used. On visual approaches, pilots may have a tendency to make flat approaches. A flat approach, however, increases landing distance and should be avoided. For example, an approach angle of 2° instead of a recommended 3° adds 500 feet to landing distance.

A more common error is excessive height over the threshold. This could be the result of an unstable approach or a stable but high approach. It also may occur during an instrument approach where the missed approach point is close to or at the runway threshold. Regardless of the cause, excessive height over the threshold most likely results in a touchdown

beyond the normal aiming point. An extra 50 feet of height over the threshold adds approximately 1,000 feet to the landing distance. It is essential that the airplane arrive at the approach threshold window exactly on altitude (50 feet above the runway).

The Flare

The flare reduces the approach rate of descent to a more acceptable rate for touchdown. Unlike light airplanes, a jet airplane should be flown onto the runway rather than "held off" the surface as speed dissipates. A jet airplane is aerodynamically clean even in the landing configuration, and its engines still produce residual thrust at idle rpm. Holding it off during the flare in an attempt to make a smooth landing greatly increases landing distance. A firm landing is normal and desirable. A firm landing does not mean a hard landing, but rather a deliberate or positive landing.

For most airports, the airplane passes over the end of the runway with the landing gear 30–45 feet above the surface, depending on the landing flap setting and the location of the touchdown zone. It takes 5–7 seconds from the time the airplane passes the end of the runway until touchdown. The flare is initiated by increasing the pitch attitude just enough to reduce the sink rate to 100–200 fpm when the landing gear is approximately 15 feet above the runway surface. In most jet airplanes, this requires a pitch attitude increase of only 1° to 3°. The thrust is smoothly reduced to idle as the flare progresses.

The normal speed bleed off during the time between passing the end of the runway and touchdown is 5 knots. Most of the decrease occurs during the flare when thrust is reduced.

If the flare is extended (held off) while an additional speed is bled off, hundreds or even thousands of feet of runway may be used up. *[Figure 15-25]* The extended flare also results in additional pitch attitude, which may lead to a tail strike. It is, therefore, essential to fly the airplane onto the runway at the target touchdown point, even if the speed is excessive. A deliberate touchdown should be planned and practiced on every flight. A positive touchdown helps prevent an extended flare.

Pilots must learn the flare characteristics of each model of airplane they fly. The visual reference cues observed from each airplane are different because window geometry and visibility are different. The geometric relationship between the pilot's eye and the landing gear is different for each make and model. It is essential that the flare maneuver be initiated at the proper height—not too high and not too low.

Beginning the flare too high or reducing the thrust too early may result in the airplane floating beyond the target touchdown point or may include a rapid pitch up as the pilot attempts to prevent a high sink rate touchdown. This can lead to a tail strike. The flare that is initiated too late may result in a hard touchdown.

Proper thrust management through the flare is also important. In many jet airplanes, the engines produce a noticeable effect on pitch trim when the thrust setting is changed. A rapid change in the thrust setting requires a quick elevator response. If the thrust levers are moved to idle too quickly during the flare, the pilot must make rapid changes in pitch control. If the thrust levers are moved more slowly, the elevator input can be more easily coordinated.

Touchdown and Rollout

A proper approach and flare positions the airplane to touch down in the touchdown target zone, which is usually about 1,000 feet beyond the runway threshold. Once the main wheels have contacted the runway, the pilot must maintain directional control and initiate the stopping process. The

Figure 15-25. *Extended flare.*

stop must be made on the runway that remains in front of the airplane. The runway distance available to stop is longest if the touchdown was on target. The energy to be dissipated is least if there is no excess speed. The stop that begins with a touchdown that is on the numbers is the easiest stop to make for any set of conditions.

At the point of touchdown, the airplane represents a very large mass that is moving at a relatively high speed. The large total energy must be dissipated by the brakes, the aerodynamic drag, and the thrust reversers. The nose wheel should be flown onto the ground immediately after touchdown because a jet airplane decelerates poorly when held in a nose-high attitude. Placing the nose wheel tire(s) on the ground assists in maintaining directional control. Also, lowering the nose gear decreases the wing AOA, decreasing the lift, placing more load onto the tires, thereby increasing tire-to-ground friction. Landing distance charts for jet airplanes assume that the nose wheel is lowered onto the runway within 4 seconds of touchdown.

There are only three forces available for stopping the airplane: wheel braking, reverse thrust, and aerodynamic braking. Of the three, the brakes are most effective and therefore the most important stopping force for most landings. When the runway is very slippery, reverse thrust and drag may be the dominant forces. Both reverse thrust and aerodynamic drag are most effective at high speeds. Neither is affected by runway surface condition. Brakes, on the other hand, are most effective at low speed. The landing rollout distance depends on the touchdown speed, what forces are applied, and when they are applied. The pilot controls the what and when factors, but the maximum braking force may be limited by tire-to-ground friction.

The pilot should begin braking as soon after touchdown and wheel spin-up as possible, and to smoothly continue the braking until stopped or a safe taxi speed is reached. However, caution should be used if the airplane is not equipped with a functioning anti-skid system. In such a case, heavy braking can cause the wheels to lock and the tires to skid.

Both directional control and braking utilize tire ground friction. They share the maximum friction force the tires can provide. Increasing either subtracts from the other. Understanding tire ground friction, how runway contamination affects it, and how to use the friction available to maximum advantage is important to a jet pilot.

Spoilers should be deployed immediately after touchdown because they are most effective at high speed. Timely deployment of spoilers increases drag by 50 to 60 percent, but more importantly, they spoil much of the lift the wing is creating, thereby causing more of the weight of the airplane to be loaded onto the wheels. The spoilers increase wheel loading by as much as 200 percent in the landing flap configuration. This increases the tire ground friction force making the maximum tire braking and cornering forces available.

Like spoilers, thrust reversers are most effective at high speeds and should be deployed quickly after touchdown. However, the pilot should not command significant reverse thrust until the nose wheel is on the ground. Otherwise, the reversers might deploy asymmetrically resulting in an uncontrollable yaw towards the side on which the most reverse thrust is being developed, in which case the pilot needs whatever nose-wheel steering is available to maintain directional control.

Key Points

Many LSAs have airframe designs that are conducive to high drag which, when combined with their low mass, results in low inertia. When attempting a crosswind landing in a high drag LSA, a rapid reduction in airspeed prior to touchdown may result in a loss of rudder and/or aileron control, which may push the aircraft off of the runway heading. This is because as the air slows across the control surfaces, the LSA's controls become ineffective. To avoid loss of control, maintain airspeed during the approach to keep the air moving over the control surfaces until the aircraft is on the ground.

LSAs with an open cockpit, easy build characteristics, low cost, and simplicity of operation and maintenance tend to be less aerodynamic and, therefore, incur more drag. The powerplant in these aircraft usually provide excess power and exhibit desirable performance. However, when power is reduced, it may be necessary to lower the nose of the aircraft to a fairly low pitch attitude in order to maintain airspeed, especially during landings and engine failure.

If the pilot makes a power off approach to landing, the approach angle will be high and the landing flare will need to be close to the ground with minimum float. This is because the aircraft will lose airspeed quickly in the flare and will not float like a more efficiently designed aircraft. Too low of an airspeed during the landing flare may lead to insufficient energy to arrest the decent which may result in a hard landing. Maintaining power during the approach will result in a reduced angle of attack and will extend the landing flare allowing more time to make adjustments to the aircraft during the landing. Always remember that rapid power reductions require an equally rapid reduction in pitch attitude to maintain airspeed.

In the event of an engine failure in an LSA, quickly transition to the required nose-down flight attitude in order to maintain airspeed. For example, if the aircraft has a power-off glide angle of 30 degrees below the horizon, position the aircraft to a nose-down 30 degree attitude as quickly as possible. The higher the pitch attitude is when the engine failure occurs, the quicker the aircraft will lose airspeed and the more likely the aircraft is to stall. Should a stall occur, decrease the aircraft's pitch attitude rapidly in order to increase airspeed to allow for a recovery. Stalls that occur at low altitudes are especially dangerous because the closer to the ground the stall occurs, the less time there is to recover. For this reason, when climbing at a low altitude, excessive pitch attitude is discouraged.

Chapter Summary

There are many considerations for a pilot when transitioning to jet powered airplanes. In addition to the information found in this chapter and type specific information that will be found in an FAA-approved Airplane Flight Manual, a pilot can find basic aerodynamic information for swept-wing jets, considerations for operating at high altitudes, and airplane upset causes and general recovery procedures in the Airplane Upset Recovery Training Aid, Supplement, pages 1-14, and all of Section 2 found at www.faa.gov/other_visit/aviation_industry/airline_operators/training/media/ap_upsetrecovery_book.pdf.

Chapter 16
Transition to Light Sport Airplanes (LSA)

Introduction

Transitioning into a light sport airplane (LSA) requires the same methodical training approach as transitioning into any other airplane. A pilot should never attempt to fly another airplane that is different than the pilot's current certification, experience, training, proficiency, or currency without proper training. Some pilots may be lulled into a false sense of security because LSAs seem to be simple. However, a pilot seeking a transition into light sport flying should follow a systematic, structured LSA training course under the guidance of a competent instructor with recent experience in the specific training airplane.

The light sport category is not a new type of airplane. It is a classification that intends to broaden the access of flight to more people. LSA has been defined as a simple-to-operate, easy-to-fly aircraft; however, "simple-to-operate" and "easy-to-fly" does not negate the need for proper and effective training. This chapter introduces the light sport category of aircraft and places emphasis on LSA transition.

Maximum gross weight of 1,320 pounds (1,430 pounds for seaplanes)

Maximum stall speed of 45 knots (51 mph)

Unpressurized cabin

Fixed or ground adjustable propeller

Single, reciprocating engine

Fixed landing gear (repositionable landing gear for seaplanes)

Maximum speed in level flight with maximum continuous power of 120 knots (138 mph)

CESSNA
Model 162
Garmin G300

INTRODUCTION

EMERGENCY PR

Section 3 provides checklist and amplified procedures for c emergencies that may occur. Emergencies caused by air engine malfunctions are extremely rare if proper preflight ins, and maintenance are practiced. Enroute weather emergencies minimized or eliminated by careful flight planning and good ju when unexpected weather is encountered. However, sho emergency arise, the basic guidelines described in this section s nsidered and applied as necessary to correct the problem. I ncy situation, the most important task is continued control c airplane and maneuver to execute a successful landing.

Emergency procedures associated with optional or supplemental equipment are found in Section 9, Supplements.

AIRSPEEDS FOR EMERGENCY OPERATIONS

ENGINE FAILURE AFTER TAKEOFF
Wing flaps UP.
Wing flaps 10° - FULL .... 70 KIAS

MAXIMUM OPERATING MANEUVERING SPEED
1320 pounds .... 65 KIAS
1200 pounds
1100 pounds

GN MANEUVERING SPEED .... 89 KIAS
.... 85 KIAS
MAXIMUM GLIDE .... 80 KIAS

PRECAUTIONARY LANDING WITH EN .... 102 KIAS

LANDING WITH

Light Sport Airplane (LSA) Background

Several groups were instrumental in the development and success of the LSA concept. These included the Federal Aviation Administration (FAA), Light Aircraft Manufacturers Association, American Society for Testing and Materials (ASTM) International, and countless individuals who promoted the concept since the early 1990s. In 2004, the FAA released a rule that created the LSA category, which covers a wide variety of aircraft including: airplane, gyroplane, lighter-than-air, weight-shift-control, glider, and powered parachute. *[Figure 16-1]*

The primary concept of the LSA is built around a defined set of standards:

- Powered (if powered) by single reciprocating engine
- Fixed landing gear, seaplanes are excluded
- Fixed pitch or ground adjustable propeller
- Maximum takeoff weight of 1,320 pounds for landplane, 1,430 for seaplane
- Maximum of two occupants
- Non-pressurized cabin

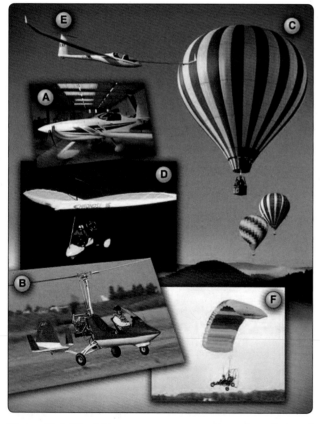

Figure 16-1. *The LSA category covers a wide variety of aircraft including: A) airplane, B) gyroplane, C) lighter-than-air, D) weight-shift-control, E) glider, and F) powered parachute.*

- Maximum speed in level flight at maximum continuous power of 120 knots calibrated airspeed (CAS)
- Maximum stall speed of 45 knots. *[Figure 16-2]*

The LSA category includes standard, special, and experimental designations. Some standard airworthiness certificated aircraft (i.e., a Piper J-2 or J-3) may meet Title 14 of the Code of Federal Regulation (14 CFR) 1.1 definition of LSA. Type certificated aircraft that continue to meet the CFR 1.1 definition of LSA allows for that type certificated aircraft to be flown by a pilot who holds a Sport Pilot certificate. The Sport Pilot certificate is discussed later in this chapter. Aircraft that are specifically manufactured for the LSA market are included in either the Special (S-LSA) or Experimental (E-LSA) designations. An approved S-LSA is manufactured in a ready-to-fly condition and an E-LSA is either a kit or plans-built aircraft based on an approved S-LSA model.

It is important to note that S-LSAs or E-LSAs are not type certificated by the FAA and are not required to meet any airworthiness requirements of 14 CFR part 23. Instead, S-LSA and E-LSA aircraft are designed and manufactured in accordance with ASTM Committee F-37 Industry Consensus Standards. Therefore, LSA aircraft designs are not subjected to the scrutiny, demands, and testing of FAA standard airworthiness certification. Industry Consensus Standards are intended to be less costly and less restrictive than 14 CFR part 23 certification requirements and, as a result, manufacturers have greater latitude with their designs. ASTM Industry Consensus Standards were accepted by the FAA in 2005, which established for the first time that the FAA accepted industry-developed standards rather than its own standards for the design and manufacture of aircraft.

ASTM Industry Consensus Standards for LSA airplanes covers the following areas:

- Design and performance
- Required equipment
- Quality assurance
- Production acceptance tests
- Aircraft operating instructions
- Maintenance and inspection procedures
- Identification and recording of major repairs and major alterations
- Continued airworthiness
- Manufacturers assembly instructions (E-LSA aircraft)

Using the ASTM Industry Consensus Standards, an LSA manufacturer can design and manufacture their aircraft and assess its compliance to the consensus standards. The

Figure 16-2. *Light sport airplane.*

The labels in the figure read:
- Maximum gross weight of 1,320 pounds (1,430 pounds for seaplanes)
- Maximum stall speed of 45 knots (51 mph)
- Unpressurized cabin
- Fixed or ground adjustable propeller
- Single, reciprocating engine
- Fixed landing gear (repositionable landing gear for seaplanes)
- Maximum speed in level flight with maximum continuous power of 120 knots (138 mph)

manufacturer then, through evaluation services offered by a designated airworthiness representative, completes the process by submitting the required paperwork to the FAA. Upon approval, an LSA manufacturer is permitted to sell ready-to-fly S-LSA aircraft.

LSA Synopsis

- The airplane must meet the weight, speed, and other criteria as described in this chapter.

- Airplanes under the S-LSA certification may be used for sport and recreation, flight training, and aircraft rental.

- Airplanes under the E-LSA certification may be used only for sport and recreation and flight instruction for the owner of the airplane. E-LSA certification is not the same as Experimental Amateur-Built. E-LSA certification is based on an approved S-LSA airplane.

- Airplanes with a standard airworthiness type certificate (i.e., a Piper J-2 or J-3) that continue to meet the 14 CFR 1.1 LSA definition may be flown by a pilot with a Sport Pilot certificate.

- Must have an FAA registration and N-number.

- United States or foreign manufacturers can be authorized.

- May be operated at night if the aircraft is equipped per 14 CFR part 91, section 91.205, if night operations are allowed by the airplane's operating limitations, and the pilot holds at least a Private Pilot certificate and a minimum of a third-class medical.

- LSAs can be flown by holders of a Sport Pilot certificate or higher level pilot certificate (recreational, private, etc.)

Sport Pilot Certificat

In addition to the LSA rules, the FAA created a new Sport Pilot certificate in 2004 that lowered the minimum training time requirements, in comparison to other pilot certificates, for newly certificated pilots wishing to exercise privileges only in LSA aircraft. A pilot that already holds a recreational, private, commercial, or airline transport pilot certificate and a current medical certificate is permitted to pilot LSA airplanes provided that he or she has the appropriate category and class ratings. For example, a commercial pilot with exclusively a rotorcraft rating cannot pilot an LSA airplane.

Pilots who hold a recreational, private, commercial, or airline transport pilot certificate with the appropriate category and class ratings but do not hold a current medical certificate may fly LSAs as long as the pilot holds a valid U.S. driver's license as evidence of medical eligibility; however, if the pilot's most recent medical certificate was denied, revoked, suspended, or withdrawn, a U.S. driver's license is not sufficient for medical eligibility. In this case, the pilot would be prohibited from flying an LSA until the pilot could be issued a third class medical.

Transition Training Considerations

Flight School

The LSA category has created new business opportunities for flight school operators. Many owners and operators of flight schools have embraced the concept of LSA aircraft and have LSAs available on their flight line for flight instruction and rental. An S-LSA may be rented to students for flight training and rented to rated pilots for pleasure flying. While S-LSAs cannot be used for compensation or hire (such as charter—however, there are some exceptions), their low cost of operation, frugal fuel usage, reliability, and low maintenance costs have made them a favorite of many students, pilots, and flight school owners. E-LSAs are not eligible for flight training and rental except when flight instruction is given to the owner of the E-LSA airplane.

When considering a transition to LSA, a potential pilot should exercise due diligence in searching for a quality flight school. Considerations should be given as in any flight training selection. First, locate a flight school that has a verifiable experience in LSA instruction and can provide the LSA academic framework. Consider if the flight school can match your needs. Some questions to be asked are the following: how many pilots the flight school has transitioned into LSAs; how many LSAs are available for instruction and rental; what are the flight school's rental and insurance policies; how is maintenance accomplished and by whom; how is scheduling accomplished; how are records maintained; what are the school's safety policies; and, take the time to personally tour the school before starting flight training. Finally, if possible, solicit feedback from other pilots that have transition into LSAs.

Flight Instructors

The flight school provides the organization for the transitioning pilot; however, it is the flight instructor that is the critical link in a successful LSA transition. Flight instructors are at first teachers of flight, so it should be considered vital that a pilot wishing to transition into LSA locate a flight instructor that has verifiable experience in LSA instruction. Considerations for selecting a flight instructor are similar to any other flight training; however, some clarity around selecting a flight instructor is needed. The Sport Pilot rule allows for a new flight instructor certificate, the CFI-S. The CFRs limit a CFI-S to instruction only in LSAs—a CFI-S cannot give instruction in a non-LSA airplane (i.e., a Cessna 150). However, a flight instructor certificated as a CFI-A can give instruction in LSA, as well as instruction in non-LSA airplanes for which the flight instructor is rated. It is important to note that a CFI-S or a CFI-A should not be the criteria for selecting an LSA flight instructor. A CFI-S with teaching experience in LSA is the correct choice compared to a CFI-A, which has minimal teaching experience in LSA airplanes.

A transitioning LSA pilot should ask the flight instructor to make available for review their LSA curriculum, syllabus, lesson plans, as well the process for tracking a pilot's progress though the transition training program. Depending on the transitioning pilot's experience, currency, and type of airplane typically flown, the flight instructor should make adjustments, as appropriate, to the LSA training curriculum. A suggested LSA transition training outline is presented:

- CFR review as pertaining to LSAs and Sport Pilots
- Pilot's Operating Handbook (POH) review
- LSA maintenance
- LSA weather considerations
- Wake turbulence avoidance
- Performance and limitations
- Operation of systems
- Ground operations
- Preflight inspection
- Before takeoff check
- Normal and crosswind takeoff/climb
- Normal and crosswind approach/landing
- Soft-field takeoff and climb
- Soft-field approach and landing
- Short-field takeoff
- Go-around/rejected landing
- Steep turns
- Power-off stalls
- Power-on stalls
- Spin awareness
- Emergency approach and landing
- Systems and equipment malfunctions

- After landing, parking, and securing

LSA Maintenance

Proper airplane maintenance is required to maximize flight safety. LSAs are no different and must be treated with the same level of care as any standard airworthiness certificated airplane. S-LSAs have greater latitude pertaining to who may conduct maintenance as compared to standard airworthiness certificated airplanes. S-LSAs may be maintained and inspected by:

- An LSA Repairman with a Maintenance rating; or,

- An FAA-certificated Airframe and Powerplant Mechanic (A&P); or,

- As specified by the aircraft manufacturer; or

- As permitted, owners performing limited maintenance on their S-LSA

The airplane maintenance manual includes the specific requirements for repair and maintenance, such as information on inspections, repair, and authorization for repairs and maintenance. Most often, S-LSA inspections can be signed off by an FAA-certificated A&P or LSA repairman with a Maintenance rating rather than an A&P with Inspection Authorization (IA); however, the aircraft maintenance manual provides the specific requirements which must be followed. The FAA does not issue Airworthiness Directives (ADs) for S-LSAs or E-LSAs. If an FAA-certified component is installed on an LSA, the FAA issues any pertaining ADs for that specific component. Manufacturer safety directives are not distributed by the FAA. S-LSA owners must comply with:

- Safety directives (alerts, bulletins, and notifications) issued by the LSA manufacturer

- ADs if any FAA-certificated components are installed

- Safety alerts (immediate action)

- Service bulletins (recommending future action)

- Safety notifications (informational)

S-LSA compliance with maintenance requirements provides greater latitude for owners and operators of these airplanes. Because of the options in complying with the maintenance requirements, pilots who are transitioning to LSAs must understand how maintenance is accomplished; who is providing the maintenance services; and verify that all compliance requirements have been met.

Airframe and Systems

Construction

LSAs may be constructed using wood, tube and fabric, metal, composite, or any combination of materials. In general, a primary effort by the manufacturer is to keep the airplane lightweight while maintaining the structural requirements. Composite LSAs tend to be sleek and modern looking with clean lines as molding of the various components allows designers great flexibility shaping the airframe. Other LSAs are authentic-looking renditions of early aviation airplanes with fabric covering a framework of steel tubes. Of course, LSAs may be anything in between using both metal and composite construction. *[Figure 16-3]* A pilot transitioning into LSA should understand the type of construction and what are typical concerns for each type of construction:

- Steel tube and fabric—while the techniques of steel tube and fabric construction hails back to the early days of aviation, this construction method has proven to be lightweight, strong, and inexpensive to build and maintain. Advances in fabric technology continue to make this method of covering airframes an excellent choice. Fabric can be limited in its life span if not properly maintained. Fabric should be free from tears, well-painted with little to no fading, and should easily spring back when lightly pressed.

- Aluminum—an aluminum-fabricated airplane has been a favorite choice for decades. Pilots should be quite familiar with this type of construction. Generally, airframes tend to be lightly rounded structures dotted with rivets and fasteners. This construction is easily inspected due to the wide-spread experience with aluminum structures. Conditions such as corrosion, working rivets, dents, and cracks should be a part of a pilot's preflight inspection.

- Composite—a composite airplane is principally made from structural epoxies and cloth-like fabrics, such as bi-directional and uni-directional fiberglass cloths, and specialty cloths like carbon fiber. Airframe components, such as wing and fuselage halves, are made in molds that result in a sculpted, mirror-like

Figure 16-3. *LSA can be constructed using both metal and composites.*

finish. Generally, composite construction has few fasteners, such as protruding rivets and bolts. Pilots should become acquainted with inspection concerns such as looking for hair-line cracks and delaminations.

Engines

LSAs use a variety of engines that range from FAA-certificated to non-FAA-certificated. Engine technology varies significantly from conventional air-cooled to high revolutions per minute (rpm)/water-cooled designs. *[Figure 16-4]* These different technologies present a transitioning pilot new training opportunities and challenges. Since most LSAs use non-FAA-certificated engines, a transitioning pilot should fully understand the engine controls, procedures, and limitations. In most LSA airplanes, engines are water-cooled, 4-cycle, carbureted with a gear reduction drive. Engines such as these have much higher operating rpms and require a gear-box to reduce the propeller rpms to the proper range. Because of the higher operating rpms, vibration and noise signatures are quite different in most LSAs when compared to most standard type certificated designs.

Instrumentation

In addition to advanced airframe and engine technology, LSAs often have advanced flight and engine instrumentation. Often installed are electronic flight instrumentation systems (EFIS) that provide attitude, airspeed, altimeter, vertical speed, direction, moving map, navigation, terrain awareness, traffic, weather, engine data, etc., all on one or two liquid crystal displays. *[Figure 16-5]* EFIS has become a cost-effective replacement for traditional mechanical gyros and instruments. Compared to mechanical instrumentation systems, EFIS requires almost no maintenance. There are tremendous advantages to EFIS systems as long as the pilot is correctly trained in its use. EFIS systems can cause a "heads down" syndrome and loss of situation awareness if the pilot is not

Figure 16-4. *A water-cooled 4-cycle engine.*

trained to quickly and properly configure, access, program, and interpret the information provided. Transition training must include, if EFIS is installed, instruction in the use of the specific EFIS installed in the training airplane. In some cases, EFIS manufacturers or third party products are available for the pilot to practice EFIS operations on a personal computer as opposed to learning their functions in flight.

Weather Considerations

Managing weather factors is important for all aircraft but becomes more significant as the weight of the airplane decreases. Smaller, lighter weight airplanes are more affected by adverse weather such as stronger winds (especially crosswinds), turbulence, terrain influences, and other hazardous conditions. *[Figures 16-6* and *16-7]* LSA Pilots should carefully consider any hazardous weather conditions and effectively use an appropriate set of personal minimums to mitigate flight risk. Some LSAs have a maximum recommend wind velocity regardless of wind direction.

Figure 16-5. *An electronic flight instrumentation system provides attitude, airspeed, altimeter, vertical speed, direction, moving map, navigation, terrain awareness, traffic, weather, and engine data all on one or two liquid crystal displays.*

[Figure 16-8] While this is not a limitation, it would be prudent to heed any factory recommendations.

Due to an LSA's lighter weight, even greater distances from convective weather should be given. Low level winds that enter and exit a thunderstorm should be avoided not only by all airplanes but operations in the vicinity of convection should not be attempted in lightweight airplanes. Weather accidents continue to plague general aviation and, while it is not possible to always fly in clear, blue, calm skies, pilots of lighter weight LSAs should carefully manage weather-related risks. For example, some consideration should be given to flight activity that crosses varying terrain boundaries, such as grass or water to hard surfaces. Differential heating can cause lighter weight airplanes to experience sinking and lift to a greater degree than heavier airplanes. Careful planning, knowledge and experience, and an understanding of the flying environment assists in mitigating weather-related risks.

Flight Environment

The stick and rudder skills required for LSAs are the same stick and rudder skills required for any airplane. This section outlines areas that are unique to LSA airplanes – most skills learned in a standard airworthiness type certificated airplane are transferrable to LSAs; however, since LSAs can vary significantly in performance, equipment and systems, and construction, pilots must seek competent flight instruction and refer to the airplane's POH for detailed and specific information prior to flight.

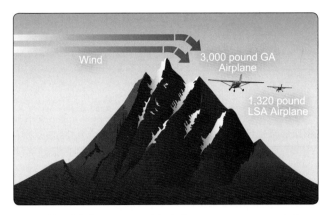

Figure 16-7. *Moderate mountain winds can create severe turbulence for LSA.*

Preflight

The preflight inspection of any airplane is critical to mitigating flight risks. A pilot transitioning into an LSA should allow adequate time to become familiar with the airplane prior to a first flight. First, the pilot and flight instructor should review the POH and cover the airplane's

Maximum Demonstrated Crosswind Velocity
Takeoff or landing .... 12 knots

Maximum Recommended Wind Velocity
All operations .... 22 knots

Figure 16-8. *Example of wind limitations that a LSA may have.*

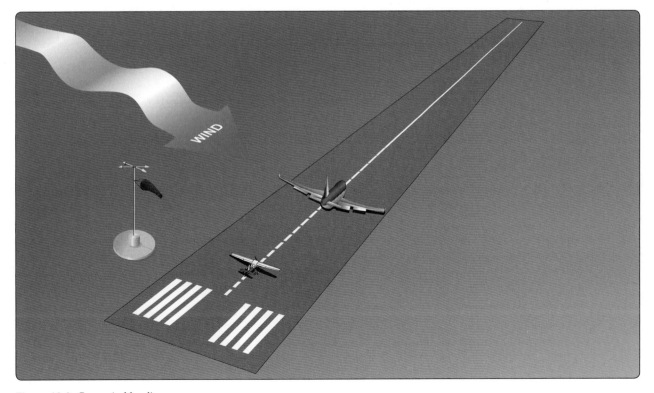

Figure 16-6. *Crosswind landing.*

limitations, systems, performance, weight and balance, normal procedures, emergency procedures, and handling requirements. *[Figure 16-9]*

Inside of the Airplane

Transitioning pilots find an LSA very familiar when conducting a preflight inspection; however, some preflight differences are worth pointing out. For example, many LSAs do not have adjustable seats but rather adjustable rudder pedals. *[Figure 16-10]* Often, LSA seats are in a fixed position. There are varied methods that LSA manufacturers have implemented for rudder pedal position adjustment. Some manufacturers use a simple removable pin while others use a knob near the rudder pedals for position adjustment. Shorter pilots may find that the adjustment range may not be sufficient for certain heights and an appropriate seat cushion may be required to have the proper range of rudder pedal movement. In addition, seats in some LSAs are in a semi-reclined position. The first time a pilot sits in a semi-reclining seat, it may seem somewhat unusual. A pilot should take time to get comfortable.

Another area that transitioning pilots require familiarity is with the flight and engine controls. These may vary significantly from airplane model to airplane model. Some

Rudder pedal position adjustment

Figure 16-10. *Adjustment lever for the rudder pedal position.*

LSA airplanes use conventional control stick while others use a yoke. One manufacturer has combined the two types of controls in what has been termed a "stoke." While this control may seem unique, it provides a completely natural feel for flight control. *[Figure 16-11]* Regardless of the flight controls, a full range of motion check of the flight controls is required. This means full forward to full forward left to full aft left to full aft right and then full forward right. Verify that each control surface moves freely and smoothly. On some LSAs, aileron control geometry, in an attempt to minimize adverse yaw, moves ailerons in a highly differential manner;

Figure 16-9. *Pilot's Operating Handbook for a LSA.*

Figure 16-11. *Stoke flight control with conventional engine controls.*

a pilot may see very little "down" aileron when compared to the "up" aileron. Pilots should always verify the direction of control surface movement.

Elevator trim on many LSAs is electrically actuated with no mechanical trim adjustment available. *[Figure 16-12]* Depending on the airplane, trim position indication may be displayed on the EFIS or an LED or mechanical indicator. On electric trim systems, as it is with any airplane, it is important to ensure that the trim position is correctly set prior to takeoff. Because trim positioning/indicting systems vary widely in LSA airplanes, pilots should fully understand not only how to position the trim, but also how to respond to a trim-run-away condition. Part of the preflight inspection should include actuating the trim switch in both nose-up and nose-down directions, verifying that the trim disconnect (if equipped) is properly functioning, ensure that the trim system circuit breaker can disconnect the trim motor from operating, and then properly setting the takeoff trim position.

Depending on the engine manufacturer, the engine controls may be completely familiar to a transitioning pilot (throttle, mixture, and carburetor heat); however, some engines have no mixture control or carburetor heat. Instead, there could be a throttle, a choke control, and carburetor preheater.

Figure 16-12. *Trim control.*

Regardless, a pilot must become familiar with the specific engine installed and its operation. A transitioning pilot also needs to become comfortable with difference between conventional engine control knobs and LSAs. In standard airworthiness airplanes, control knobs are reasonably standardized; however, LSAs may use controls that are much larger or smaller in size.

If the LSA is equipped with an EFIS, the manufacturer's EFIS Pilot Guide should be available for reference. In addition, the airplane POH likely has specific EFIS preflight procedures that must be completed. These checks are to verify that all internal tests are passed, that no red "Xs" are displayed, and that appropriate annunciators are illuminated. Some systems have a "reversionary" mode where the information from one display can be sent to another display. For example, should the Primary Flight Display (PFD) fail, information can be routed to the Multi-Function Display (MFD). Not all LSA EFIS systems are equipped with a MFD or reversionary capability, so it is important for a transitioning pilot to understand the system and limitations.

Fuel level in any airplane should be checked both visually and via the fuel level instrument or sight gauges. In LSAs, fuel level quantities can be shown on a wide range of technologies. Some models may have conventional float activated indictors while other may have the fuel level display on the EFIS with low-fuel alarm capability. It is not uncommon for an LSA airplane to have advanced EFIS technology for attitude and navigation information but have a simple sight gauge for fuel level indication. Fuel tank selection can also vary from simple on/off valves to a left/right selector. Fuel starvation remains a leading factor in aircraft accidents, which should be a reminder that when transitioning into a new airplane, time spent understanding the fuel system is time well spent.

A popular safety feature of some LSAs is a ballistic parachute. *[Figure 16-13]* These devices have been shown to be well worth their cost in the remote case of a catastrophic failure or some other unsurvivable emergency. This system rockets a parachute into deployment and then the parachute slowly lowers the aircraft. The preflight inspections of these systems require a check of the mounts, safety pin and flag, and the activation handle and cable. Because most standard airworthiness type certificated airplanes do not have these systems installed, LSA training should cover the operation and limitations of the system.

Outside of the Airplane

Transitioning pilots should feel comfortable and in a familiar setting when preflighting the outside of an LSA. Some unique areas worthy of notation are presented below.

Figure 16-13. *A ballistic recovery parachute is a popular safety feature available on some LSA.*

Propellers of LSAs may range from a conventional metal propeller to composite or wood. The preflight inspection is similar regardless of the type of propeller; however, if a transitioning pilot is principally familiar with metal propellers, time should be spent with the LSA flight instructor covering the type of propeller installed. Many LSA propellers are composite and have a ground adjustable pitch adjustment. As a result, there may be more areas to check with these types of propellers. For example, on ground adjustable propellers, ensure that the blades are tight against the hub by snugly twisting the blade at the root to verify that there is no rotation of the blade at the hub.

Many LSAs are equipped with engines that have a water cooling system. LSAs may be tightly cowled, which reduces drag, and with liquid-cooled engines, this minimizes the need for cylinder cooling inlets, which further reduces drag and improves performance. This does present a new system for a transitioning pilot to check. Preflighting this system requires that the radiator, coolant hoses, and expansion tank are checked for condition, freedom from leaks, and coolant level requirements. Most standard type certificated airplanes do not have coolant systems.

Split flaps may be used on some LSA designs. *[Figure 16-14]* These flaps hinge down from underneath the wing and inspecting these flaps require the pilot to crouch and twist low for inspection. A suitable handheld mirror can facilitate inspection without undue twisting and bending. In an attempt to keep complexity to a minimum, flap control is typically a handle that actuates the flaps. A pilot should verify that the flaps extend and retract smoothly.

Figure 16-14. *Split flap.*

Before Start and Starting Engine

Once a pilot has completed the preflight inspection of the LSA, the pilot should properly seat themselves in the airplane ensuring that the rudder pedals can be exercised with full-range movement without over-reaching. Seat belts should be checked for proper position and security. The pilot must continue to use the POH for all required checklists. Starting newer generation LSA engines can be quite simple only requiring the pull of the choke and a twist of the ignition switch. If the LSA is equipped with a standard certificated engine, starting procedures are normal and routine. The canopy or doors of an LSA may have quite different latching mechanisms than standard airworthiness airplanes. Practice latching and unlatching the doors or canopy to ensure that understanding is complete. Having a gull-wing door or sliding canopy "pop" open in flight can become an emergency in seconds.

Taxi

Like standard certificated airplanes, LSAs may have a full-castoring or steerable nosewheel or, if conventional gear, a tailwheel. In order to taxi a full-castoring nosewheel equipped airplane, the use of differential brakes is required. This type of nosewheel can require practice to develop the skill necessary to keep the airplane on the centerline while minimizing brake application or damage to the tires. The balance is just enough taxi speed so that only light taps of brake pressure in the desired direction of turn or correction is required to make a turn or correction without carrying excessive taxi speed. If the speed is too slow, application of a brake can cause the aircraft to pivot to a stop, rather than an adjustment in direction, resulting in excessive brake and tire wear. If the speed is too fast, excessive brake wear is likely.

An LSA with conventional gear (tailwheel) should be initially transitioned into during no-wind conditions. The airplane, due to its light weight, requires the development of the proper flight control responses prior to operations in any substantial wind.

Takeoff and Climb

Takeoff and climb performance of LSA can be spirited as it typically has a high horsepower to weight ratio and accelerates quickly. Due to design requirement for low stall speeds, LSAs typically have low rotation and climb speeds with impressive climb rates. Like other airplanes, the pilot should be flying the published speeds as given the airplane's POH. Stick (yoke or stoke) forces tend to be light, which may lead a transitioning pilot to initially over-control as a result of flight control deflections being greater than required. The key is to relax, have reasonable patience, and input only appropriate flight control pressures needed to get the required response. If a transitioning pilot is inducing excessive control inputs, they should minimize flight control pressures, set attitudes based on outside references, and allow the airplane to settle.

During climbs, visibility over the nose may be difficult in some LSAs. As always, it is important to properly clear the airspace for traffic and other hazards. Occasionally lowering the airplane's nose to get a good look out toward the horizon is important for managing flight safety. Shallow banked turns in both directions of 10° to 20° also allow for clearing. Trim should be used to relieve climb flight control pressures that are generally light. Because flight control pressures tend to be light, it is easy to get in the habit of flying with an LSA airplane out of trim. This is to be avoided. Trim off any flight control pressures. This allows the pilot to focus as much time as possible looking outside.

Cruise

After leveling off at cruise altitude, the airplane should be allowed to accelerate to cruise speed, reduce power to cruise rpm, adjust pitch, and then trim off any flight control pressures. *[Figure 16-15]* The first time a transitioning pilot sees cruise rpm setting of 4,800 rpm (or as recommended), they may have a sense that the engine is turning too fast; however, remember that the engine has gear-reduction

Figure 16-15. *EFIS indication of level cruise flight.*

drive and the propeller is turning much slower. If the LSA is equipped with a standard aircraft engine, rpms are in a range that the transitioning pilot is immediately comfortable. The pilot should refer to the Cruise Checklist to ensure that the airplane is properly configured.

In slower cruise flight, stick forces are likely to be light; therefore, correction to pitch and roll attitudes should be made with light pressures. Excessive pressures result in the pilot inducing excessive correction causing a chasing effect. Only enough pressure needed to correct a deviation is required. This is best accomplished with fingertip pressures only and not with a wrapped palm of the hand. Stick forces can change dramatically as airspeed changes; for example, what could be considered light control pressures at 80 knots may become quite stiff at 100 knots. A CFI-S or CFI-A experienced in the LSA airplane is able to demonstrate this effect. This effect is dependent on the specific model of LSA and any significance or relevance varies from manufacturer to manufacturer.

LSA maneuvers such as steep turns, slow flight, and stalls are typically conventional. These maneuvers should be practiced as part of a good transition training program. Steep turns in LSA airplanes tend to be quite easy to perform precisely. With light flight control pressures, stick mounted trim (if installed), and highly differential ailerons (if part of the airplane's design), makes the performance of the maneuver simpler than heavier airplanes. Basic aerodynamics applies to any airplane and factors, such as over-banking tendency, are still prevalent and must be compensated.

Slow flight in LSAs is accomplished at slower airspeeds than standard airworthiness airplanes since stall speeds tend to be well below the 45-knot limit. The first time practicing slow flight demonstrates the unique capability of LSAs. Power off stalls are typically of no particular significance as simply unloading the wing and the application of power immediately puts the airplane back flying. However, a pilot should understand that control pressures tend to be light so an aggressive forward movement of the elevator is generally not required. In addition, proper application of rudder to compensate for propeller forces is required, and retraction of any flap should be completed prior to reaching V_{FE}, which comes very quickly if full power and nose down pitch attitude are maintained. Power on stalls can result in a very high nose-up attitude unless the airplane is adequately slowed down prior to the maneuver. In addition, some manufacturers limit pitch attitudes to 30° during power on stalls. If aggressive pitch attitudes are coupled with uncoordinated rudder inputs, spin entry is likely to be quick and aggressive.

Depending on the LSA design, especially those airplanes which use control tubes rather than wires and pulleys, flight in turbulence may couple motion to the stick rather distinctively. If a transitioning pilot's flight experience is only with airplanes that have control cables and pulleys, the first flight in turbulence may be disconcerting; however, once the pilot becomes familiar with the control sensations induced by the turbulence, it only becomes another sign for the pilot to feel the airplane.

Approach and Landing

Approach and landing in an LSA is routine and comfortable. Speeds in the pattern tend to be in the 60-knot range, which makes for reasonable airspeeds to assess landing conditions. Flap limit airspeeds tend to be lower in LSAs than standard airworthiness airplanes so managing airspeed is important. Light control forces require smooth application of control pressures without over-controlling. Pitch and power are the same in an LSA as in a standard airworthiness airplane.

Crosswinds and gusty conditions can represent hazards for all airplanes; however, the lighter weights of LSA airplanes should place an emphasis in this area. Control application does not change for crosswind technique in an LSA. Manufacturers' place a maximum demonstrated crosswind speed in the POH and, until sufficient practice and experience is gained in the airplane, a transitioning pilot should have personal minimums that do not approach the manufacturer's demonstrated crosswind speed. The LSA's light weight, slow landing speeds, and light control forces can result in a pilot inducing rapid control deflections that exceed the requirements to compensate for the crosswind. However, prompt and positive control inputs are necessary in strong winds. In addition, strong gusty crosswind conditions may exceed the airplane's control capability resulting in loss of control during the landing.

Emergencies

LSAs can be advanced airplanes in regard to its engines, airframes, and instrumentation. This environment requires that a transitioning pilot thoroughly understand and be able to effectively respond to emergency requirements. While LSA are designed to be simple, a strong respect for system knowledge is required.

The airplane's POH describes the appropriate responses to the various emergency situations that may be encountered. *[Figure 16-16]* Consider a few examples; the EFIS is displaying a "red X" across the airspeed tape, electric trim runaway, or control system failure. The pilot must be able to respond to immediate actions items from memory and locate emergency procedures quickly. In the example of trim runaway, the pilot needs to quickly assess the trim runaway condition, locate and depress the trim disconnect (if installed), or pull the trim power circuit breaker. Then depending on

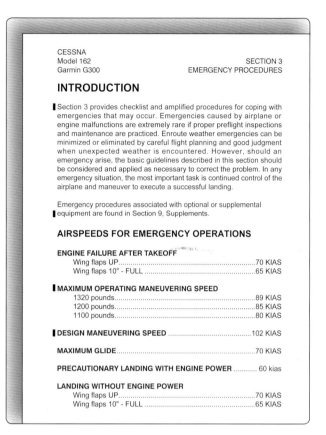

Figure 16-16. *Example of a POH Emergency Procedures section.*

control forces required to maintain pitch attitude, the pilot may need to make a no-flap landing due to the flap pitching moments. Another example is failure of the EFIS. If the EFIS "blanks" out and POH recovery procedures do not reset the EFIS, an LSA pilot may have to be prepared to land without airspeed, altitude, or vertical speed information. An effective training program covers emergencies procedures.

Post-Flight

After the airplane has been shut-down, tied-down, and secured, the pilot should conduct a complete post-flight inspection. Any squawks or discrepancies should be noted and reported to maintenance. Transitioning pilots should insist on a training debriefing where critique and planning for the next lesson takes place. Documentation of the pilot's progress should be noted on the student's records.

Key Points

Many LSA's have airframe designs that are conducive to high drag which, when combined with their low mass, results in low inertia. When attempting a crosswind landing in a high drag LSA, a rapid reduction in airspeed prior to touchdown may result in a loss of rudder and/or aileron control, which may push the aircraft off of the runway heading. This is because as the air slows across the control surfaces, the

LSA's controls become ineffective. To avoid loss of control, maintain airspeed during the approach to keep the air moving over the control surfaces until the aircraft is on the ground. LSAs with an open cockpit, easy build characteristics, low cost, and simplicity of operation and maintenance tend to be less aerodynamic and, therefore, incur more drag. The powerplant in these aircraft usually provide excess power and exhibit desirable performance. However, when power is reduced, it may be necessary to lower the nose of the aircraft to a fairly low pitch attitude in order to maintain airspeed, especially during landings and engine failure.

If the pilot makes a power off approach to landing, the approach angle will be high and the landing flare will need to be close to the ground with minimum float. This is because the aircraft will lose airspeed quickly in the flare and will not float like a more efficiently designed aircraft. Too low of an airspeed during the landing flare may lead to insufficient energy to arrest the decent which may result in a hard landing. Maintaining power during the approach will result in a reduced angle of attack and will extend the landing flare allowing more time to make adjustments to the aircraft during the landing. Always remember that rapid power reductions require an equally rapid reduction in pitch attitude to maintain airspeed.

In the event of an engine failure in an LSA, quickly transition to the required nose-down flight attitude in order to maintain airspeed. For example, if the aircraft has a power-off glide angle of 30 degrees below the horizon, position the aircraft to a nose-down 30-degree attitude as quickly as possible. The higher the pitch attitude is when the engine failure occurs, the quicker the aircraft will lose airspeed and the more likely the aircraft is to stall. Should a stall occur, decrease the aircraft's pitch attitude rapidly in order to increase airspeed to allow for a recovery. Stalls that occur at low altitudes are especially dangerous because the closer to the ground the stall occurs, the less time there is to recover. For this reason, when climbing at a low altitude, excessive pitch attitude is discouraged.

Chapter Summary

LSAs are a new category of small, lightweight aircraft that may include advanced systems, such a parachutes, EFIS, and composite construction. While the transition is not difficult, LSA does require a properly designed transition training program led by a competent CFI-S or CFI-A. Safety is of utmost importance when it comes to any flight activity. In order to properly assess the hazards of flight and mitigate flight risk, a pilot must develop the skill, judgment, and experience in order to effectively and safely pilot a LSA.

Chapter 17
Emergency Procedures

Emergency Situations

This chapter contains information on dealing with non-normal and emergency situations that may occur in flight. The key to successful management of an emergency situation, and/or preventing a non-normal situation from progressing into a true emergency, is a thorough familiarity with, and adherence to, the procedures developed by the airplane manufacturer and contained in the Federal Aviation Administration (FAA) approved Airplane Flight Manual and/or Pilot's Operating Handbook (AFM/POH). The following guidelines are generic and are not meant to replace the airplane manufacturer's recommended procedures. Rather, they are meant to enhance the pilot's general knowledge in the area of non-normal and emergency operations. If any of the guidance in this chapter conflicts in any way with the manufacturer's recommended procedures for a particular make and model airplane, the manufacturer's recommended procedures take precedence.

Emergency Landings

This section contains information on emergency landing techniques in small fixed-wing airplanes. The guidelines that are presented apply to the more adverse terrain conditions for which no practical training is possible. The objective is to instill in the pilot the knowledge that almost any terrain can be considered "suitable" for a survivable crash landing if the pilot knows how to use the airplane structure for self-protection and the protection of passengers.

Types of Emergency Landings

The different types of emergency landings are defined as follows:

- Forced landing—an immediate landing, on or off an airport, necessitated by the inability to continue further flight. A typical example of which is an airplane forced down by engine failure.

- Precautionary landing—a premeditated landing, on or off an airport, when further flight is possible but inadvisable. Examples of conditions that may call for a precautionary landing include deteriorating weather, being lost, fuel shortage, and gradually developing engine trouble.

- Ditching—a forced or precautionary landing on water.

A precautionary landing, generally, is less hazardous than a forced landing because the pilot has more time for terrain selection and the planning of the approach. In addition, the pilot can use power to compensate for errors in judgment or technique. The pilot should be aware that too many situations calling for a precautionary landing are allowed to develop into immediate forced landings, when the pilot uses wishful thinking instead of reason, especially when dealing with a self-inflicted predicament. The non-instrument-rated pilot trapped by weather, or the pilot facing imminent fuel exhaustion who does not give any thought to the feasibility of a precautionary landing, accepts an extremely hazardous alternative.

Psychological Hazards

There are several factors that may interfere with a pilot's ability to act promptly and properly when faced with an emergency. Some of these factors are listed below.

- Reluctance to accept the emergency situation—a pilot who allows the mind to become paralyzed at the thought that the airplane will be on the ground in a very short time, regardless of the pilot's actions or hopes, is severely handicapped in the handling of the emergency. An unconscious desire to delay the dreaded moment may lead to such errors as: failure to lower the nose to maintain flying speed, delay in the selection of the most suitable landing area within

reach, and indecision in general. Desperate attempts to correct whatever went wrong at the expense of airplane control fall into the same category.

- Desire to save the airplane—the pilot who has been conditioned during training to expect to find a relatively safe landing area, whenever the flight instructor closed the throttle for a simulated forced landing, may ignore all basic rules of airmanship to avoid a touchdown in terrain where airplane damage is unavoidable. Typical consequences are: making a 180° turn back to the runway when available altitude is insufficient; stretching the glide without regard for minimum control speed in order to reach a more appealing field; accepting an approach and touchdown situation that leaves no margin for error. The desire to save the airplane, regardless of the risks involved, may be influenced by two other factors: the pilot's financial stake in the airplane and the certainty that an undamaged airplane implies no bodily harm. There are times, however, when a pilot should be more interested in sacrificing the airplane so that the occupants can safely walk away from it.

- Undue concern about getting hurt—fear is a vital part of the self-preservation mechanism. However, when fear leads to panic, we invite that which we want most to avoid. The survival records favor pilots who maintain their composure and know how to apply the general concepts and procedures that have been developed through the years. The success of an emergency landing is as much a matter of the mind as of skills.

Basic Safety Concepts

General

A pilot who is faced with an emergency landing in terrain that makes extensive airplane damage inevitable should keep in mind that the avoidance of crash injuries is largely a matter of: (1) keeping the vital structure (cabin area) relatively intact by using dispensable structure (i.e., wings, landing gear, fuselage bottom) to absorb the violence of the stopping process before it affects the occupants (2) avoiding forceful bodily contact with interior structure.

The advantage of sacrificing dispensable structure is demonstrated daily on the highways. A head-on car impact against a tree at 20 miles per hour (mph) is less hazardous for a properly restrained driver than a similar impact against the driver's door. Accident experience shows that the extent of crushable structure between the occupants and the principal point of impact on the airplane has a direct bearing on the severity of the transmitted crash forces and, therefore, on survivability.

Avoiding forcible contact with interior structure is a matter of seat and body security. Unless the occupant decelerates at the same rate as the surrounding structure, no benefit is realized from its relative intactness. The occupant is brought to a stop violently in the form of a secondary collision.

Dispensable airplane structure is not the only available energy absorbing medium in an emergency situation. Vegetation, trees, and even manmade structures may be used for this purpose. Cultivated fields with dense crops, such as mature corn and grain, are almost as effective in bringing an airplane to a stop with repairable damage as an emergency arresting device on a runway. *[Figure 17-1]* Brush and small trees provide considerable cushioning and braking effect without destroying the airplane. When dealing with natural and manmade obstacles with greater strength than the dispensable airplane structure, the pilot must plan the touchdown in such a manner that only nonessential structure is "used up" in the principal slowing-down process.

The overall severity of a deceleration process is governed by speed (groundspeed) and stopping distance. The most critical of these is speed; doubling the groundspeed means quadrupling the total destructive energy and vice versa. Even a small change in groundspeed at touchdown—be it as a result of wind or pilot technique—affects the outcome of a controlled crash. It is important that the actual touchdown during an emergency landing be made at the lowest possible controllable airspeed, using all available aerodynamic devices.

Most pilots instinctively—and correctly—look for the largest available flat and open field for an emergency landing.

Actually, very little stopping distance is required if the speed can be dissipated uniformly; that is, if the deceleration forces can be spread evenly over the available distance. This concept is designed into the arresting gear of aircraft carriers that provides a nearly constant stopping force from the moment of hookup.

The typical light airplane is designed to provide protection in crash landings that expose the occupants to nine times the acceleration of gravity (9G) in a forward direction. Assuming a uniform 9G deceleration, at 50 mph the required stopping distance is about 9.4 feet. While at 100 mph, the stopping distance is about 37.6 feet—about four times as great. *[Figure 17-2]* Although these figures are based on an ideal deceleration process, it is interesting to note what

Figure 17-2. *Stopping distance vs. groundspeed.*

Figure 17-1. *Using vegetation to absorb energy.*

can be accomplished in an effectively used short stopping distance. Understanding the need for a firm but uniform deceleration process in very poor terrain enables the pilot to select touchdown conditions that spread the breakup of dispensable structure over a short distance, thereby reducing the peak deceleration of the cabin area.

Attitude and Sink Rate Control

The most critical and often the most inexcusable error that can be made in the planning and execution of an emergency landing, even in ideal terrain, is the loss of initiative over the airplane's attitude and sink rate at touchdown. When the touchdown is made on flat, open terrain, an excessive nose-low pitch attitude brings the risk of "sticking" the nose in the ground. Steep bank angles just before touchdown should also be avoided, as they increase the stalling speed and the likelihood of a wingtip strike.

Since the airplane's vertical component of velocity is immediately reduced to zero upon ground contact, it must be kept well under control. A flat touchdown at a high sink rate (well in excess of 500 feet per minute (fpm)) on a hard surface can be injurious without destroying the cabin structure, especially during gear up landings in low-wing airplanes. A rigid bottom construction of these airplanes may preclude adequate cushioning by structural deformation. Similar impact conditions may cause structural collapse of the overhead structure in high-wing airplanes. On soft terrain, an excessive sink rate may cause digging in of the lower nose structure and severe forward deceleration.

Terrain Selection

A pilot's choice of emergency landing sites is governed by:

- The route selected during preflight planning
- The height above the ground when the emergency occurs
- Excess airspeed (excess airspeed can be converted into distance and/or altitude)

The only time the pilot has a very limited choice is during the low and slow portion of the takeoff. However, even under these conditions, the ability to change the impact heading only a few degrees may ensure a survivable crash.

If beyond gliding distance of a suitable open area, the pilot should judge the available terrain for its energy absorbing capability. If the emergency starts at a considerable height above the ground, the pilot should be more concerned about first selecting the desired general area than a specific spot. Terrain appearances from altitude can be very misleading and considerable altitude may be lost before the best spot can be pinpointed. For this reason, the pilot should not hesitate

to discard the original plan for one that is obviously better. However, as a general rule, the pilot should not change his or her mind more than once; a well-executed crash landing in poor terrain can be less hazardous than an uncontrolled touchdown on an established field.

Airplane Configuratio

Since flaps improve maneuverability at slow speed, and lower the stalling speed, their use during final approach is recommended when time and circumstances permit. However, the associated increase in drag and decrease in gliding distance call for caution in the timing and the extent of their application; premature use of flap and dissipation of altitude may jeopardize an otherwise sound plan.

A hard and fast rule concerning the position of a retractable landing gear at touchdown cannot be given. In rugged terrain and trees, or during impacts at high sink rate, an extended gear would definitely have a protective effect on the cabin area. However, this advantage has to be weighed against the possible side effects of a collapsing gear, such as a ruptured fuel tank. As always, the manufacturer's recommendations as outlined in the AFM/POH should be followed.

When a normal touchdown is assured, and ample stopping distance is available, a gear-up landing on level, but soft terrain or across a plowed field may result in less airplane damage than a gear-down landing. *[Figure 17-3]* Deactivation of the airplane's electrical system before touchdown reduces the likelihood of a post-crash fire.

However, the battery master switch should not be turned off until the pilot no longer has any need for electrical power to operate vital airplane systems. Positive airplane control during the final part of the approach has priority over all other considerations, including airplane configuration and checklist tasks. The pilot should attempt to exploit the power available

Figure 17-3. *Intentional gear-up landing.*

from an irregularly running engine; however, it is generally better to switch the engine and fuel off just before touchdown. This not only ensures the pilot's initiative over the situation, but a cooled-down engine reduces the fire hazard considerably.

Approach

When the pilot has time to maneuver, the planning of the approach should be governed by the following three factors:

- Wind direction and velocity
- Dimensions and slope of the chosen field
- Obstacles in the final approach path

These three factors are seldom compatible. When compromises have to be made, the pilot should aim for a wind/obstacle/ terrain combination that permits a final approach with some margin for error in judgment or technique. A pilot who overestimates the gliding range may be tempted to stretch the glide across obstacles in the approach path. For this reason, it is sometimes better to plan the approach over an unobstructed area, regardless of wind direction. Experience shows that a collision with obstacles at the end of a ground roll or slide is much less hazardous than striking an obstacle at flying speed before the touchdown point is reached.

Terrain Types

Since an emergency landing on suitable terrain resembles a situation in which the pilot should be familiar through training, only the more unusual situations are discussed.

Confined Area

The natural preference to set the airplane down on the ground should not lead to the selection of an open spot between trees or obstacles where the ground cannot be reached without making a steep descent.

Once the intended touchdown point is reached, and the remaining open and unobstructed space is very limited, it may be better to force the airplane down on the ground than to delay touchdown until it stalls (settles). An airplane decelerates faster after it is on the ground than while airborne. Thought may also be given to the desirability of ground-looping or retracting the landing gear in certain conditions.

A river or creek can be an inviting alternative in otherwise rugged terrain. The pilot should ensure that the water or creek bed can be reached without snagging the wings. The same concept applies to road landings with one additional reason for caution: manmade obstacles on either side of a road may not be visible until the final portion of the approach.

When planning the approach across a road, it should be remembered that most highways and even rural dirt roads are paralleled by power or telephone lines. Only a sharp lookout for the supporting structures or poles may provide timely warning.

Trees (Forest)

Although a tree landing is not an attractive prospect, the following general guidelines help to make the experience survivable.

- Use the normal landing configuration (full flaps, gear down).

- Keep the groundspeed low by heading into the wind.

- Make contact at minimum indicated airspeed, but not below stall speed, and "hang" the airplane in the tree branches in a nose-high landing attitude. Involving the underside of the fuselage and both wings in the initial tree contact provides a more even and positive cushioning effect, while preventing penetration of the windshield. *[Figure 17-4]*

- Avoid direct contact of the fuselage with heavy tree trunks.

- Low, closely spaced trees with wide, dense crowns (branches) close to the ground are much better than tall trees with thin tops; the latter allow too much free fall height (a free fall from 75 feet results in an impact speed of about 40 knots, or about 4,000 fpm).

- Ideally, initial tree contact should be symmetrical; that is, both wings should meet equal resistance in the tree branches. This distribution of the load helps to maintain proper airplane attitude. It may also preclude the loss of one wing, which invariably leads to a more rapid and less predictable descent to the ground.

Figure 17-4. *Tree landing.*

- If heavy tree trunk contact is unavoidable once the airplane is on the ground, it is best to involve both wings simultaneously by directing the airplane between two properly spaced trees. Do not attempt this maneuver, however, while still airborne.

Water (Ditching) and Snow

A well-executed water landing normally involves less deceleration violence than a poor tree landing or a touchdown on extremely rough terrain. Also, an airplane that is ditched at minimum speed and in a normal landing attitude does not immediately sink upon touchdown. Intact wings and fuel tanks (especially when empty) provide floatation for at least several minutes, even if the cabin may be just below the water line in a high-wing airplane.

Loss of depth perception may occur when landing on a wide expanse of smooth water with the risk of flying into the water or stalling in from excessive altitude. To avoid this hazard, the airplane should be "dragged in" when possible. Use no more than intermediate flaps on low-wing airplanes. The water resistance of fully extended flaps may result in asymmetrical flap failure and slowing of the airplane. Keep a retractable gear up unless the AFM/POH advises otherwise.

A landing in snow should be executed like a ditching, in the same configuration and with the same regard for loss of depth perception (white out) in reduced visibility and on wide-open terrain.

Engine Failure After Takeoff (Single-Engine)

The altitude available is, in many ways, the controlling factor in the successful accomplishment of an emergency landing. If an actual engine failure should occur immediately after takeoff and before a safe maneuvering altitude is attained, it is usually inadvisable to attempt to turn back to the field from where the takeoff was made. Instead, it is safer to immediately establish the proper glide attitude, and select a field directly ahead or slightly too either side of the takeoff path.

The decision to continue straight ahead is often difficult to make unless the problems involved in attempting to turn back are seriously considered. In the first place, the takeoff was in all probability made into the wind. To get back to the takeoff field, a downwind turn must be made. This increases the groundspeed and rushes the pilot even more in the performance of procedures and in planning the landing approach. Secondly, the airplane is losing considerable altitude during the turn and might still be in a bank when the ground is contacted, resulting in the airplane cartwheeling (which would be a catastrophe for the occupants, as well as the airplane). After turning downwind, the apparent increase

in groundspeed could mislead the pilot into attempting to prematurely slow down the airplane and cause it to stall. On the other hand, continuing straight ahead or making a slight turn allows the pilot more time to establish a safe landing attitude, and the landing can be made as slowly as possible, but more importantly, the airplane can be landed while under control.

Concerning the subject of turning back to the runway following an engine failure on takeoff, the pilot should determine the minimum altitude an attempt of such a maneuver should be made in a particular airplane. Experimentation at a safe altitude should give the pilot an approximation of height lost in a descending 180° turn at idle power. By adding a safety factor of about 25 percent, the pilot should arrive at a practical decision height. The ability to make a 180° turn does not necessarily mean that the departure runway can be reached in a power-off glide; this depends on the wind, the distance traveled during the climb, the height reached, and the glide distance of the airplane without power. The pilot should also remember that a turn back to the departure runway may in fact require more than a 180° change in direction.

Consider the following example of an airplane which has taken off and climbed to an altitude of 300 feet above ground level (AGL) when the engine fails. *[Figure 17-5]* After a typical 4 second reaction time, the pilot elects to turn back to the runway. Using a standard rate (3° change in direction per second) turn, it takes 1 minute to turn 180°. At a glide speed of 65 knots, the radius of the turn is 2,100 feet, so at the completion of the turn, the airplane is 4,200 feet to one side of the runway. The pilot must turn another 45° to head the airplane toward the runway. By this time, the total change in direction is 225° equating to 75 seconds plus the 4 second reaction time. If the airplane in a poweroff glide descends at approximately 1,000 fpm, it has descended 1,316, feet placing it 1,016 feet below the runway.

Emergency Descents

An emergency descent is a maneuver for descending as rapidly as possible to a lower altitude or to the ground for an emergency landing. *[Figure 17-6]* The need for this maneuver may result from an uncontrollable fire, a sudden loss of cabin pressurization, or any other situation demanding an immediate and rapid descent. The objective is to descend the airplane as soon and as rapidly as possible within the structural limitations of the airplane. Simulated emergency descents should be made in a turn to check for other air traffic below and to look around for a possible emergency landing area. A radio call announcing descent intentions may be appropriate to alert other aircraft in the area. When initiating the descent, a bank of approximately 30 to 45° should be established to maintain positive load factors (G forces) on the airplane.

Figure 17-5. *Turning back to the runway after engine failure.*

Emergency descent training should be performed as recommended by the manufacturer, including the configuration and airspeeds. Except when prohibited by the manufacturer, the power should be reduced to idle, and the propeller control (if equipped) should be placed in the low pitch (or high revolutions per minute (rpm)) position. This allows the propeller to act as an aerodynamic brake to help prevent an excessive airspeed buildup during the descent. The landing gear and flaps should be extended as recommended by the manufacturer. This provides maximum drag so that the descent can be made as rapidly as possible, without excessive airspeed. The pilot should not allow the airplane's airspeed

to pass the never-exceed speed (V_{NE}), the maximum landing gear extended speed (V_{LE}), or the maximum flap extended speed (V_{FE}), as applicable. In the case of an engine fire, a high airspeed descent could blow out the fire. However, the weakening of the airplane structure is a major concern and descent at low airspeed would place less stress on the airplane. If the descent is conducted in turbulent conditions, the pilot must also comply with the design maneuvering speed (V_A) limitations. The descent should be made at the maximum allowable airspeed consistent with the procedure used. This provides increased drag and, therefore, the loss of altitude as quickly as possible. The recovery from an emergency descent should be initiated at a high enough altitude to ensure a safe recovery back to level flight or a precautionary landing.

When the descent is established and stabilized during training and practice, the descent should be terminated. In airplanes with piston engines, prolonged practice of emergency descents should be avoided to prevent excessive cooling of the engine cylinders.

In-Flight Fire

A fire in-flight demands immediate and decisive action. The pilot therefore must be familiar with the procedures outlined to meet this emergency contained in the AFM/POH for the

Figure 17-6. *Emergency descent.*

particular airplane. For the purposes of this handbook, in-flight fires are classified as in-flight engine fires, electrical fires, and cabin fires.

Engine Fire

An in-flight engine compartment fire is usually caused by a failure that allows a flammable substance, such as fuel, oil, or hydraulic fluid, to come in contact with a hot surface. This may be caused by a mechanical failure of the engine itself, an engine-driven accessory, a defective induction or exhaust system, or a broken line. Engine compartment fires may also result from maintenance errors, such as improperly installed/fastened lines and/or fittings resulting in leaks.

Engine compartment fires can be indicated by smoke and/or flames coming from the engine cowling area. They can also be indicated by discoloration, bubbling, and/or melting of the engine cowling skin in cases where flames and/or smoke are not visible to the pilot. By the time a pilot becomes aware of an in-flight engine compartment fire, it usually is well developed. Unless the airplane manufacturer directs otherwise in the AFM/POH, the first step on discovering a fire should be to shut off the fuel supply to the engine by placing the mixture control in the idle cut off position and the fuel selector shutoff valve to the OFF position. The ignition switch should be left ON in order to use up the fuel that remains in the fuel lines and components between the fuel selector/shutoff valve and the engine. This procedure may starve the engine compartment of fuel and cause the fire to die naturally. If the flames are snuffed out, no attempt should be made to restart the engine.

If the engine compartment fire is oil-fed, as evidenced by thick black smoke, as opposed to a fuel-fed fire, which produces bright orange flames, the pilot should consider stopping the propeller rotation by feathering or other means, such as (with constant-speed propellers) placing the pitch control lever to the minimum rpm position and raising the nose to reduce airspeed until the propeller stops rotating. This procedure stops an engine-driven oil (or hydraulic) pump from continuing to pump the flammable fluid that is feeding the fire.

Some light airplane emergency checklists direct the pilot to shut off the electrical master switch. However, the pilot should consider that unless the fire is electrical in nature, or a crash landing is imminent, deactivating the electrical system prevents the use of panel radios for transmitting distress messages and also causes air traffic control (ATC) to lose transponder returns.

Pilots of powerless single-engine airplanes are left with no choice but to make a forced landing. Pilots of twin-engine airplanes may elect to continue the flight to the nearest airport.

However, consideration must be given to the possibility that a wing could be seriously impaired and lead to structural failure. Even a brief but intense fire could cause dangerous structural damage. In some cases, the fire could continue to burn under the wing (or engine cowling in the case of a single-engine airplane) out of view of the pilot. Engine compartment fires that appear to have been extinguished have been known to rekindle with changes in airflow pattern and airspeed.

The pilot must be familiar with the airplane's emergency descent procedures. The pilot must bear in mind the following:

- The airplane may be severely structurally damaged to the point that its ability to remain under control could be lost at any moment.

- The airplane may still be on fire and susceptible to explosion.

- The airplane is expendable and the only thing that matters is the safety of those on board.

Electrical Fires

The initial indication of an electrical fire is usually the distinct odor of burning insulation. Once an electrical fire is detected, the pilot should attempt to identify the faulty circuit by checking circuit breakers, instruments, avionics, and lights. If the faulty circuit cannot be readily detected and isolated, and flight conditions permit, the battery master switch and alternator/generator switches should be turned off to remove the possible source of the fire. However, any materials that have been ignited may continue to burn.

If electrical power is absolutely essential for the flight, an attempt may be made to identify and isolate the faulty circuit by:

1. Turning the electrical master switch OFF.

2. Turning all individual electrical switches OFF.

3. Turning the master switch back ON.

4. Selecting electrical switches that were ON before the fire indication one at a time, permitting a short time lapse after each switch is turned on to check for signs of odor, smoke, or sparks.

This procedure, however, has the effect of recreating the original problem. The most prudent course of action is to land as soon as possible.

Cabin Fire

Cabin fires generally result from one of three sources: (1) careless smoking on the part of the pilot and/or passengers; (2) electrical system malfunctions; (3) heating system malfunctions. A fire in the cabin presents the pilot with two immediate demands: attacking the fire and getting the airplane safely on the ground as quickly as possible. A fire or smoke in the cabin should be controlled by identifying and shutting down the faulty system. In many cases, smoke may be removed from the cabin by opening the cabin air vents. This should be done only after the fire extinguisher (if available) is used. Then the cabin air control can be opened to purge the cabin of both smoke and fumes. If smoke increases in intensity when the cabin air vents are opened, they should be immediately closed. This indicates a possible fire in the heating system, nose compartment baggage area (if so equipped), or that the increase in airflow is feeding the fire.

On pressurized airplanes, the pressurization air system removes smoke from the cabin; however, if the smoke is intense, it may be necessary to either depressurize at altitude, if oxygen is available for all occupants, or execute an emergency descent.

In unpressurized single-engine and light twin-engine airplanes, the pilot can attempt to expel the smoke from the cabin by opening the foul weather windows. These windows should be closed immediately if the fire becomes more intense. If the smoke is severe, the passengers and crew should use oxygen masks if available, and the pilot should initiate an immediate descent. The pilot should also be aware that on some airplanes, lowering the landing gear and/or wing flaps can aggravate a cabin smoke problem.

Flight Control Malfunction/Failure

Total Flap Failure
The inability to extend the wing flaps necessitates a no-flap approach and landing. In light airplanes, a no-flap approach and landing is not particularly difficult or dangerous. However, there are certain factors that must be considered in the execution of this maneuver. A no-flap landing requires substantially more runway than normal. The increase in required landing distance could be as much as 50 percent.

When flying in the traffic pattern with the wing flaps retracted, the airplane must be flown in a relatively nose-high attitude to maintain altitude, as compared to flight with flaps extended. Losing altitude can be more of a problem without the benefit of the drag normally provided by flaps. A wider, longer traffic pattern may be required in order to avoid the necessity of diving to lose altitude and consequently building up excessive airspeed.

On final approach, a nose-high attitude can make it difficult to see the runway. This situation, if not anticipated, can result in serious errors in judgment of height and distance.

Approaching the runway in a relatively nose-high attitude can also cause the perception that the airplane is close to a stall. This may cause the pilot to lower the nose abruptly and risk touching down on the nosewheel.

With the flaps retracted and the power reduced for landing, the airplane is slightly less stable in the pitch and roll axes. Without flaps, the airplane tends to float considerably during roundout. The pilot should avoid the temptation to force the airplane onto the runway at an excessively, high speed. Neither should the pilot flare excessively because without flaps, this might cause the tail to strike the runway.

Asymmetric (Split) Flap
An asymmetric "split" flap situation is one in which one flap deploys or retracts while the other remains in position. The problem is indicated by a pronounced roll toward the wing with the least flap deflection when wing flaps are extended/retracted.

The roll encountered in a split flap situation is countered with opposite aileron. The yaw caused by the additional drag created by the extended flap requires substantial opposite rudder resulting in a cross-control condition. Almost full aileron may be required to maintain a wings-level attitude, especially at the reduced airspeed necessary for approach and landing. The pilot should not attempt to land with a crosswind from the side of the deployed flap because the additional roll control required to counteract the crosswind may not be available.

The approach to landing with a split flap condition should be flown at a higher than normal airspeed. The pilot should not risk an asymmetric stall and subsequent loss of control by flaring excessively. Rather, the airplane should be flown onto the runway so that the touchdown occurs at an airspeed consistent with a safe margin above flaps-up stall speed.

Loss of Elevator Control
In many airplanes, the elevator is controlled by two cables: a "down" cable and an "up" cable. Normally, a break or disconnect in only one of these cables does not result in a total loss of elevator control. In most airplanes, a failed cable results in a partial loss of pitch control. In the failure of the "up" elevator cable (the "down" elevator being intact and functional), the control yoke moves aft easily but produces no response. Forward yoke movement, however, beyond the neutral position produces a nosedown attitude. Conversely, a failure of the "down" elevator cable, forward movement of the control yoke produces no effect. The pilot, however, has partial control of pitch attitude with aft movement.

When experiencing a loss of up-elevator control, the pilot can retain pitch control by:

- Applying considerable nose-up trim
- Pushing the control yoke forward to attain and maintain desired attitude
- Increasing forward pressure to lower the nose and relaxing forward pressure to raise the nose
- Releasing forward pressure to flare for landing

When experiencing a loss of down-elevator control, the pilot can retain pitch control by:

- Applying considerable nosedown trim
- Pulling the control yoke aft to attain and maintain attitude
- Releasing back pressure to lower the nose and increasing back pressure to raise the nose
- Increasing back pressure to flare for landing

Trim mechanisms can be useful in the event of an in-flight primary control failure. For example, if the linkage between the cabin and the elevator fails in flight, leaving the elevator free to weathervane in the wind, the trim tab can be used to raise or lower the elevator within limits. The trim tabs are not as effective as normal linkage control in conditions such as low airspeed, but they do have some positive effect—usually enough to bring about a safe landing.

If an elevator becomes jammed, resulting in a total loss of elevator control movement, various combinations of power and flap extension offer a limited amount of pitch control. A successful landing under these conditions, however, is problematical.

Landing Gear Malfunction

Once the pilot has confirmed that the landing gear has in fact malfunctioned and that one or more gear legs refuses to respond to the conventional or alternate methods of gear extension contained in the AFM/POH, there are several methods that may be useful in attempting to force the gear down. One method is to dive the airplane (in smooth air only) to V_{NE} speed (red line on the airspeed indicator) and (within the limits of safety) execute a rapid pull up. In normal category airplanes, this procedure creates a 3.8G load on the structure, in effect making the landing gear weigh 3.8 times normal. In some cases, this may force the landing gear into the down and locked position. This procedure requires a fine control touch and good feel for the airplane. Careful consideration should be given to the fact that if the pull up is too abrupt, it may result in an accelerated stall, possible loss of control, and cause excessive structural stress to be imposed on the aircraft.

The design maneuvering speed (V_A) is a structural design airspeed used in determining the strength requirements for the airplane and its control surfaces. The structural design requirements do not cover multiple control inputs in one axis or control inputs in more than one axis at a time at any speed, even below V_A. Combined control inputs cause additional bending and twisting forces. Any airspeed above the maneuvering speed provides a positive life capability that may cause structural damage if excessive G forces are exerted on the aircraft. V_A is based on the actual gross weight of the airplane and the wing's response to a 50 foot per second wind gust or movement of the elevator. The combination of turbulence and high G loading induces even greater stress on the aircraft. Because wind gusts are not symmetrical, the total additional stress that is added to the aircraft due to turbulence is difficult to determine. Each element of the airframe and each flight control component have their own design structural load limit. Maneuvering speed is primarily determined for the wings; the elevator may be structurally damaged below this speed.

An alternative method that has proven useful in dislodging stuck landing gear (in some cases) is to induce rapid yawing. After stabilizing below V_A, the pilot should alternately and aggressively apply rudder in one direction and then the other in rapid sequence. However, be advised that operating at or below maneuvering speed does not provide structural protection against multiple full control inputs in one axis or full control inputs in more than one axis at the same time. The resulting yawing action may cause the landing gear to fall into place. The pilot must be aware that moving the rudder from stop to stop is not a load limit certification requirement for normal category airplanes. Only aircraft designed for certain high G load flight maneuvers must have a vertical fin and rudder capable to withstand abrupt pedal control application to the limits in both directions.

If all efforts to extend the landing gear have failed and a gear-up landing is inevitable, the pilot should select an airport with crash and rescue facilities. The pilot should not hesitate to request that emergency equipment is standing by.

When selecting a landing surface, the pilot should consider that a smooth, hard-surface runway usually causes less damage than rough, unimproved grass strips. A hard surface does, however, create sparks that can ignite fuel. If the airport is so equipped, the pilot can request that the runway surface be foamed. The pilot should consider burning off excess fuel. This reduces landing speed and fire potential.

If the landing gear malfunction is limited to one main landing gear leg, the pilot should consume as much fuel from that side of the airplane as practicable, thereby reducing the weight of the wing on that side. The reduced weight makes it possible to delay the unsupported wing from contacting the surface during the landing roll until the last possible moment. Reduced impact speeds result in less damage.

If only one landing gear leg fails to extend, the pilot has the option of landing on the available gear legs or landing with all the gear legs retracted. Landing on only one main gear usually causes the airplane to veer strongly in the direction of the faulty gear leg after touchdown. If the landing runway is narrow and/or ditches and obstacles line the runway edge, maximum directional control after touchdown is a necessity. In this situation, a landing with all three gear retracted may be the safest course of action.

If the pilot elects to land with one main gear retracted (and the other main gear and nose gear down and locked), the landing should be made in a nose-high attitude with the wings level. As airspeed decays, the pilot should apply whatever aileron control is necessary to keep the unsupported wing airborne as long as possible. *[Figure 17-7]* Once the wing contacts the surface, the pilot can anticipate a strong yaw in that direction. The pilot must be prepared to use full opposite rudder and aggressive braking to maintain some degree of directional control.

When landing with a retracted nosewheel (and the main gear extended and locked), the pilot should hold the nose off the ground until almost full up-elevator has been applied. *[Figure 17-8]* The pilot should then release back pressure in such a manner that the nose settles slowly to the surface. Applying and holding full up-elevator results in the nose abruptly dropping to the surface as airspeed decays, possibly resulting in burrowing and/or additional damage. Brake pressure should not be applied during the landing roll unless absolutely necessary to avoid a collision with obstacles.

If the landing must be made with only the nose gear extended, the initial contact should be made on the aft

Figure 17-7. *Landing with one main gear retracted.*

Figure 17-8. *Landing with nosewheel retracted.*

fuselage structure with a nose-high attitude. This procedure helps prevent porpoising and/or wheelbarrowing. The pilot should then allow the nosewheel to gradually touchdown, using nosewheel steering as necessary for directional control.

Systems Malfunctions

Electrical System

The loss of electrical power can deprive the pilot of numerous critical systems, and therefore should not be taken lightly even in day/visual flight rules (VFR) conditions. Most in-flight failures of the electrical system are located in the generator or alternator. Once the generator or alternator system goes off line, the electrical source in a typical light airplane is a battery. If a warning light or ammeter indicates the probability of an alternator or generator failure in an airplane with only one generating system, however, the pilot may have very little time available from the battery.

The rating of the airplane battery provides a clue to how long it may last. With batteries, the higher the amperage load, the less the usable total amperage. Thus, a 25-amp hour battery could produce 5 amps per hour for 5 hours, but if the load were increased to 10 amps, it might last only 2 hours. A 40-amp load might discharge the battery fully in about 10 or 15 minutes. Much depends on the battery condition at the time of the system failure. If the battery has been in service for a few years, its power may be reduced substantially because of internal resistance. Or if the system failure was not detected immediately, much of the stored energy may have already been used. It is essential, therefore, that the pilot immediately shed non-essential loads when the generating source fails. *[Figure 17-9]* The pilot should then plan to land at the nearest suitable airport.

What constitutes an "emergency" load following a generating system failure cannot be predetermined because the actual circumstances are always somewhat different—for example, whether the flight is VFR or instrument flight rules (IFR),

Electrical Loads for Light Single	Number of units	Total Amperes
A. Continuous Load		
Pitot Heating (Operating)	1	3.30
Wingtip Lights	4	3.00
Heater Igniter	1	1-20
**Navigation Receivers	1-4	1-2 each
**Communications Receivers	1-2	1-2 each
Fuel Indicator	1	0.40
Instrument Lights (overhead)	2	0.60
Engine Indicator	1	0.30
Compass Light	1	0.20
Landing Gear Indicator	1	0.17
Flap Indicator	1	0.17
B. Intermittent Load		
Starter	1	100.00
Landing Lights	2	17.80
Heater Blower Motor	1	14.00
Flap Motor	1	13.00
Landing Gear Motor	1	10.00
Cigarette Lighter	1	7.50
Transceiver (keyed)	1	5-7
Fuel Boost Pump	1	2.00
Cowl Flap Motor	1	1.00
Stall Warning Horn	1	1.50

** Amperage for radios varies with equipment.
In general, the more recent the model, the less amperage required.
NOTE: Panel and indicator lights usually draw less than one amp.

Figure 17-9. *Electrical load for light single.*

conducted in day or at night, in clouds or in the clear. Distance to nearest suitable airport can also be a factor.

The pilot should remember that the electrically-powered (or electrically-selected) landing gear and flaps do not function properly on the power left in a partially-depleted battery. Landing gear and flap motors use up power at rates much greater than most other types of electrical equipment. The result of selecting these motors on a partially-depleted battery may well result in an immediate total loss of electrical power.

If the pilot should experience a complete in-flight loss of electrical power, the following steps should be taken:

- Shed all but the most necessary electrically-driven equipment.

- Understand that any loss of electrical power is critical in a small airplane—notify ATC of the situation immediately. Request radar vectors for a landing at the nearest suitable airport.

- If landing gear or flaps are electrically controlled or operated, plan the arrival well ahead of time. Expect to make a no-flap landing and anticipate a manual landing gear extension.

Pitot-Static System

The source of the pressure for operating the airspeed indicator, the vertical speed indicator (VSI), and the altimeter is the pitot-static system. The major components of the pitot-static system are the impact pressure chamber and lines and the static pressure chamber and lines, each of which are subject to total or partial blockage by ice, dirt, and/or other foreign matter. Blockage of the pitot-static system adversely affects instrument operation. *[Figure 17-10]*

Partial static system blockage is insidious in that it may go unrecognized until a critical phase of flight. During takeoff, climb, and level-off at cruise altitude the altimeter, airspeed indicator, and VSI may operate normally. No indication of malfunction may be present until the airplane begins a descent.

If the static reference system is severely restricted, but not entirely blocked, as the airplane descends, the static reference pressure at the instruments begins to lag behind the actual outside air pressure. While descending, the altimeter may indicate that the airplane is higher than actual because the obstruction slows the airflow from the static port to the altimeter. The VSI confirms the altimeter's information regarding rate of change because the reference pressure is not changing at the same rate as the outside air pressure. The airspeed indicator, unable to tell whether it is experiencing more airspeed pitot pressure or less static reference pressure, indicates a higher airspeed than actual. To the pilot, the instruments indicate that the airplane is too high, too fast, and descending at a rate much less than desired.

If the pilot levels off and then begins a climb, the altitude indication may still lag. The VSI indicates that the airplane is not climbing as fast as actual. The indicated airspeed, however, may begin to decrease at an alarming rate. The least amount of pitch-up attitude may cause the airspeed needle to indicate dangerously near stall speed.

Managing a static system malfunction requires that the pilot know and understand the airplane's pitot-static system. If a system malfunction is suspected, the pilot should confirm it by opening the alternate static source. This should be done while the airplane is climbing or descending. If the instrument needles move significantly when this is done, a static pressure problem exists and the alternate source should be used during the remainder of the flight.

Failure of the pitot-static system may also have serious consequences for Electronic Flight Instrument Systems (EFIS). To satisfy the requirements of Title 14 of the Code

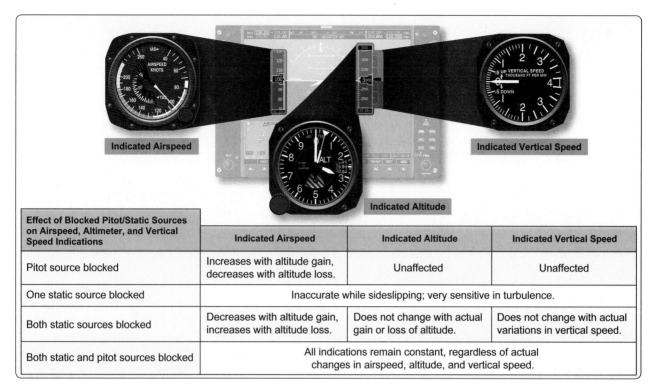

Effect of Blocked Pitot/Static Sources on Airspeed, Altimeter, and Vertical Speed Indications	Indicated Airspeed	Indicated Altitude	Indicated Vertical Speed
Pitot source blocked	Increases with altitude gain, decreases with altitude loss.	Unaffected	Unaffected
One static source blocked	Inaccurate while sideslipping; very sensitive in turbulence.		
Both static sources blocked	Decreases with altitude gain, increases with altitude loss.	Does not change with actual gain or loss of altitude.	Does not change with actual variations in vertical speed.
Both static and pitot sources blocked	All indications remain constant, regardless of actual changes in airspeed, altitude, and vertical speed.		

Figure 17-10. *Effects of blocked pitot-static sources.*

of Federal Regulations (14 CFR) part 23 for IFR flight, redundant instrumentation is required for electronic displays of airspeed, altitude, and attitude indications. Dedicated standby instruments or dual independent pilot flight displays (PFD) must be installed in the aircraft. Many of the light aircraft equipped with glass cockpits typically share the same pitot-static inputs for the backup instrumentation. Since both systems are receiving the same input signals, both may be adversely affected by obstructed or blocked pitot tubes and static ports making redundancy much less than desired. Some manufacturers combine both the air data computer (ADC) and the attitude and heading reference system (AHRS) functions so that a blockage of the input system may also affect the attitude display.

With conventional instrumentation, the design and operation are similar regardless of aircraft or manufacturer. By comparing information between the six conventional instruments, pilots are able to diagnose common failure modes. Instrument failure indications of conventional instruments and electronic flight displays may be entirely different, and electronic systems failure indications are not standardized. With the wide diversity in system design of glass cockpits, the primary display and the backup display may respond differently to any interruption of data input, and both displays may function differently than conventional instruments under the same conditions.

It is imperative for pilots to obtain equipment-specific information in reference to both the aircraft and the avionics that fully prepare them to interpret and properly respond to equipment malfunctions of electronic flight instrument displays. Rapidly changing equipment, complex systems, and the difficulty or inability to simulate failure modes and functions can impose training limitations. Pilots still must be able to respond to equipment malfunctions in a timely manner without impairing other critical flight tasks should the need arise.

Abnormal Engine Instrument Indication

The AFM/POH for the specific airplane contains information that should be followed in the event of any abnormal engine instrument indications. The table shown in *Figure 17-11* offers generic information on some of the more commonly experienced in-flight abnormal engine instrument indications, their possible causes, and corrective actions.

Door Opening In-Flight

In most instances, the occurrence of an inadvertent door opening is not of great concern to the safety of a flight, but rather, the pilot's reaction at the moment the incident happens. A door opening in flight may be accompanied by a sudden loud noise, sustained noise level, and possible vibration or buffeting. If a pilot allows himself or herself to become distracted to the

Malfunction	Probable Cause	Corrective Action
Loss of rpm during cruise flight (non-altitude engines)	Carburetor or induction icing or air filter clogging	Apply carburetor heat. If dirty filter is suspected and non-filtered air is available, switch selector to unfiltered position.
Loss of manifold pressure during cruise flight	Same as above	Same as above.
	Turbocharger failure	Possible exhaust leak. Shut down engine or use lowest practicable power setting. Land as soon as possible.
Gain of manifold pressure during cruise flight	Throttle has opened, propeller control has decreased rpm, or improper method of power reduction	Readjust throttle and tighten friction lock. Reduce manifold pressure prior to reducing rpm.
High oil temperature	Oil congealed in cooler	Reduce power. Land. Preheat engine.
	Inadequate engine cooling	Reduce power. Increase airspeed.
	Detonation or preignition	Observe cylinder head temperatures for high reading. Reduce manifold pressure. Enrich mixture.
	Forthcoming internal engine failure	Land as soon as possible or feather propeller and stop engine.
	Defective thermostatic oil cooler control	Land as soon as possible. Consult maintenance personnel.
Low oil temperature	Engine not warmed up to operating temperature	Warm engine in prescribed manner.
High oil pressure	Cold oil	Same as above.
	Possible internal plugging	Reduce power. Land as soon as possible.
Low oil pressure	Broken pressure relief valve	Land as soon as possible or feather propeller and stop engine.
	Insufficient oil	Same as above.
	Burned out bearings	Same as above.
Fluctuating oil pressure	Low oil supply, loose oil lines, defective pressure relief valve	Same as above.
High cylinder head temperature	Improper cowl flap adjustment	Adjust cowl flaps.
	Insufficient airspeed for cooling	Increase airspeed.
	Improper mixture adjustment	Adjust mixture.
	Detonation or preignition	Reduce power, enrich mixture, increase cooling airflow.
Low cylinder head temperature	Excessive cowl flap opening	Adjust cowl flaps.
	Excessively rich mixture	Adjust mixture control.
	Extended glides without clearing engine	Clear engine long enough to keep temperatures at minimum range.
Ammeter indicating discharge	Alternator or generator failure	Shed unnecessary electrical load. Land as soon as practicable.
Load meter indicating zero	Same as above	Same as above
Surging rpm and overspeeding	Defective propeller	Adjust propeller rpm.
	Defective engine	Consult maintenance.
	Defective propeller governor	Adjust propeller control. Attempt to restore normal operation.
	Defective tachometer	Consult maintenance.
	Improper mixture setting	Readjust mixture for smooth operation.
Loss of airspeed in cruise flight with manifold pressure and rpm constant	Possible loss of one or more cylinders	Land as soon as possible.
Rough running engine	Improper mixture control setting	Adjust mixture for smooth operation
	Defective ignition or valves	Consult maintenance personnel.
	Detonation or preignition	Reduce power, enrich mixture, open cowl flaps to reduce cylinder head temp. Land as soon as practicable.
	Induction air leak	Reduce power. Consult maintenance.
	Plugged fuel nozzle (fuel injection)	Same as above.
	Excessive fuel pressure or fuel flow	Lean mixture control.
Loss of fuel pressure	Engine-driven pump failure No fuel	Turn on boost pumps. Switch tanks, turn on fuel.

Figure 17-11. *Commonly experienced in-flight abnormal engine instrument indications, their possible causes, and corrective actions.*

point where attention is focused on the open door rather than maintaining control of the airplane, loss of control may result even though disruption of airflow by the door is minimal.

In the event of an inadvertent door opening in flight or on takeoff, the pilot should adhere to the following.

- Concentrate on flying the airplane. Particularly in light single and twin-engine airplanes; a cabin door that opens in flight seldom if ever compromises the airplane's ability to fly. There may be some handling effects, such as roll and/or yaw, but in most instances these can be easily overcome.

- If the door opens after lift-off, do not rush to land. Climb to normal traffic pattern altitude, fly a normal traffic pattern, and make a normal landing.

- Do not release the seat belt and shoulder harness in an attempt to reach the door. Leave the door alone. Land as soon as practicable, and close the door once safely on the ground.

- Remember that most doors do not stay wide open. They usually bang open and then settle partly closed. A slip towards the door may cause it to open wider; a slip away from the door may push it closed.

- Do not panic. Try to ignore the unfamiliar noise and vibration. Also, do not rush. Attempting to get the airplane on the ground as quickly as possible may result in steep turns at low altitude.

- Complete all items on the landing checklist.

- Remember that accidents are almost never caused by an open door. Rather, an open door accident is caused by the pilot's distraction or failure to maintain control of the airplane.

Inadvertent VFR Flight Into IMC

It is beyond the scope of this handbook to incorporate a course of training in basic attitude instrument flying. This information is contained in FAA-H-8083-15, Instrument Flying Handbook. Certain pilot certificates and/or associated ratings require training in instrument flying and a demonstration of specific instrument flying tasks on the practical test.

Pilots and flight instructors should refer to FAA-H-8083-15 for guidance in the performance of these tasks and to the appropriate practical test standards (PTS) for information on the standards to which these required tasks must be performed for the particular certificate level and/or rating. The pilot should remember, however, that unless these tasks are practiced on a continuing and regular basis, skill erosion begins almost immediately. In a very short time, the pilot's assumed level of confidence is much higher than the performance he or she is actually able to demonstrate should the need arise.

Accident statistics show that the pilot who has not been trained in attitude instrument flying, or one whose instrument skills have eroded, lose control of the airplane in about 10 minutes once forced to rely solely on instrument reference. The purpose of this section is to provide guidance on practical emergency measures to maintain airplane control for a limited period of time in the event a VFR pilot encounters instrument meteorological conditions (IMC). The main goal is not precision instrument flying; rather, it is to help the VFR pilot keep the airplane under adequate control until suitable visual references are regained.

The first steps necessary for surviving an encounter with IMC by a VFR pilot are as follows:

- Recognition and acceptance of the seriousness of the situation and the need for immediate remedial action

- Maintaining control of the airplane

- Obtaining the appropriate assistance in getting the airplane safely on the ground

Recognition

A VFR pilot is in IMC conditions anytime he or she is unable to maintain airplane attitude control by reference to the natural horizon regardless of the circumstances or the prevailing weather conditions. Additionally, the VFR pilot is, in effect, in IMC anytime he or she is inadvertently or intentionally for an indeterminate period of time unable to navigate or establish geographical position by visual reference to landmarks on the surface. These situations must be accepted by the pilot involved as a genuine emergency requiring appropriate action.

The pilot must understand that unless he or she is trained, qualified, and current in the control of an airplane solely by reference to flight instruments, he or she is unable to do so for any length of time. Many hours of VFR flying using the attitude indicator as a reference for airplane control may lull a pilot into a false sense of security based on an overestimation of his or her personal ability to control the airplane solely by instrument reference. In VFR conditions, even though the pilot thinks he or she is controlling the airplane by instrument reference, the pilot also receives an overview of the natural horizon and may subconsciously rely on it more than the attitude indicator. If the natural horizon were to suddenly disappear, the untrained instrument pilot would be subject to vertigo, spatial disorientation, and inevitable control loss.

Maintaining Airplane Control

Once the pilot recognizes and accepts the situation, he or she must understand that the only way to control the airplane safely is by using and trusting the flight instruments. Attempts to control the airplane partially by reference to flight instruments while searching outside of the airplane for visual confirmation of the information provided by those instruments results in inadequate airplane control. This may be followed by spatial disorientation and complete control loss.

The most important point to be stressed is that the pilot must not panic. The task at hand may seem overwhelming, and the situation may be compounded by extreme apprehension. The pilot therefore must make a conscious effort to relax. The pilot must understand the most important concern—in fact the only concern at this point—is to keep the wings level. An uncontrolled turn or bank usually leads to difficulty in achieving the objectives of any desired flight condition. The pilot finds that good bank control has the effect of making pitch control much easier.

The pilot should remember that a person cannot feel control pressures with a tight grip on the controls. Relaxing and learning to "control with the eyes and the brain," instead of only the muscles usually takes considerable conscious effort.

The pilot must believe what the flight instruments show about the airplane's attitude regardless of what the natural senses tell. The vestibular sense (motion sensing by the inner ear) can and will confuse the pilot. Because of inertia, the sensory areas of the inner ear cannot detect slight changes in airplane attitude, nor can they accurately sense attitude changes that occur at a uniform rate over a period of time. On the other hand, false sensations are often generated, leading the pilot to believe the attitude of the airplane has changed when, in fact, it has not. These false sensations result in the pilot experiencing spatial disorientation.

Attitude Control

An airplane is, by design, an inherently stable platform and, except in turbulent air, maintains approximately straight-and-level flight if properly trimmed and left alone. It is designed to maintain a state of equilibrium in pitch, roll, and yaw. The pilot must be aware, however, that a change about one axis affects the stability of the others. The typical light airplane exhibits a good deal of stability in the yaw axis, slightly less in the pitch axis, and even lesser still in the roll axis. The key to emergency airplane attitude control, therefore, is to:

- Trim the airplane with the elevator trim so that it maintains hands-off level flight at cruise airspeed.

- Resist the tendency to overcontrol the airplane. Fly the attitude indicator with fingertip control. No attitude changes should be made unless the flight instruments indicate a definite need for a change.

- Make all attitude changes smooth and small, yet with positive pressure. Remember that a small change

Figure 17-12. *Attitude indicator.*

as indicated on the horizon bar corresponds to a proportionately much larger change in actual airplane attitude.

- Make use of any available aid in attitude control, such as autopilot or wing leveler.

The primary instrument for attitude control is the attitude indicator. *[Figure 17-12]* Once the airplane is trimmed so that it maintains hands-off level flight at cruise airspeed, that airspeed need not vary until the airplane must be slowed for landing. All turns, climbs, and descents can and should be made at this airspeed. Straight flight is maintained by keeping the wings level using "fingertip pressure" on the control wheel. Any pitch attitude change should be made by using no more than one bar width up or down.

Turns

Turns are perhaps the most potentially dangerous maneuver for the untrained instrument pilot for two reasons:

- The normal tendency of the pilot to overcontrol, leading to steep banks and the possibility of a "graveyard spiral."

- The inability of the pilot to cope with the instability resulting from the turn.

When a turn must be made, the pilot must anticipate and cope with the relative instability of the roll axis. The smallest practical bank angle should be used—in any case no more than 10° bank angle. *[Figure 17-13]* A shallow bank takes very little vertical lift from the wings resulting in little if any deviation in altitude. It may be helpful to turn a few degrees and then return to level flight if a large change in heading must be made. Repeat the process until the desired heading is reached. This process may relieve the progressive overbanking that often results from prolonged turns.

Figure 17-13. *Level turn.*

Climbs

If a climb is necessary, the pilot should raise the miniature airplane on the attitude indicator no more than one bar width and apply power. *[Figure 17-14]* The pilot should not attempt to attain a specific climb speed but accept whatever speed results. The objective is to deviate as little as possible from level flight attitude in order to disturb the airplane's equilibrium as little as possible. If the airplane is properly trimmed, it assumes a nose-up attitude on its own commensurate with the amount of power applied. Torque and P-factor cause the airplane to have a tendency to bank and turn to the left. This must be anticipated and compensated for. If the initial power application results in an inadequate rate of climb, power should be increased in increments of 100 rpm or 1 inch of manifold pressure until the desired rate of climb is attained. Maximum available power is seldom necessary. The more power that is used, the more the airplane wants to bank and turn to the left. Resuming level flight is accomplished by first decreasing pitch attitude to level on the attitude indicator using slow but deliberate pressure, allowing airspeed to increase to near cruise value, and then decreasing power.

Figure 17-14. *Level climb.*

Descents

Descents are very much the opposite of the climb procedure if the airplane is properly trimmed for hands-off straight-and-level flight. In this configuration, the airplane requires a certain amount of thrust to maintain altitude. The pitch attitude is controlling the airspeed. The engine power, therefore, (translated into thrust by the propeller) is maintaining the selected altitude. Following a power reduction, however slight, there is an almost imperceptible decrease in airspeed. However, even a slight change in speed results in less down load on the tail, whereupon the designed nose heaviness of the airplane causes it to pitch down just enough to maintain the airspeed for which it was trimmed. The airplane then descends at a rate directly proportionate to the amount of thrust that has been removed. Power reductions should be made in increments of 100 rpm or 1 inch of manifold pressure and the resulting rate of descent should never exceed 500 fpm. The wings should be held level on the attitude indicator, and the pitch attitude should not exceed one bar width below level. *[Figure 17-15]*

Combined Maneuvers

Combined maneuvers, such as climbing or descending turns, should be avoided if at all possible by an untrained instrument pilot already under the stress of an emergency situation. Combining maneuvers only compound the problems

Figure 17-15. *Level descent.*

encountered in individual maneuvers and increase the risk of control loss. Remember that the objective is to maintain airplane control by deviating as little as possible from straight-and-level flight attitude and thereby maintaining as much of the airplane's natural equilibrium as possible.

When being assisted by ATC, the pilot may detect a sense of urgency as he or she is being directed to change heading and/or altitude. This sense of urgency reflects a normal concern for safety on the part of the controller. But the pilot must not let this prompt him or her to attempt a maneuver that could result in loss of control.

Transition to Visual Flight

One of the most difficult tasks a trained and qualified instrument pilot must contend with is the transition from instrument to visual flight prior to landing. For the untrained instrument pilot, these difficulties are magnified.

The difficulties center around acclimatization and orientation. On an instrument approach, the trained instrument pilot must prepare in advance for the transition to visual flight. The pilot must have a mental picture of what he or she expects to see once the transition to visual flight is made and quickly acclimatize to the new environment. Geographical orientation must also begin before the transition, as the pilot must visualize where the airplane is in relation to the airport/runway when the transition occurs so that the approach and landing may be completed by visual reference to the ground.

In an ideal situation, the transition to visual flight is made with ample time, at a sufficient altitude above terrain, and to visibility conditions sufficient to accommodate acclimatization and geographical orientation. This, however, is not always the case. The untrained instrument pilot may find the visibility still limited, the terrain completely unfamiliar, and altitude above terrain such that a "normal" airport traffic pattern and landing approach is not possible. Additionally, the pilot is most likely under considerable self-induced psychological pressure to get the airplane on the ground. The pilot must take this into account and, if possible, allow time to become acclimatized and geographically oriented before attempting an approach and landing, even if it means flying straight and level for a time or circling the airport. This is especially true at night.

Chapter Summary

This chapter provided general guidance and recommended procedures that may apply to light single-engine airplanes involved in certain emergency situations. The information presented is intended to enhance the general knowledge of emergency operations with the clear understanding that the manufacturer's recommended emergency procedures take precedence. The chapter offers explanation concerning design structural load damage that may be imposed on the aircraft while performing emergency gear extension techniques. Rapid and abrupt pitch attitude changes executed at high forward airspeed may impose structural damage on the aircraft and flight controls. Normal category aircraft may not be designed to withstand abrupt pedal applications necessary to dislodge the landing gear.

Additional information is provided addressing failure of the pitot-static system in aircraft with EFIS. The redundancy of these systems as required by 14 CFR part 23 for IFR flight may be less than desired because both the primary and backup instrumentation may be receiving signal data input from the same pitot-static source. The failure indications of EFIS may be entirely different from conventional instruments making recognition of system malfunction much more difficult for the pilot. Lack of system standardization compounds the problem making equipment specific information and knowledge imperative to determine electronic display malfunctions. The inability to simulate certain failure modes during training and evaluation makes the pilot less prepared for an actual emergency. As electronic avionics become more technically advanced, the training and proficiency needed to safely operate these systems must keep pace.

Glossary

Numbers and Symbols

100-hour Inspection. An inspection, identical in scope to an annual inspection. Must be conducted every 100 hours of flight on aircraft of under 12,500 pounds that are used for hire.

A

Absolute altitude. The vertical distance of an airplane above the terrain, or above ground level (AGL).

Absolute ceiling. The altitude at which a climb is no longer possible.

Accelerate-go distance. The distance required to accelerate to V1 with all engines at takeoff power, experience an engine failure at V1 and continue the takeoff on the remaining engine(s). The runway required includes the distance required to climb to 35 feet by which time V2 speed must be attained.

Accelerate-stop distance. The distance required to accelerate to V1 with all engines at takeoff power, experience an engine failure at V1, and abort the takeoff and bring the airplane to a stop using braking action only (use of thrust reversing is not considered).

Acceleration. Force involved in overcoming inertia, and which may be defined as a change in velocity per unit of time.

Accessories. Components that are used with an engine, but are not a part of the engine itself. Units such as magnetos, carburetors, generators, and fuel pumps are commonly installed engine accessories.

Adjustable stabilizer. A stabilizer that can be adjusted in flight to trim the airplane, thereby allowing the airplane to fly hands-off at any given airspeed.

Adverse yaw. A condition of flight in which the nose of an airplane tends to yaw toward the outside of the turn. This is caused by the higher induced drag on the outside wing, which is also producing more lift. Induced drag is a by-product of the lift associated with the outside wing.

Aerodynamic ceiling. The point (altitude) at which, as the indicated airspeed decreases with altitude, it progressively merges with the low speed buffet boundary where prestall buffet occurs for the airplane at a load factor of 1.0 G.

Aerodynamics. The science of the action of air on an object, and with the motion of air on other gases. Aerodynamics deals with the production of lift by the aircraft, the relative wind, and the atmosphere.

Ailerons. Primary flight control surfaces mounted on the trailing edge of an airplane wing, near the tip. Ailerons control roll about the longitudinal axis.

Air start. The act or instance of starting an aircraft's engine while in flight, especially a jet engine after flameout.

Aircraft logbooks. Journals containing a record of total operating time, repairs, alterations or inspections performed, and all Airworthiness Directive (AD) notes complied with. A maintenance logbook should be kept for the airframe, each engine, and each propeller.

Airfoil. An airfoil is any surface, such as a wing, propeller, rudder, or even a trim tab, which provides aerodynamic force when it interacts with a moving stream of air.

Airmanship skills. The skills of coordination, timing, control touch, and speed sense in addition to the motor skills required to fly an aircraft.

Airmanship. A sound acquaintance with the principles of flight, the ability to operate an airplane with competence and precision both on the ground and in the air, and the exercise of sound judgment that results in optimal operational safety and efficiency.

Airplane Flight Manual (AFM). A document developed by the airplane manufacturer and approved by the Federal Aviation Administration (FAA). It is specific to a particular make and model airplane by serial number and it contains operating procedures and limitations.

Airplane Owner/Information Manual. A document developed by the airplane manufacturer containing general information about the make and model of an airplane. The airplane owner's manual is not FAA-approved and is not specific to a particular serial numbered airplane. This manual is not kept current, and therefore cannot be substituted for the AFM/POH.

Airport/Facility Directory. A publication designed primarily as a pilot's operational manual containing all airports, seaplane bases, and heliports open to the public including communications data, navigational facilities, and certain special notices and procedures. This publication is issued in seven volumes according to geographic area.

Airworthiness. A condition in which the aircraft conforms to its type certificated design including supplemental type certificates, and field approved alterations. The aircraft must also be in a condition for safe flight as determined by annual, 100 hour, preflight and any other required inspections.

Airworthiness Certificate. A certificate issued by the FAA to all aircraft that have been proven to meet the minimum standards set down by the Code of Federal Regulations.

Airworthiness Directive. A regulatory notice sent out by the FAA to the registered owner of an aircraft informing the owner of a condition that prevents the aircraft from continuing to meet its conditions for airworthiness. Airworthiness Directives (AD notes) must be complied with within the required time limit, and the fact of compliance, the date of compliance, and the method of compliance must be recorded in the aircraft's maintenance records.

Alpha mode of operation. The operation of a turboprop engine that includes all of the flight operations, from takeoff to landing. Alpha operation is typically between 95 percent to 100 percent of the engine operating speed.

Alternate air. A device which opens, either automatically or manually, to allow induction airflow to continue should the primary induction air opening become blocked.

Alternate static source. A manual port that when opened allows the pitot static instruments to sense static pressure from an alternate location should the primary static port become blocked.

Alternator/generator. A device that uses engine power to generate electrical power.

Altimeter. A flight instrument that indicates altitude by sensing pressure changes.

Altitude (AGL). The actual height above ground level (AGL) at which the aircraft is flying.

Altitude (MSL). The actual height above mean sea level (MSL) at which the aircraft is flying.

Altitude chamber. A device that simulates high altitude conditions by reducing the interior pressure. The occupants will suffer from the same physiological conditions as flight at high altitude in an unpressurized aircraft.

Altitude engine. A reciprocating aircraft engine having a rated takeoff power that is producible from sea level to an established higher altitude.

Angle of attack. The acute angle between the chord line of the airfoil and the direction of the relative wind.

Angle of incidence. The angle formed by the chord line of the wing and a line parallel to the longitudinal axis of the airplane.

Annual inspection. A complete inspection of an aircraft and engine, required by the Code of Federal Regulations, to be accomplished every 12 calendar months on all certificated aircraft. Only an A&P technician holding an Inspection Authorization can conduct an annual inspection.

Anti-icing. The prevention of the formation of ice on a surface. Ice may be prevented by using heat or by covering the surface with a chemical that prevents water from reaching the surface. Anti-icing should not be confused with deicing, which is the removal of ice after it has formed on the surface.

Attitude indicator. An instrument which uses an artificial horizon and miniature airplane to depict the position of the airplane in relation to the true horizon. The attitude indicator senses roll as well as pitch, which is the up and down movement of the airplane's nose.

Attitude. The position of an aircraft as determined by the relationship of its axes and a reference, usually the earth's horizon.

Autokinesis. This is caused by staring at a single point of light against a dark background for more than a few seconds. After a few moments, the light appears to move on its own.

Autopilot. An automatic flight control system which keeps an aircraft in level flight or on a set course. Automatic pilots can be directed by the pilot, or they may be coupled to a radio navigation signal.

Axes of an aircraft. Three imaginary lines that pass through an aircraft's center of gravity. The axes can be considered as imaginary axles around which the aircraft turns. The three axes pass through the center of gravity at 90° angles to each other. The axis from nose to tail is the longitudinal axis, the axis that passes from wingtip to wingtip is the lateral axis, and the axis that passes vertically through the center of gravity is the vertical axis.

Axial flow compressor. A type of compressor used in a turbine engine in which the airflow through the compressor is essentially linear. An axial-flow compressor is made up of several stages of alternate rotors and stators. The compressor ratio is determined by the decrease in area of the succeeding stages.

B

Back side of the power curve. Flight regime in which flight at a higher airspeed requires a lower power setting and a lower airspeed requires a higher power setting in order to maintain altitude.

Balked landing. A go-around.

Ballast. Removable or permanently installed weight in an aircraft used to bring the center of gravity into the allowable range.

Balloon. The result of a too aggressive flare during landing causing the aircraft to climb.

Basic empty weight (GAMA). Basic empty weight includes the standard empty weight plus optional and special equipment that has been installed.

Best angle of climb (V_X). The speed at which the aircraft will produce the most gain in altitude in a given distance.

Best glide. The airspeed in which the aircraft glides the furthest for the least altitude lost when in non-powered flight.

Best rate of climb (V_Y). The speed at which the aircraft will produce the most gain in altitude in the least amount of time.

Blade face. The flat portion of a propeller blade, resembling the bottom portion of an airfoil.

Bleed air. Compressed air tapped from the compressor stages of a turbine engine by use of ducts and tubing. Bleed air can be used for deice, anti-ice, cabin pressurization, heating, and cooling systems.

Bleed valve. In a turbine engine, a flapper valve, a popoff valve, or a bleed band designed to bleed off a portion of the compressor air to the atmosphere. Used to maintain blade angle of attack and provide stall-free engine acceleration and deceleration.

Boost pump. An electrically driven fuel pump, usually of the centrifugal type, located in one of the fuel tanks. It is used to provide fuel to the engine for starting and providing fuel pressure in the event of failure of the engine driven pump. It also pressurizes the fuel lines to prevent vapor lock.

Buffeting. The beating of an aerodynamic structure or surface by unsteady flow, gusts, etc.; the irregular shaking or oscillation of a vehicle component owing to turbulent air or separated flow.

Bus bar. An electrical power distribution point to which several circuits may be connected. It is often a solid metal strip having a number of terminals installed on it.

Bus tie. A switch that connects two or more bus bars. It is usually used when one generator fails and power is lost to its bus. By closing the switch, the operating generator powers both busses.

Bypass air. The part of a turbofan's induction air that bypasses the engine core.

Bypass ratio. The ratio of the mass airflow in pounds per second through the fan section of a turbofan engine to the mass airflow that passes through the gas generator portion of the engine. Or, the ratio between fan mass airflow (lb/sec.) and core engine mass airflow (lb/sec.).

C

Cabin pressurization. A condition where pressurized air is forced into the cabin simulating pressure conditions at a much lower altitude and increasing the aircraft occupants comfort.

Calibrated airspeed (CAS). Indicated airspeed corrected for installation error and instrument error. Although manufacturers attempt to keep airspeed errors to a minimum, it is not possible to eliminate all errors throughout the airspeed operating range. At certain airspeeds and with certain flap settings, the installation and instrument errors may total several knots. This error is generally greatest at low airspeeds. In the cruising and higher airspeed ranges, indicated airspeed and calibrated airspeed are approximately the same. Refer to the airspeed calibration chart to correct for possible airspeed errors.

Cambered. The camber of an airfoil is the characteristic curve of its upper and lower surfaces. The upper camber is more pronounced, while the lower camber is comparatively flat. This causes the velocity of the airflow immediately above the wing to be much higher than that below the wing.

Carburetor ice. Ice that forms inside the carburetor due to the temperature drop caused by the vaporization of the fuel. Induction system icing is an operational hazard because it can cut off the flow of the fuel/air charge or vary the fuel/air ratio.

Carburetor. 1. Pressure: A hydromechanical device employing a closed feed system from the fuel pump to the discharge nozzle. It meters fuel through fixed jets according to the mass airflow through the throttle body and discharges it under a positive pressure. Pressure carburetors are distinctly different from float-type carburetors, as they do not incorporate a vented float chamber or suction pickup from a discharge nozzle located in the venturi tube. 2. Float-type: Consists essentially of a main air passage through which the engine draws its supply of air, a mechanism to control the quantity of fuel discharged in relation to the flow of air, and a means of regulating the quantity of fuel/air mixture delivered to the engine cylinders.

Cascade reverser. A thrust reverser normally found on turbofan engines in which a blocker door and a series of cascade vanes are used to redirect exhaust gases in a forward direction.

Center of gravity (CG). The point at which an airplane would balance if it were possible to suspend it at that point. It is the mass center of the airplane, or the theoretical point at which the entire weight of the airplane is assumed to be concentrated. It may be expressed in inches from the reference datum, or in percent of mean aerodynamic chord (MAC). The location depends on the distribution of weight in the airplane.

Center-of-gravity limits. The specified forward and aft points within which the CG must be located during flight. These limits are indicated on pertinent airplane specifications.

Center-of-gravity range. The distance between the forward and aft CG limits indicated on pertinent airplane specifications.

Centrifugal flow compressor. An impeller-shaped device that receives air at its center and slings air outward at high velocity into a diffuser for increased pressure. Also referred to as a radial outflow compressor.

Chord line. An imaginary straight line drawn through an airfoil from the leading edge to the trailing edge.

Circuit breaker. A circuit-protecting device that opens the circuit in case of excess current flow. A circuit breakers differs from a fuse in that it can be reset without having to be replaced.

Clear air turbulence. Turbulence not associated with any visible moisture.

Climb gradient. The ratio between distance traveled and altitude gained.

Cockpit resource management. Techniques designed to reduce pilot errors and manage errors that do occur utilizing cockpit human resources. The assumption is that errors are going to happen in a complex system with error-prone humans.

Coefficient of lift. See lift coefficient.

Coffin corner. The flight regime where any increase in airspeed will induce high speed Mach buffet and any decrease in airspeed will induce low speed Mach buffet.

Combustion chamber. The section of the engine into which fuel is injected and burned.

Common traffic advisory frequency. The common frequency used by airport traffic to announce position reports in the vicinity of the airport.

Complex aircraft. An aircraft with retractable landing gear, flaps, and a controllable-pitch propeller, or is turbine powered.

Compression ratio. 1. In a reciprocating engine, the ratio of the volume of an engine cylinder with the piston at the bottom center to the volume with the piston at top center. 2. In a turbine engine, the ratio of the pressure of the air at the discharge to the pressure of air at the inlet.

Compressor bleed air. See bleed air.

Compressor bleed valves. See bleed valve.

Compressor section. The section of a turbine engine that increases the pressure and density of the air flowing through the engine.

Compressor stall. In gas turbine engines, a condition in an axial-flow compressor in which one or more stages of rotor blades fail to pass air smoothly to the succeeding stages. A stall condition is caused by a pressure ratio that is incompatible with the engine rpm Compressor stall will be indicated by a rise in exhaust temperature or rpm fluctuation, and if allowed to continue, may result in flameout and physical damage to the engine.

Compressor surge. A severe compressor stall across the entire compressor that can result in severe damage if not quickly corrected. This condition occurs with a complete stoppage of airflow or a reversal of airflow.

Condition lever. In a turbine engine, a powerplant control that controls the flow of fuel to the engine. The condition lever sets the desired engine rpm within a narrow range between that appropriate for ground and flight operations.

Configuration. This is a general term, which normally refers to the position of the landing gear and flaps.

Constant speed propeller. A controllable pitch propeller whose pitch is automatically varied in flight by a governor to maintain a constant rpm in spite of varying air loads.

Control touch. The ability to sense the action of the airplane and its probable actions in the immediate future, with regard to attitude and speed variations, by sensing and evaluation of varying pressures and resistance of the control surfaces transmitted through the cockpit flight controls.

Controllability. A measure of the response of an aircraft relative to the pilot's flight control inputs.

Controllable pitch propeller. A propeller in which the blade angle can be changed during flight by a control in the cockpit.

Conventional landing gear. Landing gear employing a third rear-mounted wheel. These airplanes are also sometimes referred to as tailwheel airplanes.

Coordinated flight. Application of all appropriate flight and power controls to prevent slipping or skidding in any flight condition.

Coordination. The ability to use the hands and feet together subconsciously and in the proper relationship to produce desired results in the airplane.

Core airflow. Air drawn into the engine for the gas generator.

Cowl flaps. Devices arranged around certain air-cooled engine cowlings which may be opened or closed to regulate the flow of air around the engine.

Crab. A flight condition in which the nose of the airplane is pointed into the wind a sufficient amount to counteract a crosswind and maintain a desired track over the ground.

Crazing. Small fractures in aircraft windshields and windows caused from being exposed to the ultraviolet rays of the sun and temperature extremes.

Critical altitude. The maximum altitude under standard atmospheric conditions at which a turbocharged engine can produce its rated horsepower.

Critical angle of attack. The angle of attack at which a wing stalls regardless of airspeed, flight attitude, or weight.

Critical engine. The engine whose failure has the most adverse effect on directional control.

Cross controlled. A condition where aileron deflection is in the opposite direction of rudder deflection.

Crossfeed. A system that allows either engine on a twin-engine airplane to draw fuel from any fuel tank.

Crosswind component. The wind component, measured in knots, at 90° to the longitudinal axis of the runway.

Current limiter. A device that limits the generator output to a level within that rated by the generator manufacturer.

D

Datum (reference datum). An imaginary vertical plane or line from which all measurements of moment arm are taken. The datum is established by the manufacturer. Once the datum has been selected, all moment arms and the location of CG range are measured from this point.

Decompression sickness. A condition where the low pressure at high altitudes allows bubbles of nitrogen to form in the blood and joints causing severe pain. Also known as the bends.

Deicer boots. Inflatable rubber boots attached to the leading edge of an airfoil. They can be sequentially inflated and deflated to break away ice that has formed over their surface.

Deicing. Removing ice after it has formed.

Delamination. The separation of layers.

Density altitude. This altitude is pressure altitude corrected for variations from standard temperature. When conditions are standard, pressure altitude and density altitude are the same. If the temperature is above standard, the density altitude is higher than pressure altitude. If the temperature is below standard, the density altitude is lower than pressure altitude. This is an important altitude because it is directly related to the airplane's performance.

Designated pilot examiner (DPE). An individual designated by the FAA to administer practical tests to pilot applicants.

Detonation. The sudden release of heat energy from fuel in an aircraft engine caused by the fuel-air mixture reaching its critical pressure and temperature. Detonation occurs as a violent explosion rather than a smooth burning process.

Dewpoint. The temperature at which air can hold no more water.

Differential ailerons. Control surface rigged such that the aileron moving up moves a greater distance than the aileron moving down. The up aileron produces extra parasite drag to compensate for the additional induced drag caused by the down aileron. This balancing of the drag forces helps minimize adverse yaw.

Diffusion. Reducing the velocity of air causing the pressure to increase.

Directional stability. Stability about the vertical axis of an aircraft, whereby an aircraft tends to return, on its own, to flight aligned with the relative wind when disturbed from that equilibrium state. The vertical tail is the primary contributor to directional stability, causing an airplane in flight to align with the relative wind.

Ditching. Emergency landing in water.

Downwash. Air deflected perpendicular to the motion of the airfoil.

Drag. An aerodynamic force on a body acting parallel and opposite to the relative wind. The resistance of the atmosphere to the relative motion of an aircraft. Drag opposes thrust and limits the speed of the airplane.

Drag curve. A visual representation of the amount of drag of an aircraft at various airspeeds.

Drift angle. Angle between heading and track.

Ducted-fan engine. An engine-propeller combination that has the propeller enclosed in a radial shroud. Enclosing the propeller improves the efficiency of the propeller.

Dutch roll. A combination of rolling and yawing oscillations that normally occurs when the dihedral effects of an aircraft are more powerful than the directional stability. Usually dynamically stable but objectionable in an airplane because of the oscillatory nature.

Dynamic hydroplaning. A condition that exists when landing on a surface with standing water deeper than the tread depth of the tires. When the brakes are applied, there is a possibility that the brake will lock up and the tire will ride on the surface of the water, much like a water ski. When the tires are hydroplaning, directional control and braking action are virtually impossible. An effective anti-skid system can minimize the effects of hydroplaning.

Dynamic stability. The property of an aircraft that causes it, when disturbed from straight-and-level flight, to develop forces or moments that restore the original condition of straight and level.

E

Electrical bus. See bus bar.

Electrohydraulic. Hydraulic control which is electrically actuated.

Elevator. The horizontal, movable primary control surface in the tail section, or empennage, of an airplane. The elevator is hinged to the trailing edge of the fixed horizontal stabilizer.

Emergency locator transmitter. A small, self-contained radio transmitter that will automatically, upon the impact of a crash, transmit an emergency signal on 121.5, 243.0, or 406.0 MHz.

Empennage. The section of the airplane that consists of the vertical stabilizer, the horizontal stabilizer, and the associated control surfaces.

Engine pressure ratio (EPR). The ratio of turbine discharge pressure divided by compressor inlet pressure that is used as an indication of the amount of thrust being developed by a turbine engine.

Environmental systems. In an aircraft, the systems, including the supplemental oxygen systems, air conditioning systems, heaters, and pressurization systems, which make it possible for an occupant to function at high altitude.

Equilibrium. A condition that exists within a body when the sum of the moments of all of the forces acting on the body is equal to zero. In aerodynamics, equilibrium is when all opposing forces acting on an aircraft are balanced (steady, unaccelerated flight conditions).

Equivalent shaft horsepower (ESHP). A measurement of the total horsepower of a turboprop engine, including that provided by jet thrust.

Exhaust gas temperature (EGT). The temperature of the exhaust gases as they leave the cylinders of a reciprocating engine or the turbine section of a turbine engine.

Exhaust manifold. The part of the engine that collects exhaust gases leaving the cylinders.

Exhaust. The rear opening of a turbine engine exhaust duct. The nozzle acts as an orifice, the size of which determines the density and velocity of the gases as they emerge from the engine.

F

False horizon. An optical illusion where the pilot confuses a row of lights along a road or other straight line as the horizon.

False start. See hung start.

Feathering propeller (feathered). A controllable pitch propeller with a pitch range sufficient to allow the blades to be turned parallel to the line of flight to reduce drag and prevent further damage to an engine that has been shut down after a malfunction.

Fixation. A psychological condition where the pilot fixes attention on a single source of information and ignores all other sources.

Fixed shaft turboprop engine. A turboprop engine where the gas producer spool is directly connected to the output shaft.

Fixed-pitch propellers. Propellers with fixed blade angles. Fixed-pitch propellers are designed as climb propellers, cruise propellers, or standard propellers.

Flaps. Hinged portion of the trailing edge between the ailerons and fuselage. In some aircraft, ailerons and flaps are interconnected to produce full-span "flaperons." In either case, flaps change the lift and drag on the wing.

Flat pitch. A propeller configuration when the blade chord is aligned with the direction of rotation.

Flicker vertigo. A disorienting condition caused from flickering light off the blades of the propeller.

Flight director. An automatic flight control system in which the commands needed to fly the airplane are electronically computed and displayed on a flight instrument. The commands are followed by the human pilot with manual control inputs or, in the case of an autopilot system, sent to servos that move the flight controls.

Flight idle. Engine speed, usually in the 70-80 percent range, for minimum flight thrust.

Floating. A condition when landing where the airplane does not settle to the runway due to excessive airspeed.

Force (F). The energy applied to an object that attempts to cause the object to change its direction, speed, or motion. In aerodynamics, it is expressed as F, T (thrust), L (lift), W (weight), or D (drag), usually in pounds.

Form drag. The part of parasite drag on a body resulting from the integrated effect of the static pressure acting normal to its surface resolved in the drag direction.

Forward slip. A slip in which the airplane's direction of motion continues the same as before the slip was begun. In a forward slip, the airplane's longitudinal axis is at an angle to its flightpath.

Free power turbine engine. A turboprop engine where the gas producer spool is on a separate shaft from the output shaft. The free power turbine spins independently of the gas producer and drives the output shaft.

Friction drag. The part of parasitic drag on a body resulting from viscous shearing stresses over its wetted surface.

Frise-type aileron. Aileron having the nose portion projecting ahead of the hinge line. When the trailing edge of the aileron moves up, the nose projects below the wing's lower surface and produces some parasite drag, decreasing the amount of adverse yaw.

Fuel control unit. The fuel-metering device used on a turbine engine that meters the proper quantity of fuel to be fed into the burners of the engine. It integrates the parameters of inlet air temperature, compressor speed, compressor discharge pressure, and exhaust gas temperature with the position of the cockpit power control lever.

Fuel efficiency. Defined as the amount of fuel used to produce a specific thrust or horsepower divided by the total potential power contained in the same amount of fuel.

Fuel heaters. A radiator-like device which has fuel passing through the core. A heat exchange occurs to keep the fuel temperature above the freezing point of water so that entrained water does not form ice crystals, which could block fuel flow.

Fuel injection. A fuel metering system used on some aircraft reciprocating engines in which a constant flow of fuel is fed to injection nozzles in the heads of all cylinders just outside of the intake valve. It differs from sequential fuel injection in which a timed charge of high-pressure fuel is sprayed directly into the combustion chamber of the cylinder.

Fuel load. The expendable part of the load of the airplane. It includes only usable fuel, not fuel required to fill the lines or that which remains trapped in the tank sumps.

Fuel tank sump. A sampling port in the lowest part of the fuel tank that the pilot can utilize to check for contaminants in the fuel.

Fuselage. The section of the airplane that consists of the cabin and/or cockpit, containing seats for the occupants and the controls for the airplane.

G

Gas generator. The basic power producing portion of a gas turbine engine and excluding such sections as the inlet duct, the fan section, free power turbines, and tailpipe. Each manufacturer designates what is included as the gas generator, but generally consists of the compressor, diffuser, combustor, and turbine.

Gas turbine engine. A form of heat engine in which burning fuel adds energy to compressed air and accelerates the air through the remainder of the engine. Some of the energy is extracted to turn the air compressor, and the remainder accelerates the air to produce thrust. Some of this energy can be converted into torque to drive a propeller or a system of rotors for a helicopter.

Glide ratio. The ratio between distance traveled and altitude lost during non-powered flight.

Glidepath. The path of an aircraft relative to the ground while approaching a landing.

Global position system (GPS). A satellite-based radio positioning, navigation, and time-transfer system.

Go-around. Terminating a landing approach.

Governing range. The range of pitch a propeller governor can control during flight.

Governor. A control which limits the maximum rotational speed of a device.

Gross weight. The total weight of a fully loaded aircraft including the fuel, oil, crew, passengers, and cargo.

Ground adjustable trim tab. A metal trim tab on a control surface that is not adjustable in flight. Bent in one direction or another while on the ground to apply trim forces to the control surface.

Ground effect. A condition of improved performance encountered when an airplane is operating very close to the ground. When an airplane's wing is under the influence of ground effect, there is a reduction in upwash, downwash, and wingtip vortices. As a result of the reduced wingtip vortices, induced drag is reduced.

Ground idle. Gas turbine engine speed usually 60-70 percent of the maximum rpm range, used as a minimum thrust setting for ground operations.

Ground loop. A sharp, uncontrolled change of direction of an airplane on the ground.

Ground power unit (GPU). A type of small gas turbine whose purpose is to provide electrical power, and/or air pressure for starting aircraft engines. Aground unit is connected to the aircraft when needed. Similar to an aircraft-installed auxiliary power unit.

Groundspeed (GS). The actual speed of the airplane over the ground. It is true airspeed adjusted for wind. Groundspeed decreases with a headwind, and increases with a tailwind.

Ground track. The aircraft's path over the ground when in flight.

Gust penetration speed. The speed that gives the greatest margin between the high and low Mach speed buffets.

Gyroscopic precession. An inherent quality of rotating bodies, which causes an applied force to be manifested 90° in the direction of rotation from the point where the force is applied.

H

Hand propping. Starting an engine by rotating the propeller by hand.

Heading. The direction in which the nose of the aircraft is pointing during flight.

Heading bug. A marker on the heading indicator that can be rotated to a specific heading for reference purposes, or to command an autopilot to fly that heading.

Heading indicator. An instrument which senses airplane movement and displays heading based on a 360° azimuth, with the final zero omitted. The heading indicator, also called a directional gyro, is fundamentally a mechanical instrument designed to facilitate the use of the magnetic compass. The heading indicator is not affected by the forces that make the magnetic compass difficult to interpret.

Headwind component. The component of atmospheric winds that acts opposite to the aircraft's flightpath.

High performance aircraft. An aircraft with an engine of more than 200 horsepower.

Horizon. The line of sight boundary between the earth and the sky.

Horsepower. The term, originated by inventor James Watt, means the amount of work a horse could do in one second. One horsepower equals 550 foot-pounds per second, or 33,000 foot-pounds per minute.

Hot start. In gas turbine engines, a start which occurs with normal engine rotation, but exhaust temperature exceeds prescribed limits. This is usually caused by an excessively rich mixture in the combustor. The fuel to the engine must be terminated immediately to prevent engine damage.

Hung start. In gas turbine engines, a condition of normal light off but with rpm remaining at some low value rather than increasing to the normal idle rpm This is often the result of insufficient power to the engine from the starter. In the event of a hung start, the engine should be shut down.

Hydraulics. The branch of science that deals with the transmission of power by incompressible fluids under pressure.

Hydroplaning. A condition that exists when landing on a surface with standing water deeper than the tread depth of the tires. When the brakes are applied, there is a possibility that the brake will lock up and the tire will ride on the surface of the water, much like a water ski. When the tires are hydroplaning, directional control and braking action are virtually impossible. An effective anti-skid system can minimize the effects of hydroplaning.

Hypoxia. A lack of sufficient oxygen reaching the body tissues.

I

Igniter plugs. The electrical device used to provide the spark for starting combustion in a turbine engine. Some igniters resemble spark plugs, while others, called glow plugs, have a coil of resistance wire that glows red hot when electrical current flows through the coil.

Impact ice. Ice that forms on the wings and control surfaces or on the carburetor heat valve, the walls of the air scoop, or the carburetor units during flight. Impact ice collecting on the metering elements of the carburetor may upset fuel metering or stop carburetor fuel flow.

Inclinometer. An instrument consisting of a curved glass tube, housing a glass ball, and damped with a fluid similar to kerosene. It may be used to indicate inclination, as a level, or, as used in the turn indicators, to show the relationship between gravity and centrifugal force in a turn.

Indicated airspeed (IAS). The direct instrument reading obtained from the airspeed indicator, uncorrected for variations in atmospheric density, installation error, or instrument error. Manufacturers use this airspeed as the basis for determining airplane performance. Takeoff, landing, and stall speeds listed in the AFM or POH are indicated airspeeds and do not normally vary with altitude or temperature.

Indicated altitude. The altitude read directly from the altimeter (uncorrected) when it is set to the current altimeter setting.

Induced drag. That part of total drag which is created by the production of lift. Induced drag increases with a decrease in airspeed.

Induction manifold. The part of the engine that distributes intake air to the cylinders.

Inertia. The opposition which a body offers to a change of motion.

Initial climb. This stage of the climb begins when the airplane leaves the ground, and a pitch attitude has been established to climb away from the takeoff area.

Instrument Flight Rules (IFR). Rules that govern the procedure for conducting flight in weather conditions below VFR weather minimums. The term "IFR" also is used to define weather conditions and the type of flight plan under which an aircraft is operating.

Integral fuel tank. A portion of the aircraft structure, usually a wing, which is sealed off and used as a fuel tank. When a wing is used as an integral fuel tank, it is called a "wet wing."

Intercooler. A device used to reduce the temperature of the compressed air before it enters the fuel metering device. The resulting cooler air has a higher density, which permits the engine to be operated with a higher power setting.

Internal combustion engines. An engine that produces power as a result of expanding hot gases from the combustion of fuel and air within the engine itself. A steam engine where coal is burned to heat up water inside the engine is an example of an external combustion engine.

International Standard Atmosphere (ISA). Standard atmospheric conditions consisting of a temperature of 59 °F (15 °C), and a barometric pressure of 29.92 "Hg. (1013.2 mb) at sea level. ISA values can be calculated for various altitudes using a standard lapse rate of approximately 2 °C per 1,000 feet.

Interstage turbine temperature (ITT). The temperature of the gases between the high pressure and low pressure turbines.

Inverter. An electrical device that changes DC to AC power.

J

Jet powered airplane. An aircraft powered by a turbojet or turbofan engine.

K

Kinesthesia. The sensing of movements by feel.

L

Lateral axis. An imaginary line passing through the center of gravity of an airplane and extending across the airplane from wingtip to wingtip.

Lateral stability (rolling). The stability about the longitudinal axis of an aircraft. Rolling stability or the ability of an airplane to return to level flight due to a disturbance that causes one of the wings to drop.

Lead-acid battery. A commonly used secondary cell having lead as its negative plate and lead peroxide as its positive plate. Sulfuric acid and water serve as the electrolyte.

Leading edge devices. High lift devices which are found on the leading edge of the airfoil. The most common types are fixed slots, movable slats, and leading edge flaps.

Leading edge. The part of an airfoil that meets the airflow first.

Leading edge flap. A portion of the leading edge of an airplane wing that folds downward to increase the camber, lift, and drag of the wing. The leading-edge flaps are extended for takeoffs and landings to increase the amount of aerodynamic lift that is produced at any given airspeed.

Licensed empty weight. The empty weight that consists of the airframe, engine(s), unusable fuel, and undrainable oil plus standard and optional equipment as specified in the equipment list. Some manufacturers used this term prior to GAMA standardization.

Lift. One of the four main forces acting on an aircraft. On a fixed-wing aircraft, an upward force created by the effect of airflow as it passes over and under the wing.

Lift coefficient. A coefficient representing the lift of a given airfoil. Lift coefficient is obtained by dividing the lift by the free-stream dynamic pressure and the representative area under consideration.

Lift/drag ratio (L/D). The efficiency of an airfoil section. It is the ratio of the coefficient of lift to the coefficient of drag for any given angle of attack.

Lift-off. The act of becoming airborne as a result of the wings lifting the airplane off the ground, or the pilot rotating the nose up, increasing the angle of attack to start a climb.

Limit load factor. Amount of stress, or load factor, that an aircraft can withstand before structural damage or failure occurs.

Load factor. The ratio of the load supported by the airplane's wings to the actual weight of the aircraft and its contents. Also referred to as G-loading.

Longitudinal axis. An imaginary line through an aircraft from nose to tail, passing through its center of gravity. The longitudinal axis is also called the roll axis of the aircraft. Movement of the ailerons rotates an airplane about its longitudinal axis.

Longitudinal stability (pitching). Stability about the lateral axis. A desirable characteristic of an airplane whereby it tends to return to its trimmed angle of attack after displacement.

M

Mach. Speed relative to the speed of sound. Mach 1 is the speed of sound.

Mach buffet. Airflow separation behind a shock-wave pressure barrier caused by airflow over flight surfaces exceeding the speed of sound.

Mach compensating device. A device to alert the pilot of inadvertent excursions beyond its certified maximum operating speed.

Mach critical. The Mach speed at which some portion of the airflow over the wing first equals Mach 1.0. This is also the speed at which a shock wave first appears on the airplane.

Mach tuck. A condition that can occur when operating a swept-wing airplane in the transonic speed range. A shock wave could form in the root portion of the wing and cause the air behind it to separate. This shock-induced separation causes the center of pressure to move aft. This, combined with the increasing amount of nose down force at higher speeds to maintain left flight, causes the nose to "tuck." If not corrected, the airplane could enter a steep, sometimes unrecoverable dive.

Magnetic compass. A device for determining direction measured from magnetic north.

Main gear. The wheels of an aircraft's landing gear that supports the major part of the aircraft's weight.

Maneuverability. Ability of an aircraft to change directions along a flightpath and withstand the stresses imposed upon it.

Maneuvering speed (V_A). The maximum speed where full, abrupt control movement can be used without overstressing the airframe.

Manifold pressure (MP). The absolute pressure of the fuel/air mixture within the intake manifold, usually indicated in inches of mercury.

Maximum allowable takeoff power. The maximum power an engine is allowed to develop for a limited period of time; usually about one minute.

Maximum landing weight. The greatest weight that an airplane normally is allowed to have at landing.

Maximum ramp weight. The total weight of a loaded aircraft, including all fuel. It is greater than the takeoff weight due to the fuel that will be burned during the taxi and runup operations. Ramp weight may also be referred to as taxi weight.

Maximum takeoff weight. The maximum allowable weight for takeoff.

Maximum weight. The maximum authorized weight of the aircraft and all of its equipment as specified in the Type Certificate Data Sheets (TCDS) for the aircraft.

Maximum zero fuel weight (GAMA). The maximum weight, exclusive of usable fuel.

Minimum controllable airspeed. An airspeed at which any further increase in angle of attack, increase in load factor, or reduction in power, would result in an immediate stall.

Minimum drag speed (L/D_{MAX}). The point on the total drag curve where the lift-to-drag ratio is the greatest. At this speed, total drag is minimized.

Mixture. The ratio of fuel to air entering the engine's cylinders.

M_{MO}. Maximum operating speed expressed in terms of a decimal of Mach speed.

Moment arm. The distance from a datum to the applied force.

Moment index (or index). A moment divided by a constant such as 100, 1,000, or 10,000. The purpose of using a moment index is to simplify weight and balance computations of airplanes where heavy items and long arms result in large, unmanageable numbers.

Moment. The product of the weight of an item multiplied by its arm. Moments are expressed in pound-inches (lb-in). Total moment is the weight of the airplane multiplied by the distance between the datum and the CG.

Movable slat. A movable auxiliary airfoil on the leading edge of a wing. It is closed in normal flight but extends at high angles of attack. This allows air to continue flowing over the top of the wing and delays airflow separation.

Mushing. A flight condition caused by slow speed where the control surfaces are marginally effective.

N

N_1, N_2, N_3. Spool speed expressed in percent rpm. N_1 on a turboprop is the gas producer speed. N_1 on a turbofan or turbojet engine is the fan speed or low pressure spool speed. N_2 is the high pressure spool speed on engine with 2 spools and medium pressure spool on engines with 3 spools with N_3 being the high pressure spool.

Nacelle. A streamlined enclosure on an aircraft in which an engine is mounted. On multiengine propeller-driven airplanes, the nacelle is normally mounted on the leading edge of the wing.

Negative static stability. The initial tendency of an aircraft to continue away from the original state of equilibrium after being disturbed.

Negative torque sensing (NTS). A system in a turboprop engine that prevents the engine from being driven by the propeller. The NTS increases the blade angle when the propellers try to drive the engine.

Neutral static stability. The initial tendency of an aircraft to remain in a new condition after its equilibrium has been disturbed.

Nickel-cadmium battery (NiCad). A battery made up of alkaline secondary cells. The positive plates are nickel hydroxide, the negative plates are cadmium hydroxide, and potassium hydroxide is used as the electrolyte.

Normal category. An airplane that has a seating configuration, excluding pilot seats, of nine or less, a maximum certificated takeoff weight of 12,500 pounds or less, and intended for nonacrobatic operation.

Normalizing (turbonormalizing). A turbocharger that maintains sea level pressure in the induction manifold at altitude.

O

Octane. The rating system of aviation gasoline with regard to its antidetonating qualities.

Overboost. A condition in which a reciprocating engine has exceeded the maximum manifold pressure allowed by the manufacturer. Can cause damage to engine components.

Overspeed. A condition in which an engine has produced more rpm than the manufacturer recommends, or a condition in which the actual engine speed is higher than the desired engine speed as set on the propeller control.

Overtemp. A condition in which a device has reached a temperature above that approved by the manufacturer or any exhaust temperature that exceeds the maximum allowable for a given operating condition or time limit. Can cause internal damage to an engine.

Overtorque. A condition in which an engine has produced more torque (power) than the manufacturer recommends, or a condition in a turboprop or turboshaft engine where the engine power has exceeded the maximum allowable for a given operating condition or time limit. Can cause internal damage to an engine.

P

Parasite drag. That part of total drag created by the design or shape of airplane parts. Parasite drag increases with an increase in airspeed.

Payload (GAMA). The weight of occupants, cargo, and baggage.

P-factor. A tendency for an aircraft to yaw to the left due to the descending propeller blade on the right producing more thrust than the ascending blade on the left. This occurs when the aircraft's longitudinal axis is in a climbing attitude in relation to the relative wind. The P-factor would be to the right if the aircraft had a counterclockwise rotating propeller.

Pilot's Operating Handbook (POH). A document developed by the airplane manufacturer and contains the FAA approved Airplane Flight Manual (AFM) information.

Piston engine. A reciprocating engine.

Pitch. The rotation of an airplane about its lateral axis, or on a propeller, the blade angle as measured from plane of rotation.

Pivotal altitude. A specific altitude at which, when an airplane turns at a given groundspeed, a projecting of the sighting reference line to a selected point on the ground will appear to pivot on that point.

Pneumatic systems. The power system in an aircraft used for operating such items as landing gear, brakes, and wing flaps with compressed air as the operating fluid.

Porpoising. Oscillating around the lateral axis of the aircraft during landing.

Position lights. Lights on an aircraft consisting of a red light on the left wing, a green light on the right wing, and a white light on the tail. CFRs require that these lights be displayed in flight from sunset to sunrise.

Positive static stability. The initial tendency to return to a state of equilibrium when disturbed from that state.

Power distribution bus. See bus bar.

Power lever. The cockpit lever connected to the fuel control unit for scheduling fuel flow to the combustion chambers of a turbine engine.

Power. Implies work rate or units of work per unit of time, and as such, it is a function of the speed at which the force is developed. The term "power required" is generally associated with reciprocating engines.

Powerplant. A complete engine and propeller combination with accessories.

Practical slip limit. The maximum slip an aircraft is capable of performing due to rudder travel limits.

Precession. The tilting or turning of a gyro in response to deflective forces causing slow drifting and erroneous indications in gyroscopic instruments.

Preignition. Ignition occurring in the cylinder before the time of normal ignition. Preignition is often caused by a local hot spot in the combustion chamber igniting the fuel/air mixture.

Pressure altitude. The altitude indicated when the altimeter setting window (barometric scale) is adjusted to 29.92. This is the altitude above the standard datum plane, which is a theoretical plane where air pressure (corrected to 15 °C) equals 29.92 "Hg. Pressure altitude is used to compute density altitude, true altitude, true airspeed, and other performance data.

Profile drag. The total of the skin friction drag and form drag for a two-dimensional airfoil section.

Propeller blade angle. The angle between the propeller chord and the propeller plane of rotation.

Propeller lever. The control on a free power turbine turboprop that controls propeller speed and the selection for propeller feathering.

Propeller slipstream. The volume of air accelerated behind a propeller producing thrust.

Propeller synchronization. A condition in which all of the propellers have their pitch automatically adjusted to maintain a constant rpm among all of the engines of a multiengine aircraft.

Propeller. A device for propelling an aircraft that, when rotated, produces by its action on the air, a thrust approximately perpendicular to its plane of rotation. It includes the control components normally supplied by its manufacturer.

R

Ramp weight. The total weight of the aircraft while on the ramp. It differs from takeoff weight by the weight of the fuel that will be consumed in taxiing to the point of takeoff.

Rate of turn. The rate in degrees/second of a turn.

Reciprocating engine. An engine that converts the heat energy from burning fuel into the reciprocating movement of the pistons. This movement is converted into a rotary motion by the connecting rods and crankshaft.

Reduction gear. The gear arrangement in an aircraft engine that allows the engine to turn at a faster speed than the propeller.

Region of reverse command. Flight regime in which flight at a higher airspeed requires a lower power setting and a lower airspeed requires a higher power setting in order to maintain altitude.

Registration certificate. A State and Federal certificate that documents aircraft ownership.

Relative wind. The direction of the airflow with respect to the wing. If a wing moves forward horizontally, the relative wind moves backward horizontally. Relative wind is parallel to and opposite the flightpath of the airplane.

Reverse thrust. A condition where jet thrust is directed forward during landing to increase the rate of deceleration.

Reversing propeller. A propeller system with a pitch change mechanism that includes full reversing capability. When the pilot moves the throttle controls to reverse, the blade angle changes to a pitch angle and produces a reverse thrust, which slows the airplane down during a landing.

Roll. The motion of the aircraft about the longitudinal axis. It is controlled by the ailerons.

Roundout (flare). A pitch-up during landing approach to reduce rate of descent and forward speed prior to touchdown.

Rudder. The movable primary control surface mounted on the trailing edge of the vertical fin of an airplane. Movement of the rudder rotates the airplane about its vertical axis.

Ruddervator. A pair of control surfaces on the tail of an aircraft arranged in the form of a V. These surfaces, when moved together by the control wheel, serve as elevators, and when moved differentially by the rudder pedals, serve as a rudder.

Runway centerline lights. Runway centerline lights are installed on some precision approach runways to facilitate landing under adverse visibility conditions. They are located along the runway centerline and are spaced at 50-foot intervals. When viewed from the landing threshold, the runway centerline lights are white until the last 3,000 feet of the runway. The white lights begin to alternate with red for the next 2,000 feet, and for the last 1,000 feet of the runway, all centerline lights are red.

Runway centerline markings. The runway centerline identifies the center of the runway and provides alignment guidance during takeoff and landings. The centerline consists of a line of uniformly spaced stripes and gaps.

Runway edge lights. Runway edge lights are used to outline the edges of runways during periods of darkness or restricted visibility conditions. These light systems are classified according to the intensity or brightness they are capable of producing: they are the High Intensity Runway Lights (HIRL), Medium Intensity Runway Lights (MIRL), and the Low Intensity Runway Lights (LIRL). The HIRL and MIRL systems have variable intensity controls, whereas the LIRLs normally have one intensity setting.

Runway end identifier lights (REIL). One component of the runway lighting system. These lights are installed at many airfields to provide rapid and positive identification of the approach end of a particular runway.

Runway incursion. Any occurrence at an airport involving an aircraft, vehicle, person, or object on the ground that creates a collision hazard or results in loss of separation with an aircraft taking off, intending to takeoff, landing, or intending to land.

Runway threshold markings. Runway threshold markings come in two configurations. They either consist of eight longitudinal stripes of uniform dimensions disposed symmetrically about the runway centerline, or the number of stripes is related to the runway width. A threshold marking helps identify the beginning of the runway that is available for landing. In some instances, the landing threshold may be displaced.

S

Safety (SQUAT) switch. An electrical switch mounted on one of the landing gear struts. It is used to sense when the weight of the aircraft is on the wheels.

Scan. A procedure used by the pilot to visually identify all resources of information in flight.

Sea level. A reference height used to determine standard atmospheric conditions and altitude measurements.

Segmented circle. A visual ground based structure to provide traffic pattern information.

Service ceiling. The maximum density altitude where the best rate-of-climb airspeed will produce a 100 feet-per-minute climb at maximum weight while in a clean configuration with maximum continuous power.

Servo tab. An auxiliary control mounted on a primary control surface, which automatically moves in the direction opposite the primary control to provide an aerodynamic assist in the movement of the control.

Shaft horse power (SHP). Turboshaft engines are rated in shaft horsepower and calculated by use of a dynamometer device. Shaft horsepower is exhaust thrust converted to a rotating shaft.

Shock waves. A compression wave formed when a body moves through the air at a speed greater than the speed of sound.

Sideslip. A slip in which the airplane's longitudinal axis remains parallel to the original flightpath, but the airplane no longer flies straight ahead. Instead, the horizontal component of wing lift forces the airplane to move sideways toward the low wing.

Single engine absolute ceiling. The altitude that a twin engine airplane can no longer climb with one engine inoperative.

Single engine service ceiling. The altitude that a twin engine airplane can no longer climb at a rate greater then 50 fpm with one engine inoperative.

Skid. A condition where the tail of the airplane follows a path outside the path of the nose during a turn.

Slip. An intentional maneuver to decrease airspeed or increase rate of descent, and to compensate for a crosswind on landing. A slip can also be unintentional when the pilot fails to maintain the aircraft in coordinated flight.

Specific fuel consumption. Number of pounds of fuel consumed in 1 hour to produce 1 HP.

Speed. The distance traveled in a given time.

Speed brakes. A control system that extends from the airplane structure into the airstream to produce drag and slow the airplane.

Speed instability. A condition in the region of reverse command where a disturbance that causes the airspeed to decrease causes total drag to increase, which in turn, causes the airspeed to decrease further.

Speed sense. The ability to sense instantly and react to any reasonable variation of airspeed.

Spin. An aggravated stall that results in what is termed an "autorotation" wherein the airplane follows a downward corkscrew path. As the airplane rotates around the vertical axis, the rising wing is less stalled than the descending wing creating a rolling, yawing, and pitching motion.

Spiral instability. A condition that exists when the static directional stability of the airplane is very strong as compared to the effect of its dihedral in maintaining lateral equilibrium.

Spiraling slipstream. The slipstream of a propeller-driven airplane rotates around the airplane. This slipstream strikes the left side of the vertical fin, causing the airplane to yaw slightly. Vertical stabilizer offset is sometimes used by aircraft designers to counteract this tendency.

Split shaft turbine engine. See free power turbine engine.

Spoilers. High-drag devices that can be raised into the air flowing over an airfoil, reducing lift and increasing drag. Spoilers are used for roll control on some aircraft. Deploying spoilers on both wings at the same time allows the aircraft to descend without gaining speed. Spoilers are also used to shorten the ground roll after landing.

Spool. A shaft in a turbine engine which drives one or more compressors with the power derived from one or more turbines.

Stabilator. A single-piece horizontal tail surface on an airplane that pivots around a central hinge point. A stabilator serves the purposes of both the horizontal stabilizer and the elevator.

Stability. The inherent quality of an airplane to correct for conditions that may disturb its equilibrium, and to return or to continue on the original flightpath. It is primarily an airplane design characteristic.

Stabilized approach. A landing approach in which the pilot establishes and maintains a constant angle glidepath towards a predetermined point on the landing runway. It is based on the pilot's judgment of certain visual cues, and depends on the maintenance of a constant final descent airspeed and configuration.

Stall. A rapid decrease in lift caused by the separation of airflow from the wing's surface brought on by exceeding the critical angle of attack. A stall can occur at any pitch attitude or airspeed.

Stall strips. A spoiler attached to the inboard leading edge of some wings to cause the center section of the wing to stall before the tips. This assures lateral control throughout the stall.

Standard atmosphere. At sea level, the standard atmosphere consists of a barometric pressure of 29.92 inches of mercury ("Hg) or 1013.2 millibars, and a temperature of 15 °C (59 °F). Pressure and temperature normally decrease as altitude increases. The standard lapse rate in the lower atmosphere for each 1,000 feet of altitude is approximately 1 "Hg and 2 °C (3.5 °F). For example, the standard pressure and temperature at 3,000 feet mean sea level (MSL) is 26.92 "Hg (29.92 – 3) and 9 °C (15 °C – 6 °C).

Standard day. See standard atmosphere.

Standard empty weight (GAMA). This weight consists of the airframe, engines, and all items of operating equipment that have fixed locations and are permanently installed in the airplane; including fixed ballast, hydraulic fluid, unusable fuel, and full engine oil.

Standard weights. These have been established for numerous items involved in weight and balance computations. These weights should not be used if actual weights are available.

Standard-rate turn. A turn at the rate of 3° per second which enables the airplane to complete a 360° turn in 2 minutes.

Starter/generator. A combined unit used on turbine engines. The device acts as a starter for rotating the engine, and after running, internal circuits are shifted to convert the device into a generator.

Static stability. The initial tendency an aircraft displays when disturbed from a state of equilibrium.

Station. A location in the airplane that is identified by a number designating its distance in inches from the datum. The datum is, therefore, identified as station zero. An item located at station +50 would have an arm of 50 inches.

Stick puller. A device that applies aft pressure on the control column when the airplane is approaching the maximum operating speed.

Stick pusher. A device that applies an abrupt and large forward force on the control column when the airplane is nearing an angle of attack where a stall could occur.

Stick shaker. An artificial stall warning device that vibrates the control column.

Stress risers. A scratch, groove, rivet hole, forging defect or other structural discontinuity that causes a concentration of stress.

Subsonic. Speed below the speed of sound.

Supercharger. An engine- or exhaust-driven air compressor used to provide additional pressure to the induction air so the engine can produce additional power.

Supersonic. Speed above the speed of sound.

Supplemental Type Certificate (STC). A certificate authorizing an alteration to an airframe, engine, or component that has been granted an Approved Type Certificate.

Swept wing. A wing planform in which the tips of the wing are farther back than the wing root.

T

Tailwheel aircraft. See conventional landing gear.

Takeoff roll (ground roll). The total distance required for an aircraft to become airborne.

Target reverser. A thrust reverser in a jet engine in which clamshell doors swivel from the stowed position at the engine tailpipe to block all of the outflow and redirect some component of the thrust forward.

Taxiway lights. Omnidirectional lights that outline the edges of the taxiway and are blue in color.

Taxiway turnoff lights. Flush lights which emit a steady green color.

Tetrahedron. A large, triangular-shaped, kite-like object installed near the runway. Tetrahedrons are mounted on a pivot and are free to swing with the wind to show the pilot the direction of the wind as an aid in takeoffs and landings.

Throttle. The valve in a carburetor or fuel control unit that determines the amount of fuel-air mixture that is fed to the engine.

Thrust line. An imaginary line passing through the center of the propeller hub, perpendicular to the plane of the propeller rotation.

Thrust reversers. Devices which redirect the flow of jet exhaust to reverse the direction of thrust.

Thrust. The force which imparts a change in the velocity of a mass. This force is measured in pounds but has no element of time or rate. The term, thrust required, is generally associated with jet engines. A forward force which propels the airplane through the air.

Timing. The application of muscular coordination at the proper instant to make flight, and all maneuvers incident thereto, a constant smooth process.

Tire cord. Woven metal wire laminated into the tire to provide extra strength. A tire showing any cord must be replaced prior to any further flight.

Torque meter. An indicator used on some large reciprocating engines or on turboprop engines to indicate the amount of torque the engine is producing.

Torque sensor. See torque meter.

Torque. 1. A resistance to turning or twisting. 2. Forces that produce a twisting or rotating motion. 3. In an airplane, the tendency of the aircraft to turn (roll) in the opposite direction of rotation of the engine and propeller.

Total drag. The sum of the parasite and induced drag.

Touchdown zone lights. Two rows of transverse light bars disposed symmetrically about the runway centerline in the runway touchdown zone.

Track. The actual path made over the ground in flight.

Trailing edge. The portion of the airfoil where the airflow over the upper surface rejoins the lower surface airflow.

Transition liner. The portion of the combustor that directs the gases into the turbine plenum.

Transonic. At the speed of sound.

Transponder. The airborne portion of the secondary surveillance radar system. The transponder emits a reply when queried by a radar facility.

Tricycle gear. Landing gear employing a third wheel located on the nose of the aircraft.

Trim tab. A small auxiliary hinged portion of a movable control surface that can be adjusted during flight to a position resulting in a balance of control forces.

Triple spool engine. Usually a turbofan engine design where the fan is the N_1 compressor, followed by the N_2 intermediate compressor, and the N_3 high pressure compressor, all of which rotate on separate shafts at different speeds.

Tropopause. The boundary layer between the troposphere and the mesosphere which acts as a lid to confine most of the water vapor, and the associated weather, to the troposphere.

Troposphere. The layer of the atmosphere extending from the surface to a height of 20,000 to 60,000 feet depending on latitude.

True airspeed (TAS). Calibrated airspeed corrected for altitude and nonstandard temperature. Because air density decreases with an increase in altitude, an airplane has to be flown faster at higher altitudes to cause the same pressure difference between pitot impact pressure and static pressure. Therefore, for a given calibrated airspeed, true airspeed increases as altitude increases; or for a given true airspeed, calibrated airspeed decreases as altitude increases.

True altitude. The vertical distance of the airplane above sea level—the actual altitude. It is often expressed as feet above mean sea level (MSL). Airport, terrain, and obstacle elevations on aeronautical charts are true altitudes.

T-tail. An aircraft with the horizontal stabilizer mounted on the top of the vertical stabilizer, forming a T.

Turbine blades. The portion of the turbine assembly that absorbs the energy of the expanding gases and converts it into rotational energy.

Turbine outlet temperature (TOT). The temperature of the gases as they exit the turbine section.

Turbine plenum. The portion of the combustor where the gases are collected to be evenly distributed to the turbine blades.

Turbine rotors. The portion of the turbine assembly that mounts to the shaft and holds the turbine blades in place.

Turbine section. The section of the engine that converts high pressure high temperature gas into rotational energy.

Turbocharger. An air compressor driven by exhaust gases, which increases the pressure of the air going into the engine through the carburetor or fuel injection system.

Turbofan engine. A turbojet engine in which additional propulsive thrust is gained by extending a portion of the compressor or turbine blades outside the inner engine case. The extended blades propel bypass air along the engine axis but between the inner and outer casing. The air is not combusted but does provide additional thrust.

Turbojet engine. A jet engine incorporating a turbine-driven air compressor to take in and compress air for the combustion of fuel, the gases of combustion being used both to rotate the turbine and create a thrust producing jet.

Turboprop engine. A turbine engine that drives a propeller through a reduction gearing arrangement. Most of the energy in the exhaust gases is converted into torque, rather than its acceleration being used to propel the aircraft.

Turbulence. An occurrence in which a flow of fluid is unsteady.

Turn coordinator. A rate gyro that senses both roll and yaw due to the gimbal being canted. Has largely replaced the turn-and-slip indicator in modern aircraft.

Turn-and-slip indicator. A flight instrument consisting of a rate gyro to indicate the rate of yaw and a curved glass inclinometer to indicate the relationship between gravity and centrifugal force. The turn-and-slip indicator indicates the relationship between angle of bank and rate of yaw. Also called a turn-and-bank indicator.

Turning error. One of the errors inherent in a magnetic compass caused by the dip compensating weight. It shows up only on turns to or from northerly headings in the Northern Hemisphere and southerly headings in the Southern Hemisphere. Turning error causes the compass to lead turns to the north or south and lag turns away from the north or south.

U

Ultimate load factor. In stress analysis, the load that causes physical breakdown in an aircraft or aircraft component during a strength test, or the load that according to computations, should cause such a breakdown.

Unfeathering accumulator. Tanks that hold oil under pressure which can be used to unfeather a propeller.

UNICOM. A nongovernment air/ground radio communication station which may provide airport information at public use airports where there is no tower or FSS.

Unusable fuel. Fuel that cannot be consumed by the engine. This fuel is considered part of the empty weight of the aircraft.

Useful load. The weight of the pilot, copilot, passengers, baggage, usable fuel, and drainable oil. It is the basic empty weight subtracted from the maximum allowable gross weight. This term applies to general aviation aircraft only.

Utility category. An airplane that has a seating configuration, excluding pilot seats, of nine or less, a maximum certificated takeoff weight of 12,500 pounds or less, and intended for limited acrobatic operation.

V

V-bars. The flight director displays on the attitude indicator that provide control guidance to the pilot.

V-speeds. Designated speeds for a specific flight condition.

Vapor lock. A condition in which air enters the fuel system and it may be difficult, or impossible, to restart the engine. Vapor lock may occur as a result of running a fuel tank completely dry, allowing air to enter the fuel system. On fuel-injected engines, the fuel may become so hot it vaporizes in the fuel line, not allowing fuel to reach the cylinders.

V_A. The design maneuvering speed. This is the "rough air" speed and the maximum speed for abrupt maneuvers. If during flight, rough air or severe turbulence is encountered, reduce the airspeed to maneuvering speed or less to minimize stress on the airplane structure. It is important to consider weight when referencing this speed. For example, VA may be 100 knots when an airplane is heavily loaded, but only 90 knots when the load is light.

Vector. A force vector is a graphic representation of a force and shows both the magnitude and direction of the force.

Velocity. The speed or rate of movement in a certain direction.

Vertical axis. An imaginary line passing vertically through the center of gravity of an aircraft. The vertical axis is called the z-axis or the yaw axis.

Vertical card compass. A magnetic compass that consists of an azimuth on a vertical card, resembling a heading indicator with a fixed miniature airplane to accurately present the heading of the aircraft. The design uses eddy current damping to minimize lead and lag during turns.

Vertical speed indicator (VSI). An instrument that uses static pressure to display a rate of climb or descent in feet per minute. The VSI can also sometimes be called a vertical velocity indicator (VVI).

Vertical stability. Stability about an aircraft's vertical axis. Also called yawing or directional stability.

V_{FE}. The maximum speed with the flaps extended. The upper limit of the white arc.

V_{FO}. The maximum speed that the flaps can be extended or retracted.

VFR Terminal Area Charts (1:250,000). Depict Class B airspace which provides for the control or segregation of all the aircraft within the Class B airspace. The chart depicts topographic information and aeronautical information which includes visual and radio aids to navigation, airports, controlled airspace, restricted areas, obstructions, and related data.

V-G diagram. A chart that relates velocity to load factor. It is valid only for a specific weight, configuration, and altitude and shows the maximum amount of positive or negative lift the airplane is capable of generating at a given speed. Also shows the safe load factor limits and the load factor that the aircraft can sustain at various speeds.

Visual approach slope indicator (VASI). The most common visual glidepath system in use. The VASI provides obstruction clearance within 10° of the extended runway centerline, and to 4 nautical miles (NM) from the runway threshold.

Visual Flight Rules (VFR). Code of Federal Regulations that govern the procedures for conducting flight under visual conditions.

V_{LE}. Landing gear extended speed. The maximum speed at which an airplane can be safely flown with the landing gear extended.

V_{LOF}. Lift-off speed. The speed at which the aircraft departs the runway during takeoff.

V_{LO}. Landing gear operating speed. The maximum speed for extending or retracting the landing gear if using an airplane equipped with retractable landing gear.

V_{MC}. Minimum control airspeed. This is the minimum flight speed at which a twin-engine airplane can be satisfactorily controlled when an engine suddenly becomes inoperative and the remaining engine is at takeoff power.

V_{MD}. Minimum drag speed.

V_{MO}. Maximum operating speed expressed in knots.

V_{NE}. Never-exceed speed. Operating above this speed is prohibited since it may result in damage or structural failure. The red line on the airspeed indicator.

V_{NO}. Maximum structural cruising speed. Do not exceed this speed except in smooth air. The upper limit of the green arc.

V_P. Minimum dynamic hydroplaning speed. The minimum speed required to start dynamic hydroplaning.

V_R. Rotation speed. The speed that the pilot begins rotating the aircraft prior to lift-off.

V_{S0}. Stalling speed or the minimum steady flight speed in the landing configuration. In small airplanes, this is the power-off stall speed at the maximum landing weight in the landing configuration (gear and flaps down). The lower limit of the white arc.

V_{S1}. Stalling speed or the minimum steady flight speed obtained in a specified configuration. For most airplanes, this is the power-off stall speed at the maximum takeoff weight in the clean configuration (gear up, if retractable, and flaps up). The lower limit of the green arc.

V_{SSE}. Safe, intentional one-engine inoperative speed. The minimum speed to intentionally render the critical engine inoperative.

V-tail. A design which utilizes two slanted tail surfaces to perform the same functions as the surfaces of a conventional elevator and rudder configuration. The fixed surfaces act as both horizontal and vertical stabilizers.

V_X. Best angle-of-climb speed. The airspeed at which an airplane gains the greatest amount of altitude in a given distance. It is used during a short-field takeoff to clear an obstacle.

V_{XSE}. Best angle of climb speed with one engine inoperative. The airspeed at which an airplane gains the greatest amount of altitude in a given distance in a light, twin-engine airplane following an engine failure.

V_Y. Best rate-of-climb speed. This airspeed provides the most altitude gain in a given period of time.

V$_{YSE}$. Best rate-of-climb speed with one engine inoperative. This airspeed provides the most altitude gain in a given period of time in a light, twin engine airplane following an engine failure.

W

Wake turbulence. Wingtip vortices that are created when an airplane generates lift. When an airplane generates lift, air spills over the wingtips from the high pressure areas below the wings to the low pressure areas above them. This flow causes rapidly rotating whirlpools of air called wingtip vortices or wake turbulence.

Waste gate. A controllable valve in the tailpipe of an aircraft reciprocating engine equipped with a turbocharger. The valve is controlled to vary the amount of exhaust gases forced through the turbocharger turbine.

Weathervane. The tendency of the aircraft to turn into the relative wind.

Weight. A measure of the heaviness of an object. The force by which a body is attracted toward the center of the Earth (or another celestial body) by gravity. Weight is equal to the mass of the body times the local value of gravitational acceleration. One of the four main forces acting on an aircraft. Equivalent to the actual weight of the aircraft. It acts downward through the aircraft's center of gravity toward the center of the Earth. Weight opposes lift.

Weight and balance. The aircraft is said to be in weight and balance when the gross weight of the aircraft is under the max gross weight, and the center of gravity is within limits and will remain in limits for the duration of the flight.

Wheelbarrowing. A condition caused when forward yoke or stick pressure during takeoff or landing causes the aircraft to ride on the nosewheel alone.

Wind correction angle. Correction applied to the course to establish a heading so that track will coincide with course.

Wind direction indicators. Indicators that include a wind sock, wind tee, or tetrahedron. Visual reference will determine wind direction and runway in use.

Wind shear. A sudden, drastic shift in windspeed, direction, or both that may occur in the horizontal or vertical plane.

Windmilling. When the air moving through a propeller creates the rotational energy.

Windsock. A truncated cloth cone open at both ends and mounted on a freewheeling pivot that indicates the direction from which the wind is blowing.

Wing. Airfoil attached to each side of the fuselage and are the main lifting surfaces that support the airplane in flight.

Wing area. The total surface of the wing (square feet), which includes control surfaces and may include wing area covered by the fuselage (main body of the airplane), and engine nacelles.

Wing span. The maximum distance from wingtip to wingtip.

Wingtip vortices. The rapidly rotating air that spills over an airplane's wings during flight. The intensity of the turbulence depends on the airplane's weight, speed, and configuration. It is also referred to as wake turbulence. Vortices from heavy aircraft may be extremely hazardous to small aircraft.

Wing twist. A design feature incorporated into some wings to improve aileron control effectiveness at high angles of attack during an approach to a stall.

Y

Yaw. Rotation about the vertical axis of an aircraft.

Yaw string. A string on the nose or windshield of an aircraft in view of the pilot that indicates any slipping or skidding of the aircraft.

Z

Zero fuel weight. The weight of the aircraft to include all useful load except fuel.

Zero sideslip. A maneuver in a twin-engine airplane with one engine inoperative that involves a small amount of bank and slightly uncoordinated flight to align the fuselage with the direction of travel and minimize drag.

Zero thrust (simulated feather). An engine configuration with a low power setting that simulates a propeller feathered condition.

Index

A

Abnormal engine instrument indication.....................17-13
Absence of propeller
    Drag...15-7
    Effect...15-6
    Slipstream...15-6
Academic material (knowledge and risk
management)..4-20
    Prevention through ADM and risk management......4-21
    Prevention through proportional counter-response.....4-21
    Recovery..4-22
Accelerated stalls 4-10
Accelerate-go distance 12-9
Accelerate-stop distance 12-9
After-landing...2-18
After-landing roll 13-6
Airframe and systems 16-5
Airplane-based UPRT..4-22
Airplane configuration 17-4
Airplane equipment and lighting 10-4
Airport and navigation lighting aids 10-5
Airport traffic patterns and operations.........................7-2
All-attitude/all-envelope flight training methods.........4-23
All-engine service ceiling of multiengine airplanes 12-9
Alternator/generator..12-7
Angle of attack.. 4-2, 13-2
Anti-icing/deicing 12-8
Approach...17-5
Approach and landing..................................... 10-8, 16-12
    Night emergencies..10-9
Approaches to stalls (impending stalls), power-on or
power-off...4-8
Attitude and sink rate control.......................................17-4
Attitude flying...3-4

B

Ballooning during round out.......................................8-30
Bank control..3-5
Basic safety concepts 17-2

Before start and starting engine 16-10
Before-takeoff check..2-17
    Avionics 2-18
    Electrical system...2-17
    Engine operation...2-17
    Flight controls 2-17
    Flight instruments...2-18
    Fuel system...2-17
    Takeoff briefing..2-18
    Trim..2-17
    Vacuum system 2-18
Bouncing during touchdown.......................................8-31
Brakes...2-8

C

Cabin fire 17-8
Captain's briefing...15-22
Cascade reversers...15-16
Chandelle 9-5
Climb gradient 12-9
Climbs and climbing turns..3-16
    Climbing turns...3-18
    Establishing a climb 3-17
        Best angle of climb (V_X)...........................3-16
        Best rate of climb (V_Y) 3-16
        Normal climb 3-16
Combustion heater 12-6
Constant radius during turning flight............................6-4
Construction..16-5
    Aluminum...16-5
    Composite...16-5
    Steel tube and fabric 16-5
Continuous ignition...15-4
Controllable-pitch propeller...11-4
    Blade angle control...11-7
    Climb...11-6
    Constant-speed propeller.................................11-4
    Constant-speed propeller operation.................11-7
    Cruise 11-6

Fixed-pitch propellers......................................11-4
  Governing range...11-7
  Takeoff...11-6
Control touch 1-1
Coordinated flight 4-2
Coordination 1-1
Correcting drift during straight-and-level flight 6-3
Cross-control stall 4-11
Crosswind after-landing roll 13-7
Crosswind approach and landing 8-14, 12-16
  Crosswind after-landing roll...........................8-16
  Crosswind final approach...............................8-14
    Crab method...8-14
    Wing-low (sideslip) method 8-15
  Crosswind round out (flare) 8-15
  Crosswind touchdown 8-15
  Maximum safe crosswind velocities 8-17
Crosswind takeoff 5-6, 13-4
  Initial climb 5-8
  Lift-off...5-8
  Takeoff roll 5-6
Cruise 16-11

D

Defining an airplane upset 4-2
Descents and descending turns 3-19
  Descent at minimum safe airspeed................3-19
  Emergency descent.....................................3-20
  Partial power descent..................................3-19
Directional control 13-3
Door opening in-flight 17-13
Drag devices...15-14
Drift and ground track control 6-3

E

Effect and use of the flight controls
  Feel of the airplane 3-4
Electrical fires 17-8
Elementary eights...6-11
  Eights across a road....................................6-13
  Eights along a road.....................................6-11
  Eights around pylons...................................6-13
  Eights-on-pylons 6-14
Elevator trim stall...4-12
Emergencies...16-12
Emergency approaches and landings (simulated)........8-26
Emergency descents..17-6
Emergency landings..17-2
  Psychological hazards 17-2
  Types of emergency landings.......................17-2
    Ditching 17-2

    Forced landing 17-2
    Precautionary landing 17-2
Emergency situations......................................17-1
Engine and propeller 2-9
Engine failure
  After lift-off...12-19
  After takeoff 12-21
  After takeoff (single-engine)........................17-6
  During flight..12-22
Engine fire...17-8
Engine inoperative approach and landing.................12-23
Engine inoperative flight principles..................12-23
Engines..16-6
Engine shutdown...2-19
Engine starting 2-12
Environmental factors.....................................4-18
Exhaust gas temperature (EGT)........................15-3

F

False start 14-9
Faulty approaches and landings 8-27
Feathering 12-3, 12-4
Flap effectiveness..11-3
Flight control malfunction/failure....................17-9
  Asymmetric (split) flap 17-9
  Landing gear malfunction 17-10
  Loss of elevator control...............................17-9
  Total flap failure..17-9
Flight director/autopilot 12-6
Flight environment..16-7
Flight standards service...................................1-5
Floating during round out 8-30
Forward slip 8-12
Four fundamentals 3-2
  Climbs 3-2
  Descents..3-2
  Straight-and-level flight 3-2
  Turns...3-2
Fowler flap 11-3
Fuel and oil 2-6
Fuel crossfeed 12-6, 12-22
Fuel heaters 15-4
Full stalls
  Power-off...4-8
  Power-on 4-9
Function of flaps 11-2
Fundamentals of stall recovery 4-7

G

Gas turbine engine 14-2
Glides...3-20

Gliding turns..3-21
Go-around 12-18
   Rejected landings 8-12
   Attitude 8-13
   Configuration 8-13
   Ground effect..8-14
   Power..8-13
Ground loop 8-34, 13-8
Ground operation 12-12

H

Hand propping 2-13
Hard landing..8-33
High final approach....................................8-28
High-performance airplane 11-1
High round out...8-29
Human factors 4-18
   Diversion of attention............................4-18
   IMC 4-18
   Sensory overload/deprivation.................4-18
   Spatial disorientation............................4-19
   Startle response.....................................4-19
   Surprise response...................................4-19
   Task saturation 4-18
   VMC to IMC 4-18
Hydraulic pump 11-11
Hydroplaning 8-35
   Dynamic hydroplaning...........................8-35
   Reverted rubber hydroplaning.................8-35
   Viscous hydroplaning.............................8-36

I

Inadvertent VFR flight into IMC 17-15
   Attitude control.....................................17-16
   Climbs 17-17
   Combined maneuvers 17-17
   Descents..17-17
   Maintaining airplane control 17-15
   Recognition 17-15
   Transition to visual flight......................17-18
   Turns..17-16
In-flight fire..17-7
Initial climb..15-24
Inside of the airplane.................................16-8
Instrumentation 16-6
Integrated flight instruction.......................3-5
Intentional slips..8-11
Intentional spins 4-16
Interstage turbine temperature (ITT) 15-4

J

Jet airplane approach and landing..............15-25
   Approach speed 15-27
   Glidepath control..................................15-28
   Landing requirements.............................15-25
   Landing speeds.....................................15-25
      Approach climb 15-26
      Landing climb 15-26
      V_{REF} 15-25
      V_{SO}...15-25
   Stabilized approach 15-27
   The flare 15-28
   Touchdown and rollout 15-29
Jet engine basics.......................................15-2
Jet engine efficiency 15-6
Jet engine ignition....................................15-4

L

Landing 13-5, 14-10
Landing gear 2-8, 13-2
   Instability...13-2
Landing gear control selected up, single-engine
climb performance adequate 12-20
   Checklist..12-21
   Climb...12-21
   Configuration 12-21
   Control...12-20
Landing gear control selected up, single-engine
climb performance inadequate.....................12-20
Landing gear down 12-19
Late or rapid round out 8-30
Lazy eight...9-6
Level off and cruise 12-14
Level turns 3-10
   Establishing a turn 3-13
      Medium turns 3-11
      Shallow turns 3-11
      Steep turns 3-11
   Turn radius 3-12
Liftoff..13-4
   Rotation 5-2
Light sport airplane (LSA) background........16-2
Loss of control in-flight (LOC-I).................4-1
Low final approach 8-27
Low speed flight 15-10
LSA maintenance.......................................16-5
LSA synopsis 16-3

M

Mach buffet boundaries 15-9
Maneuvering by reference to ground objects.................6-2
Minimum equipment list and configuration
deviation list...15-18
Multiengine training considerations 12-28

N

Night illusions...10-3
    Black-hole approach...10-3
    Visual autokinesis 10-3
Night vision...10-2
Noise abatement..5-13
Normal and crosswind takeoff and climb 12-13
Normal approach and landing.........................8-2, 12-14
    After-landing roll..8-8
    Base leg 8-2
    Final approach 8-3
        Estimating height and movement 8-5
        Use of flaps 8-4
    Round out (flare) 8-6
    Stabilized approach concept...............................8-9
    Touchdown...8-7
Normal takeoff...5-3
    Initial climb 5-5
    Lift-off...5-4
    Takeoff roll 5-3, 13-3
Nose baggage compartment......................................12-7

O

Operating the jet engine 15-3
Operational considerations.......................................14-9
Operation of systems..12-3
Orientation and navigation.......................................10-7
Outer wing surfaces 2-5
Outside of the airplane 16-9

P

Parking 2-19
Performance and limitations 12-9
Pilot equipment 10-4
Pilot sensations in jet flying 15-17
Pitch and power...3-23
Pitch control 3-5
Plain (hinge) flap...11-3
Porpoising 8-32
Post-flight...2-19, 16-12
    Securing and servicing 2-19
Power control 3-5
Power-off accuracy approaches 8-22

    90° Power-off approach 8-22
    180° Power-off approach 8-23
Preflight...16-7
Preflight assessment of the aircraft 2-2
Preparation and preflight..10-6
Pre-takeoff procedures 15-20
Prior to takeoff 5-2
Propellers 12-3
Propeller synchronization 12-6

R

Recovery from overspeed conditions...........................15-9
Rectangular course...6-6
Rejected takeoff 12-19, 15-22
Rejected takeoff/engine failure 5-12
Retractable landing gear 11-11
    Controls and position indicators.......................11-11
    Emergency gear extension systems....................11-12
    Landing gear safety devices 11-11
    Landing gear systems 11-11
        Electrical landing gear retraction system...........11-11
        Hydraulic landing gear retraction system 11-11
    Operational procedures.....................................11-12
        Approach and landing....................................11-15
        Preflight 11-12
        Takeoff and climb 11-13
Reverse thrust and beta range operations...................14-7
Risk and resource management 2-9, 2-10
    Identifying the hazard 2-10
    Resource management 2-11
        Aeronautical decision-making (ADM) 2-11
        Flight deck resource management....................2-11
        Situational awareness 2-11
        Task management..2-11
    Risk 2-10
    Risk assessment 2-10
Role of the FAA...1-2
Role of the flight instructor...1-7
Role of the pilot examiner...1-6
Roles of FSTDs and airplanes in UPRT 4-22
Rotation and lift-off..15-24

S

Safety considerations 7-5
Secondary stall...4-10
Setting power 15-4
Short-field approach and landing.................... 8-18, 12-17
Short-field landing 13-7
Short-field takeoff..13-4
Short-field takeoff and climb 12-17
Short field takeoff and maximum performance climb ...5-10

Initial climb 5-11
Lift-off...5-10
Takeoff roll 5-10
Sideslip..8-11
Single-engine service ceiling 12-9
Slotted flap 11-3
Slow acceleration of the jet engine 15-6
Slow final approach 8-28
Slow flight.. 4-3, 12-26
Maneuvering in slow flight 4-4
Soft-field approach and landing......................8-21
Soft-field landing 13-8
Soft-field takeoff..13-4
Soft/rough-field takeoff and climb..................5-11
Initial climb 5-12
Lift-off...5-12
Sources of flight training 1-8
Airman certification standards (ACS)..............1-10
Flight safety practices..................................1-11
Collision avoidance 1-11
Positive transfer of controls 1-15
Runway incursion avoidance...................1-12
Stall awareness..1-12
Use of checklists 1-13
Practical test standards (PTS).......................1-10
Speed margins...15-7
Speed sense 1-1
Spin awareness.. 4-13, 12-28
Spin procedures..4-14
Developed phase..4-15
Entry phase 4-14
Incipient phase..4-14
Recovery phase 4-15
Spiral dive 4-23
Split flap..11-3
Sport pilot certificate.......................................16-3
Stabilized approach 14-10
Stall characteristics 4-6
Stall recognition..4-5
Angle of attack indicators..............................4-6
Feel 4-5
Hearing 4-6
Kinesthesia..4-6
Vision...4-6
Stalls.. 4-5, 12-26, 15-11
Accelerated approach to stall 12-27
Engine inoperative—loss of directional control
demonstration 12-28
Full stall.. 4-5, 12-27
Impending stall 4-5
Power-off stalls (approach and landing) 12-26

Power-on stalls (takeoff and departure) 12-27
Spin awareness 12-28
Stall training 4-8
Standard airport traffic patterns 7-2
Base leg 7-4
Crosswind leg 7-4
Departure leg 7-4
Downwind leg 7-4
Entry leg 7-3
Starting, taxiing, and runup........................10-6
Steep spiral 9-4
Steep turns...9-2
Straight-and-level flight...................................3-6
Level flight..3-8
Straight flight...3-7
S-turns across a road 6-8
Systems malfunctions 17-11
Electrical system..17-11
Pitot-static system 17-12

T

Takeoff and climb............................ 10-7, 16-11
Takeoff and departure 14-10
Takeoff checks 2-18
Takeoff roll 15-21
Ground roll 5-2
Takeoffs 14-10
Taxi 16-10
Taxiing 2-14, 13-2
Terrain selection..17-4
Terrain types 17-5
Confined areas 17-5
Trees (forest) 17-5
Water (ditching) and snow 17-6
Thrust reversers..15-15
Thrust to thrust lever relationship 15-5
Timing..1-1
Tires 2-8
Touchdown 13-5
Crosswinds 13-6
Three-point landing.....................................13-5
Wheel landing..13-6
Touchdown in a drift or crab 8-34
Tracking over and parallel to a straight line 6-6
Training considerations...................................14-11
Flight training 14-12
Ground training 14-12
Training for night flight 10-6
Transition training...11-16
Transition training considerations..................16-4
Flight instructors 16-4

Flight school 16-4

Trim control 3-5, 3-10

Turbine inlet temperature (TIT) 15-4

Turbine outlet temperature (TOT) 15-4

Turbocharging 11-8

    Ground boosting versus altitude turbocharging 11-9

    Heat management 11-10

    Operating characteristics 11-9

    Turbocharger failure 11-10

        Low manifold pressure 11-11

        Over-boost condition 11-10

Turboprop airplane electrical systems 14-8

Turboprop engines 14-2

Turboprop engine types 14-3

    Fixed shaft 14-3

    Split shaft/free turbine engine 14-5

Turbulent air approach and landing 8-18

U

Unusual attitudes versus upsets 4-17

Upset prevention and recovery 4-17

Upset prevention and recovery training (UPRT) 4-19

Use of power 8-29

V

V_1 15-23

    Maximum V_1 15-23

    Minimum V_1 15-23

    Reduced V_1 15-23

Variation of thrust with RPM 15-5

Visibility 13-3

Visual inspection of the aircraft 2-2

Visual preflight assessment 2-3

V-speeds 12-2, 15-20

    V_{LOF} 12-2

    V_{MC} 12-2

    V_R 12-2

    V_{REF} 12-2

    V_{SSE} 12-2

    V_X 12-2

    V_{XSE} 12-2

    V_Y 12-2

    V_{YSE} 12-2

W

Weather considerations 16-6

Weathervaning 13-3

Weight and balance 12-11

    Basic empty weight 12-11

    Empty weight 12-11

    Maximum landing weight 12-12

    Ramp weight 12-12

    Standard empty weight 12-11

    Zero fuel weight 12-11

Weight and balance requirements related to spins 4-17

Wheel barrowing 8-33

Wing rising after touchdown 8-35

Y

Yaw damper 12-7

Federal Aviation Administration

Airplane Flying Handbook (FAA-H-8083-3B) Addendum

Due to a technical glitch, Chapter 12 of the Airplane Flying Handbook (FAA-H-8083-3B) abruptly ends on page 12-28. The following addendum contains the remaining material that is missing from the current version of the chapter.

Airplane Flying Handbook (FAA-H-8083-3B) Addendum
Chapter 12: Transition to Multiengine Airplanes

Engine Inoperative—Loss of Directional Control Demonstration

An engine inoperative—loss of directional control demonstration, often referred to as a "V_{MC} demonstration," is a task on the practical test for a multiengine class rating. A thorough knowledge of the factors that affect V_{MC}, as well as its definition, is essential for multiengine pilots, and as such an essential part of that task. V_{MC} is a speed established by the manufacturer, published in the AFM/POH, and marked on most airspeed indicators with a red radial line. A knowledgeable multiengine pilot understands that V_{MC} is **not** a fixed airspeed under all conditions. V_{MC} is a fixed airspeed only for the very specific set of circumstances under which it was determined during aircraft certification.

In reality, V_{MC} varies with a variety of factors as outlined below. The V_{MC} noted in practice and demonstration, or in actual single-engine operation, could be less or even greater than the published value, depending on conditions and technique.

Historically, in aircraft certification, V_{MC} is the sea level calibrated airspeed at which, when the critical engine is suddenly made inoperative, it is possible to maintain control of the airplane with that engine still inoperative and then maintain straight flight at the same speed with an angle of bank not more than 5°.

The foregoing refers to the determination of V_{MC} under "dynamic" conditions. This technique is only used by highly experienced test pilots during aircraft certification. It is unsafe to be attempted outside of these circumstances.

In aircraft certification, there is also a determination of V_{MC} under "static," or steady-state conditions. If there is a difference between the dynamic and static speeds, the higher of the two is published as V_{MC}. The static determination is simply the ability to maintain straight flight at V_{MC} with a bank angle of not more than 5°. This more closely resembles the V_{MC} demonstration task in the practical test for a multiengine rating.

The AFM/POH-published V_{MC} is determined with the "critical" engine inoperative. The critical engine is the engine whose failure had the most adverse effect on directional control. On twins with each engine rotating in conventional, clockwise rotation as viewed from the pilot's seat, the critical engine will be the left engine.

Multiengine airplanes are subject to P-factor just as single-engine airplanes are. The descending propeller blade of each engine will produce greater thrust than the ascending blade when the airplane is operated under power and at positive angles of attack. The descending propeller blade of the right engine is also a greater distance from the center of gravity, and therefore has a longer moment arm than the descending propeller blade of the left engine. As a result, failure of the left engine will result in the most asymmetrical thrust (adverse yaw) as the right engine will be providing the remaining thrust. [Figure 12-20]

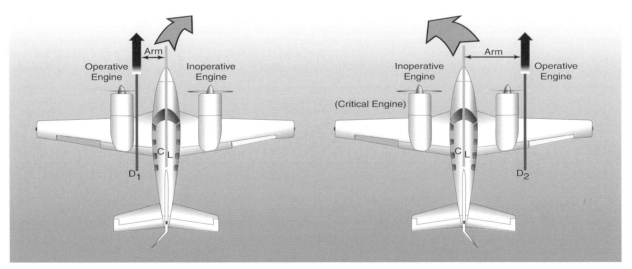

Figure 12-20. *Forces created during single-engine operation.*

Many twins are designed with a counter-rotating right engine. With this design, the degree of asymmetrical thrust is the same with either engine inoperative. No engine is more critical than the other, and a V_{MC} demonstration may be performed with either engine windmilling.

Historically, in aircraft certification, dynamic V_{MC} has been determined under the following conditions:

- **Maximum Available Takeoff Power.** V_{MC} increases as power is increased on the operating engine. With normally aspirated engines, V_{MC} is highest at takeoff power and sea level, and decreases with altitude. With turbocharged engines, takeoff power, and therefore V_{MC}, remains constant with increases in altitude up to the engine's critical altitude (the altitude where the engine can no longer maintain 100 percent power). Above the critical altitude, V_{MC} decreases just as it would with a normally aspirated engine, whose critical altitude is sea level. V_{MC} tests are conducted at a variety of altitudes. The results of those tests are then extrapolated to a single, sea level value.

- **Windmilling propeller.** V_{MC} increases with increased drag on the inoperative engine. V_{MC} is highest, therefore, when the critical engine propeller is windmilling at the low pitch, high rpm blade angle. V_{MC} is determined with the critical engine propeller windmilling in the takeoff position, unless the engine is equipped with an autofeather system.

- **Most unfavorable weight and center-of-gravity position.** V_{MC} increases as the center of gravity is moved aft. The moment arm of the rudder is reduced, and therefore its effectivity is reduced, as the center of gravity is moved aft. For a typical light twin, the aft-most CG limit is the most unfavorable CG position. Historically, 14 CFR part 23 calls for V_{MC} to be determined at the most unfavorable weight. For twins certificated under CAR 3 or early 14 CFR part 23, the weight at which V_{MC} was determined was not specified. V_{MC} increases as weight is reduced. [Figure 12-21]

Figure 12-21. *Effect of CG location on yaw.*

- **Landing gear retracted.** V_{MC} increases when the landing gear is retracted. Extended landing gear aids directional stability, which tends to decrease V_{MC}.

- **Wing flaps in the takeoff position.** For most twins, this will be 0° of flaps.

- **Cowl flaps in the takeoff position.**

- **Airplane trimmed for takeoff.**

- **Airplane airborne and the ground effect negligible.**

- **Maximum of 5° angle of bank**. V_{MC} is highly sensitive to bank angle. To prevent claims of an unrealistically low V_{MC} speed in aircraft certification, the manufacturer is permitted to use a maximum of a 5° bank angle toward the operative engine. The horizontal component of lift generated by the bank balances the side force from the rudder, rather than using sideslip to do so. Sideslip requires more rudder deflection, which in turn increases V_{MC}. The bank angle works in the manufacturer's favor in lowering V_{MC} since using high bank angles reduces required rudder deflection. However, this method may result in unsafe flight from both the large sideslip and the need to increase the angle of attack in order to maintain the vertical component of lift.

V_{MC} is reduced significantly with increases in bank angle. Conversely, V_{MC} increases significantly with decreases in bank angle. Tests have shown that V_{MC} may increase more than 3 knots for each degree of bank angle less than 5°. Loss of directional control may be experienced at speeds almost 20 knots above published V_{MC} when the wings are held level.

The 5° bank angle maximum is an historical limit imposed upon manufacturers in aircraft certification. The 5° bank does not inherently establish zero sideslip or best single-engine climb performance. Zero sideslip, and therefore best single-engine climb performance, may occur at bank angles less than 5°. The determination of V_{MC} in certification is solely concerned with the minimum speed for directional control under a very specific set of circumstances, and not the optimum airplane attitude or configuration for climb performance.

During dynamic V_{MC} determination in aircraft certification, cuts of the critical engine using the mixture control are performed by flight test pilots while gradually reducing the speed with each attempt. V_{MC} is the minimum speed at which directional control could be maintained within 20° of the original entry heading when a cut of the critical engine was made. During such tests, the climb angle with both engines operating was high, and the pitch attitude following the engine cut had to be quickly lowered to regain the initial speed. Pilots should understand that attempting to demonstrate V_{MC} with an engine cut from high power, or intentionally failing an engine at speeds less than V_{SSE} creates a high likelihood for loss of control and an accident.

The actual demonstration of V_{MC} and recovery in flight training more closely resembles static V_{MC} determination in aircraft certification. For a demonstration that avoids the hazard of unintended contact with the ground, the pilot selects an altitude that will allow performance of the maneuver at least 3,000 feet AGL. The following description assumes a twin with noncounter-rotating engines, where the left engine is critical.

With the landing gear retracted and the flaps set to the takeoff position, the pilot slows the airplane to approximately 10 knots above V_{SSE} or V_{YSE} (whichever is higher) and trims for takeoff. For the remainder of the maneuver, the trim setting remains unaltered. The pilot selects an entry heading and sets high rpm on both propeller controls. Power on the left engine is throttled back to idle as the right engine power is advanced to the takeoff setting. The landing gear warning horn will sound as long as a throttle is retarded, however the pilot listens carefully for the stall warning horn or watches for the stall warning light. The left yawing and rolling moment of the asymmetrical thrust is counteracted primarily with right rudder. A bank angle of up to 5° (a right bank in this case) may be established as appropriate for the make and model.

While maintaining entry heading, the pitch attitude is slowly increased to decelerate at a rate of 1 knot per second (no faster). As the airplane slows and control effectivity decays, the pilot counteracts the increasing yawing tendency with additional rudder pressure. Aileron displacement will also increase in order to maintain the established bank. An airspeed is soon reached where full right rudder travel and up to a 5° right bank can no longer counteract the asymmetrical thrust, and the airplane will begin to yaw uncontrollably to the left.

The moment the pilot first recognizes the uncontrollable yaw, or experiences **any** symptom associated with a stall, the pilot retards the throttle for the operating engine to stop the yaw as the pitch attitude is decreased. Recovery is made to straight flight on the entry heading at V_{SSE} or V_{YSE}, before setting symmetrical power. The recovery demonstration does not include increasing power on the windmilling engine alone.

To keep the foregoing description simple, there were several important background details that were not covered. The rudder pressure during the demonstration can be quite high. In certification, 150 pounds of force is permitted before the limiting factor becomes rudder pressure, not rudder travel. Most twins will run out of rudder travel long before 150 pounds of pressure is required. Still, it will seem considerable.

Maintaining altitude is **not** a criterion in accomplishing this maneuver. This is a demonstration of controllability, not performance. Many airplanes will lose (or gain) altitude during the demonstration. Remaining at or above a minimum of 3,000 feet AGL throughout the maneuver is considered to be effective risk mitigation of certain hazards.

V$_{MC}$ Demo Stall Avoidance

As discussed earlier, with normally aspirated engines, V$_{MC}$ decreases with altitude. Stalling speed (V$_S$), however, remains the same. Except for a few models, published V$_{MC}$ is almost always higher than V$_S$. At sea level there is usually a margin of several knots between V$_{MC}$ and V$_S$, but the margin decreases with altitude, and at some altitude, V$_{MC}$ and V$_S$ are the same. [Figure 12-22]

Figure 12-22. *Graph depicting relationship of V$_{MC}$ to V$_S$.*

Should a stall occur while the airplane is under asymmetrical power a spin entry is likely. The yawing moment induced from asymmetrical thrust is little different from that induced by full rudder in an intentional spin in the appropriate model of single-engine airplane. In this case, however, the airplane will depart controlled flight in the direction of the idle engine, not in the direction of applied rudder. Twins are not required to demonstrate recoveries from spins, and their spin recovery characteristics are generally very poor.

Where V$_S$ is encountered before V$_{MC}$, the departure from controlled flight might be quite sudden, with strong yawing and rolling tendencies to the inverted orientation, and a spin entry. Therefore, during a V$_{MC}$ demonstration, if there are **any** symptoms of an impending stall such as a stall warning light or horn, airframe or elevator buffet, or sudden loss of control effectiveness; terminate the maneuver immediately by reducing the angle of attack as the throttle is retarded and return the airplane to the entry airspeed. Note that noise within the flight deck may mask the sound of the stall warning horn.

While the V$_{MC}$ demonstration shows the earliest onset of a loss of directional control when performed in accordance with the foregoing procedures, avoid a stalled condition. Avoid stalls with asymmetrical thrust, such that the V$_{MC}$ demonstration does not degrade into a single-engine stall. A V$_{MC}$ demonstration that is allowed to degrade into a single-engine stall with high asymmetrical thrust may result in an unrecoverable loss of control and a fatal accident.

An actual demonstration of V$_{MC}$ may not be possible under certain conditions of density altitude, or with airplanes whose V$_{MC}$ is equal to or less than V$_S$. Under those circumstances, as a training technique, a demonstration of V$_{MC}$ may safely be conducted by artificially limiting rudder travel to simulate maximum available rudder. A speed well above V$_S$ (approximately 20 knots) is recommended when limiting rudder travel.

The rudder limiting technique avoids the hazards of spinning as a result of stalling with high asymmetrical power, yet is effective in demonstrating the loss of directional control.

To reduce the risk of a loss of control, avoid performing any V$_{MC}$ demonstration from a high pitch attitude with both engines operating and then reducing power on one engine.

Multiengine Training Considerations

Flight training in a multiengine airplane can be safely accomplished if both the instructor and the student are cognizant of the following factors.

- The participants conduct a preflight briefing of the objectives, maneuvers, expected student actions, and completion standards before the flight begins.

- A clear understanding exists as to how simulated emergencies will be introduced, and what action the student is expected to take.

The introduction, practice, and testing of emergency procedures has always been a sensitive subject. Surprising a multiengine student with an emergency without a thorough briefing beforehand creates a hazardous condition. Simulated engine failures, for example, can very quickly become actual emergencies or lead to loss of the airplane when approached carelessly. Stall-spin accidents in training for emergencies rival the number of stall-spin accidents from actual emergencies. The training risk normally gets mitigated by a briefing. Pulling circuit breakers is not recommended for training purposes and can lead to a subsequent gear up landing.

Many normal, abnormal, and emergency procedures can be introduced and practiced in the airplane as it sits on the ground without the engines running. In this respect, the airplane is used as a cockpit procedures trainer (CPT), ground trainer, or simulator. The value of this training may be substantial. The engines do not have to be operating for real learning to occur. Upon completion of a training session, care should be taken to restore items to their proper positions.

Pilots who do not use a checklist effectively will be at a significant disadvantage in multiengine airplanes. Use of the checklist is essential to safe operation of airplanes, and it is risky to conduct a flight without one. The manufacturer's checklist or an aftermarket checklist that conforms to the manufacturer's procedures for the specific make, model, and model year may be used. If there is a procedural discrepancy between the checklist and the AFM/POH, then the AFM/POH always takes precedence.

Certain immediate action items (such as a response to an engine failure in a critical phase of flight) are best committed to memory. After they are accomplished, and as work load permits, the pilot can compare the action taken with a checklist.

Simulated engine failures during the takeoff ground roll may be accomplished with the mixture control. The simulated failure should be introduced at a speed no greater than 50 percent of V_{MC}. If a student does not react promptly by retarding both throttles, the instructor can always pull the other mixture.

The FAA recommends that all in-flight simulated engine failures below 3,000 feet AGL, be introduced with a smooth reduction of the throttle. Thus, the engine is kept running and is available for instant use, if necessary. Smooth throttle reduction avoids abusing the engine and possibly causing damage. Simulation of inflight engine failures below V_{SSE} introduces a very high and unnecessary training risk.

If the engines are equipped with dynamic crankshaft counterweights, it is essential to make throttle reductions for simulated failures smoothly. Other areas leading to dynamic counterweight damage include high rpm and low manifold pressure combinations, overboosting, and propeller feathering. Severe damage or repetitive abuse to counterweights will eventually lead to engine failure. Dynamic counterweights are found on larger, more complex engines—instructors may check with maintenance personnel or the engine manufacturer to determine if their engines are so equipped.

When an instructor simulates an engine failure, the student responds with the appropriate memory items and retards the propeller control toward the FEATHER position. Assuming zero thrust will be set, the instructor promptly moves the propeller control forward and sets the appropriate manifold pressure and rpm. It is vital that the student be kept informed of the instructor's intentions. At this point the instructor may say words to the effect, "I have the right engine; you have the left. I have set zero thrust and the right engine is simulated feathered." Any ambiguity as to who is operating what systems or controls increases the likelihood of an unintended outcome.

Following a simulated engine failure, the instructor cares for the "failed" engine just as the student cares for the operative engine. If zero thrust is set to simulate a feathered propeller, the cowl flap is normally closed and the mixture leaned. An occasional clearing of the engine is also desirable. If possible, avoid high power applications immediately following a prolonged cool-down at a zero-thrust power setting. A competent flight instructor teaches the multiengine student pilot about the critical importance of feathering the propeller in a timely manner should an actual engine failure situation ever be encountered. A windmilling propeller, in many cases, has given the improperly trained multiengine pilot the mistaken perception that the engine is still developing useful thrust, resulting in a psychological reluctance to feather, as feathering results in cessation of propeller rotation. The flight instructor spends ample time demonstrating the difference in the performance capabilities of the airplane with a simulated feathered propeller (zero thrust) as opposed to a windmilling propeller.

Actual and safe propeller feathering for training is performed at altitudes and positions where safe landings on established airports may be readily accomplished if the propeller will not unfeather. Plan unfeathering and restart to be completed no lower than 3,000 feet AGL. At certain elevations and with many popular multiengine training airplanes, this may be above the single-engine service ceiling, and level flight will not be possible.

Repeated feathering and unfeathering is hard on the engine and airframe, and is done as necessary to ensure adequate training. The FAA's Airman Certification Standards for a multiengine class rating contains a task for feathering and unfeathering of one propeller during flight in airplanes in which it is safe to do so.

While much of this chapter has been devoted to the unique flight characteristics of a multiengine airplane with one engine inoperative, the modern well-maintained reciprocating engine is remarkably reliable. When training in an airplane, initiation of a simulated engine inoperative emergency at low altitude normally occurs at a minimum of 400 feet AGL to mitigate the risk involved and only after the student has successfully mastered engine inoperative procedures at higher altitudes. Initiating a simulated low altitude engine inoperative emergency in the airplane at extremely low altitude, immediately after liftoff, or below VSSE creates a situation where they are non-existent safety margins.

For training in maneuvers that would be hazardous in flight, or for initial and recurrent qualification in an advanced multiengine airplane, consider a simulator training center or manufacturer's training course. Comprehensive training manuals and classroom instruction are available along with system training aids, audio/visuals, and flight training devices and simulators. Training under a wide variety of environmental and aircraft conditions is available through simulation. Emergency procedures that would be either dangerous or impossible to accomplish in an airplane can be done safely and effectively in a flight training device or simulator. The flight training device or simulator need not necessarily duplicate the specific make and model of airplane to be useful. Highly effective instruction can be obtained in training devices for other makes and models as well as generic training devices.

The majority of multiengine training is conducted in four-to-six place airplanes at weights significantly less than maximum. Single-engine performance, particularly, at low density altitudes, may be deceptively good. To experience the performance expected at higher weights, altitudes and temperatures, the instructor may occasionally artificially limit the amount of manifold pressure available on the operative engine. Airport operations above the single-engine ceiling can also be simulated in this matter. Avoid loading the airplane with passengers to practice emergencies at maximum takeoff weight since this practice creates an unnecessary training hazard.

The use of the touch-and-go landing and takeoff in multiengine flight training has always been somewhat controversial. The value of the learning experience may be offset by the hazards of reconfiguring the airplane for takeoff in extremely limited time as well as the loss of the follow-through ordinarily experienced in a full stop landing. Touch-and-goes are not recommended during initial aircraft familiarization in multiengine airplanes.

If touch-and-goes are to be performed at all, the student and instructor responsibilities need to be carefully briefed prior to each flight. Following touchdown, the student will ordinarily maintain directional control while keeping the left hand on the yoke and the right hand on the throttles. The instructor resets the flaps and trim and announces when the airplane has been reconfigured. The multiengine airplane needs considerably more runway to perform a touch-and-go than a single-engine airplane. A full stop-taxi back landing is preferable during initial familiarization. Solo touch-and-goes in twins are strongly discouraged.

Chapter Summary

Small multiengine airplanes handle much like single-engine airplanes as long as both engines are functioning normally. A competent multiengine pilot, however, acquires the additional knowledge, risk mitigation strategies, and practical skills required to fly a multiengine airplane in case a loss of thrust from one engine actually occurs. In that case, the pilot will be able to take appropriate action leading to a safe outcome. Much of this chapter discussed loss of directional control. How to obtain the best performance with an inoperative engine was also described in detail. These two considerations correspond to the red radial line (V_{MC}) and the blue radial line (V_{YSE}) on the airspeed indicator. The actions a pilot takes when dealing with stalls, V_{MC}, or best performance vary greatly. Understanding these concepts, knowing how to mitigate the risks, and possessing the skills to handle an engine failure in a variety of situations, allows a pilot to enjoy the increased performance and safety provided when flying a multiengine airplane.